Computer Arithmetic and Formal Proofs

Series Editor
Valérie Berthé

Computer Arithmetic and Formal Proofs

Verifying Floating-point Algorithms with the Coq System

Sylvie Boldo
Guillaume Melquiond

ELSEVIER

First published 2017 in Great Britain and the United States by ISTE Press Ltd and Elsevier Ltd

ISTE Press Ltd
27-37 St George's Road
London SW19 4EU
UK

Elsevier Ltd
The Boulevard, Langford Lane
Kidlington, Oxford, OX5 1GB
UK

www.iste.co.uk

www.elsevier.com

Notices

Knowledge and best practice in this field are constantly changing. As new research and experience broaden our understanding, changes in research methods, professional practices, or medical treatment may become necessary.

Practitioners and researchers must always rely on their own experience and knowledge in evaluating and using any information, methods, compounds, or experiments described herein. In using such information or methods they should be mindful of their own safety and the safety of others, including parties for whom they have a professional responsibility.

To the fullest extent of the law, neither the Publisher nor the authors, contributors, or editors, assume any liability for any injury and/or damage to persons or property as a matter of products liability, negligence or otherwise, or from any use or operation of any methods, products, instructions, or ideas contained in the material herein.

For information on all our publications visit our website at http://store.elsevier.com/

British Library Cataloguing-in-Publication Data
A CIP record for this book is available from the British Library
Library of Congress Cataloging in Publication Data
A catalog record for this book is available from the Library of Congress
ISBN 978-1-78548-112-3

Printed and bound in the UK and US

Contents

Preface

Once upon a time in 2010, we began the development of the Flocq library to merge the existing disparate formalizations of floating-point arithmetic for the Coq proof assistant. It was not an easy task: we wanted to have our cake and eat it too. On the one hand, we wanted high-level properties on floating-point numbers as real numbers with a mathematical definition of rounding (such as the nearest floating-point number). On the other hand, we also wanted to be able to perform computations inside the proof assistant (which is antinomic with considering only real numbers). Finally, more than have our cake and eat it too, we wanted to sell the used flour: we wanted to explore exotic formats, their deficiencies and properties (such as fixed-point formats, IEEE-754 floating-point formats, flush-to-zero formats, and many others).

Since that time, we have changed office and have improved the Flocq library in various ways. As time flew by, we did not feel the need to change our original design choices, the only small difference being the definition of a reasonable value for $ulp(0)$. We developed more automation, more formalized high-level properties, proved algorithms, and verified programs. Moreover, we got users! Contrary to our forecast, the library has been used by other researchers for their own deeds worldwide, for teaching purposes, and even in a commercial software.

Now is the time to put everything together to offer a comprehensive lecture on Flocq and on several tools related to it. We have aimed at homogeneity, understandability, and pedagogy. We have assumed our readers to be Master's or PhD students, engineers or researchers, who are already familiar either with floating-point arithmetic or with formal verification and who would like to broaden their expertise to the other domain.

We hope you will get to learn something new about formal proofs and computer arithmetic. More importantly, we sincerely hope that, after reading this book, you will feel like formally verifying your own floating-point algorithms.

Acknowledgments

We are indebted to a large amount of people, including (by alphabetical order starting with letter p):

– Our parents. Thanks for your deep and fundamental support that has helped us to become better persons/researchers.

– Our reviewers. Thanks for reading and criticizing the first version of this book. You really helped us make it better: Valérie Berthé, François Bobot, François Clément, Florian Faissole, Stef Graillat, Philippe Langlois, Christoph Lauter, Xavier Leroy, Assia Mahboubi, Claude Marché, Érik Martin-Dorel, Jean-Michel Muller, Laurence Rideau, Pierre Roux and Laurent Théry. Special thanks to Claude-Pierre Jeannerod for his numerous comments. The remaining errors are all ours.

– Our team. Thanks for your cheerfulness, chocolate, coffee, and tea. We have spent a large amount of time in a dull white room without sunlight and we really enjoyed your support.

– The Boullis family: Nicolas, Cyril, and Cédric. Thanks for letting one of the authors sometimes work during the evenings/nights/weekends, and for having given her the best time of her life.

– Our collaborators. Thanks for working with us! You have been strictly necessary for some of the results presented in this book:

- part of Chapter 4 is based on some work done with Assia Mahboubi, Thomas Sibut-Pinote, and the members of the TaMaDi project, in particular Érik Martin-Dorel;

- parf of Chapter 5 is based on some work done with Jean-Michel Muller, Laurence Rideau, and Laurent Théry;

- part of Chapter 6 is based on some work done with Sylvain Pion;

- Chapter 7 is based on some work done with Jacques-Henri Jourdan and Xavier Leroy;

- part of Chapter 8 is based on some work done with Jean-Christophe Filliâtre, Claude Marché, and Thi Minh Tuyen Nguyen;

- Chapter 9 is based on some work done with François Clément, Jean-Christophe Filliâtre, Catherine Lelay, Micaela Mayero, and Pierre Weis.

– Our employer, Inria. Thanks for supporting us while we spent a lot of time writing this book.

Sylvie BOLDO
Guillaume MELQUIOND
August 2017

List of Algorithms

Introduction

Computer arithmetic is a field of computer science that investigates how computers should represent numbers and perform operations on them. It includes integer arithmetic, fixed-point arithmetic, and the arithmetic this book focuses on: floating-point (FP) arithmetic, which will be more thoroughly described in Chapter 1. For now, let us say that it is the common way computers approximate real numbers and that it is described in the IEEE-754 standard [IEE 08]. As in scientific notations, numbers are represented using an exponent and a significand, except that this significand has to fit on a certain amount of bits. As this number of bits (called precision) is limited, each operation may be inexact due to the rounding. This makes computer arithmetic sometimes inaccurate: the result of a long computation may be far from the mathematical result that would have been obtained if all the computations were correct. This also makes computer arithmetic unintuitive: for instance, the FP addition is not always associative. Let us see what the following two C expressions using the double type on 64 bits compute. Note that the C expression 0x1p60 stands for the value 2^{60}:

– the expression (1 + 0x1p60) – 0x1p60 returns 0;

– the expression 1 + (0x1p60 – 0x1p60) returns 1.

Many strange behaviors caused by FP operations may be found in [MUL 10]. The following example is Muller's sequence defined by

$$
\begin{cases}
u_0 &= 2, \\
u_1 &= -4, \\
u_n &= 111 - \dfrac{1130}{u_{n-1}} + \dfrac{3000}{u_{n-1}u_{n-2}}.
\end{cases}
$$

In practice, it converges toward 100, while mathematically, it should converge toward 6. This discrepancy can be explained by giving a closed form for u_n:

$$u_n = \frac{\alpha 100^{n+1} + \beta 6^{n+1} + \gamma 5^{n+1}}{\alpha 100^n + \beta 6^n + \gamma 5^n},$$

with the values of α, β, and γ depending on the initial values. The aforementioned values of u_0 and u_1 imply that $\alpha = 0$ and $\beta \neq 0$, so the sequence should converge toward 6. But the rounding error on the first values makes it as if α was small but nonzero, therefore the computed sequence converges toward 100.

Many more subtleties exist, such as very small FP numbers (called subnormal numbers), exceptional values, and so on. Guaranteeing that an algorithm is correct or nearly correct for some given inputs is therefore a difficult task, which requires trust and automation.

Formal proofs may be used to provide more trust. If an article or person tells you something is true (such as a theorem, an algorithm, a program), will you believe it? This question is more sociological than scientific [DEM 77]. You may believe it for many different reasons: because you trust the author, because the result seems reasonable, because everybody believes it, because it looks like something you already believe, because you read the proof, because you tested the program, and so on. We all have a trust base, meaning a set of persons, theorems, algorithms, and programs we believe in. But we all have a different trust base, so to convince as many people as possible, you have to reduce the trust base as much as possible.

Proofs are usually considered as an absolute matter and mathematicians are those who can discover these absolute facts and show them. Mathematicians, however, might fail. An old book of 1935 makes an inventory of 500 errors made by famous mathematicians [LEC 35]. Given the current publication rate, an updated inventory would now take several bookshelves. As humans are fallible, how can we make proofs more reliable? A solution is to rely on a less-fallible device, the computer, to carefully check the proofs. The lack of imagination of computers, however, means we need to detail the proofs much more than mathematicians usually do.

We need a "language of rules, the kind of language even a thing as stupid as a computer can use" (Porter) [MAC 04]. Formal logic studies the languages used to define abstract objects such as groups or real numbers and reason about them. Logical reasoning aims at checking every step of the proof, so as to guarantee only justified assumptions and correct inferences are used. The reasoning steps that are applied to deduce from a property believed to be true a new property believed to be true called an inference rule. They are usually handled at a syntactic level: only the form of the statements matters, their content does not. For instance, the *modus ponens* rule states that, if both properties A and "if A then B" hold, then property B holds too, whatever the meaning of A and B. To check these syntactic rules are correctly applied, a mechanical device as stupid as a computer can be used. It will be faster and more systematic than a human being, as long as the proof is comprehensive and given in a language understandable by the computer. To decrease the trust base, one may choose the LCF approach [GOR 00], meaning the only thing to trust is a small kernel that entirely checks the proof. It does not matter whether the rest of the system, such as proof search, is bug-ridden, as each proof step will ultimately be checked by the trusted kernel. In this case, the proof must be complete: there is no more skipping some part of the proof for shortening or legibility.

Even when we rely on formal systems, how sure are we of the result? This can be restated as "what trust can you put in a formal system?", or as "who checks the checker?" [POL 98]. There is a complementary and more pragmatic point of view influenced by Davis [DAV 72] with this nice title: "Fidelity in Mathematical Discourse: Is One and One Really Two?". It describes "Platonic mathematics" and its beliefs. In particular, Platonic mathematics assumes that symbol manipulation is perfect. Davis argues that this manipulation is in fact wrong with a given probability. This probability depends on whether the manipulation is performed by a human ($\approx 10^{-3}$) or a machine ($\approx 10^{-10}$). Indeed, a human may forget a corner case or lose concentration, while a cosmic ray may mislead a computer to accept an incorrect proof. Although perfection is not achievable, computers and formal proof assistants are still the best way to decrease the probability of failure, and hence increase the trust.

Among the many existing formal systems, the one we use in this book is the Coq proof assistant, which provides a higher-order logic with dependent types, as described in Chapter 2. For the sociological and historical aspects of mechanized proofs, we refer the reader to MacKenzie [MAC 04].

Formal libraries for floating-point arithmetic have been designed to help with the verification of FP properties of circuits, algorithms, and programs. Formal methods have been applied quite early to the IEEE-754 standard, firstly for a precise specification of FP arithmetic: what does this standard in English really mean? Barrett wrote such a specification in Z [BAR 89]. It was very low-level and included all the features. Later, Miner and Carreño specified the IEEE-754 and 854 standards in PVS and HOL [CAR 95a, MIN 95, CAR 95b]; the definitions were more generic and not as low-level.

Due to the high cost of an incorrect design, processor makers also turned to formal proofs. A nice survey of hardware formal verification techniques, and not just about FP arithmetic, has been done by Kern and Greenstreet [KER 99]. Regarding FP arithmetic, AMD researchers, particularly Russinoff, have proved in ACL2 many FP operators mainly for the AMD-K5 and AMD-K7 chipsmakeatletter [[MOO 98], [RUS 98], [RUS 99], [RUS 00]], and the square root of the IBM Power4 chip [SAW 02]. Specific quotient and square root algorithms have been verified in ACL2 with a focus on obtaining small and correct digit-selection tables [RUS 13]. This last work also shows that there are several mistakes in a previous work by Kornerup [KOR 05]. As for Intel, the Pentium Pro operations have been verified [OLE 99]. The Pentium 4 divider has been verified in the Forte verification environment [KAI 03]. A full proof of the exponential function, from the mathematical function to the circuit was achieved using HOL4 [AKB 10]. A development by Jacobi in PVS is composed of both a low-level specification of the standard and proofs about the correctness of FP units [JAC 02]. This has been implemented in the VAMP processor [JAC 05]. Finally, during his PhD, Harrison verified a full implementation of the exponential function [HAR 00]. He therefore developed a rather high-level formalization of FP arithmetic in HOL Light with the full range of FP values [HAR 99a]. Among others, he has developed algorithms for computing transcendental functions that take into account the hardware, namely IA-64 [HAR 99b].

As far as Coq is concerned, there have previously been three different Coq formalizations of FP arithmetic, designed by the authors of this book. The PFF library was initiated by Daumas, Rideau, and Théry [DAU 01]. Its main characteristic is the fact that roundings are axiomatically formalized. A rounding here is a relation between a real number and an FP number (and the radix, and the FP format). This library contains many proved results (theorems and algorithms) [BOL 04a] but it lacks automation and there is no computational contents. The Gappalib library, on the other hand, is designed to make it possible for Coq to verify the FP properties proved by the Gappa tool [DAU 10] and thus provides computable operations with rounding but only for radix 2. Finally, the CoqInterval library is designed for automation based on intensive radix-2 FP computations [MAR 16].

This book mostly revolves around Flocq [BOL 11b], a Coq formal library described in Chapter 3. It serves as the lower layer of both CoqInterval and Gappalib and supersedes PFF for proving algorithms as it offers genericity, both in terms of radix and formats. In particular, the exploration of properties of exotic formats was impossible in either previous formalizations. The theorem names given in this book refer to the corresponding Coq names in Flocq. The auxiliary lemmas are not named: only the lemmas dedicated to be used out of Flocq are given. Most of the applications and examples of this book, from decision procedures (Chapter 4) to compilers (Chapter 7) are based on Coq and Flocq. Note that this book is more than just formal proofs of theorems representing FP properties or algorithms. We also explain how to verify programs in Chapter 8.

The outline of this book is as follows. Chapter 1 presents some basic notions of FP arithmetic. Two aspects are covered: the first relates it to real arithmetic through formats and rounded computations, while the second anchors FP arithmetic to the

IEEE-754 standard and its handling of exceptional values such as infinities and signed zeros.

Chapter 2 then gives an overview of the Coq proof assistant: what the logic formalism is, how the user can perform proof steps, and which libraries that are useful for formalizing FP arithmetic are shipped by default with Coq.

After these two introductory chapters on FP arithmetic and the Coq system, Chapter 3 describes the core of the Flocq library, a Coq formalization of FP arithmetic. It starts by defining what FP formats are and how to merge them into a unified framework. It then gives an abstract description of rounding operators. So that FP computations can be performed within Coq or extracted as reference libraries, this chapter also gives a more computational definition of rounding operators. It finally extends this formalism so as to cover part of the IEEE-754 format.

Chapters 1 and 2, as well as sections 3.1 and 3.2 provide the prerequisites needed for the following chapters, which can mostly be read in no particular order.

Formal proofs of FP algorithms are often a long and tedious process, so any tool that might automate part of them is welcome. Chapter 4 presents a few tools dedicated to proving statements on real or FP arithmetic. Some are tactics from the standard distribution of Coq while others are shipped as external libraries: CoqInterval and Gappa.

While FP operations usually approximate real operations, in some situations the rounded values actually represent the infinitely-precise values. There are some other situations where, while the computations introduce some rounding errors, there is a way to compute some correcting terms so as to recover all the information. Chapter 5 details some of these situations.

At this point, all of the basic blocks have been laid out and one can tackle the verification of larger FP algorithms. Chapter 6 presents a few examples whose correctness was formally proved. These examples give an overview of the kind of specifications one could verify for an FP algorithm: bounds on round-off errors, exactness or robustness of computations, and so on.

Formally verifying some properties about an FP algorithm is no small work but for this work to be meaningful, the verified properties should still hold when the algorithm is eventually executed. Several things might get in the way, the compiler being one of them. This might be due to actual compiler bugs or simply because the programmers and compiler writers have different expectations regarding the compilation of FP code. To alleviate this issue, the CompCert C compiler comes with a precise semantics for FP operations and with a Coq formal proof that it actually implements this semantics. Chapter 7 also gives some details on how the support for FP arithmetic was added to this compiler.

Chapter 8 extends Chapter 6 toward the verification of actual programs. Indeed, the verification of an FP algorithm might have been performed under the assumption that its execution is safe, e.g. operations have not overflowed, arrays have not been accessed outside their bounds. This chapter covers these other aspects of program verification through several examples. It also gives some tools for verifying programs independently of the way some FP operations might be reassociated.

Finally, Chapter 9 tackles an example whose verification involves some complicated mathematical reasoning. This example is a small solver for the one-dimensional wave equation; its specification states how close the computed FP results are from the real solution to the partial differential equation. One of the components of the correctness proof is the Coquelicot formalization of real analysis.

Floating-Point Arithmetic

Roughly speaking, floating-point (FP) arithmetic is the way numerical quantities are handled by the computer. Many different programs rely on FP computations such as control software, weather forecasts, and hybrid systems (embedded systems mixing continuous and discrete behaviors). FP arithmetic corresponds to scientific notation with a limited number of digits for the integer significand. On modern processors, it is specified by the IEEE-754 standard [IEE 08] which defines formats, attributes and roundings, exceptional values, and exception handling. FP arithmetic lacks several basic properties of real arithmetic; for example, addition is not associative. FP arithmetic is therefore often considered as strange and unintuitive. This chapter presents some basic knowledge about FP arithmetic, including numbers and their encoding, and operations and rounding. Further readings about FP arithmetic include [STE 74, GOL 91, MUL 10].

This chapter is organized as follows: section 1.1 provides a high-level view of FP numbers as a discrete set, and a high-level view of rounding and FP operations. Section 1.2 shows some simple results on the round-off error of a computation. Section 1.3 presents a low-level view of FP numbers, which includes exceptional values such as NaNs or infinities and how operations behave on them. Section 1.4 gives additional definitions and results useful in the rest of the book: faithful rounding, double rounding, and rounding to odd.

1.1. FP numbers as real numbers

We begin with a high-level view of FP numbers, which will be refined in section 1.3. A definition of FP formats and numbers is given in section 1.1.1. An important quantity is the unit in the last place (ulp) of an FP number defined in section 1.1.2. A definition of rounding and the standard rounding modes are given in section 1.1.3. Some FP operations and what they should compute are given in section 1.1.4.

1.1.1. *FP formats*

Let us consider an integer β called the radix. It is usually either 2 or 10. In a simplified model, an FP number is just a real value $m\beta^e$ satisfying some constraints on m and e given below. The integers m and e are called the *integer significand* and the *exponent*. The usual FP formats only consider FP numbers $m\beta^e$ with $|m| < \beta^\varrho$, with ϱ being called the *precision*. Let us also bound the exponent: $e_{\min} \leq e \leq e_{\max}$. An FP number is therefore a real value representable in the considered FP format:

$$m\beta^e \quad \text{such that} \quad |m| < \beta^\varrho \wedge e_{\min} \leq e \leq e_{\max}.$$

With this definition of FP numbers, there might be several representations of the same real value. For instance, in radix 2 and precision 3, the real number 1 is representable by any of $(1_2, 0)$, $(10_2, -1)$, and $(100_2, -2)$. The set of representations of a given real value is called a *cohort*. We define a *canonical* representation as follows: it is the FP number with minimal exponent (or equivalently maximal integer significand). In our previous example, it is $(100_2, -2)$.

The IEEE-754 standard adopts a slightly different but equivalent point of view with m being a fixed-point number [IEE 08, section 3.3]. The FP number is represented as $d_0 \bullet d_1 d_2 \ldots d_{\varrho-1} \beta^{e_s}$. The value $d_0 \bullet d_1 d_2 \ldots d_{\varrho-1}$ is called the *mantissa*. The exponent is shifted compared to our exponent with the integer significand: $e_s = e + \varrho - 1$. The minimal and maximal exponents are also shifted accordingly. Our exponent e also corresponds to the q notation of the IEEE-754 for decimal formats. We will stick with the integer significand throughout this book.

It is common to split FP finite numbers into two classes: normal and subnormal numbers. Subnormal numbers can be represented as $m\beta^{e_{\min}}$ with $|m| < \beta^{\varrho-1}$. Equivalently, subnormal numbers are FP numbers smaller than $\beta^{\varrho-1+e_{\min}}$. Normal numbers can be represented as $m\beta^e$ with $\beta^{\varrho-1} \leq |m| < \beta^\varrho$ and $e_{\min} \leq e \leq e_{\max}$. Normal numbers are FP numbers greater than or equal to $\beta^{\varrho-1+e_{\min}}$. The numbers in an FP format are therefore a discrete finite set that may be represented on the real axis as done in Figure 1.1. As FP numbers are just a subset of real numbers in this section, there is only one zero seen as a subnormal number. Section 1.3 will later give a notion of sign for zero.

The maximal FP number is $\Omega = (\beta^\varrho - 1)\beta^{e_{\max}}$. When the canonical FP number $m\beta^e$ considered is such that $e > e_{\max}$, we have an *overflow* and may get a special value. This is ignored until section 1.3.

When the value considered is smaller than the minimal *normal* FP number (that is $\beta^{\varrho-1+e_{\min}}$), we have an *underflow* and may get a subnormal number. Some formats may ignore subnormal numbers; this is called *flush-to-zero* or *abrupt underflow*. Subnormal numbers are flushed to the value zero when they are either inputs or outputs of FP operations. One consequence of abrupt underflow is that $x = y$ is no longer equivalent to $\circ(x - y) = 0$ (with $\circ$ being rounding to nearest, see section 1.1.3).

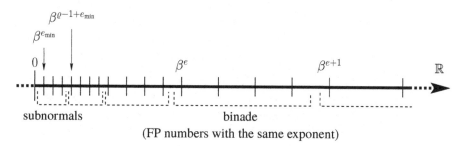

subnormals   binade
(FP numbers with the same exponent)

Figure 1.1. *Distribution of FP numbers over the real axis.*

The IEEE-754 standard defines several FP formats. The most used one in this book is *binary64* which corresponds to radix 2 with $\varrho = 53$. Two other binary formats are *binary32* with $\varrho = 24$ and *binary128* with $\varrho = 113$. The standard also defines some radix-10 formats, namely *decimal64* with $\varrho = 16$ and *decimal128* with $\varrho = 34$. More details can be found in section 3.1.2.1.

1.1.2. *Unit in the last place*

As seen in Figure 1.1, when the exponent increases, the FP numbers are further apart. For instance, there is no FP number between $\beta^{\varrho-1}$ and $\beta^{\varrho-1} + 1$; there is no FP number between β^{ϱ} and $\beta^{\varrho} + \beta$. For a positive FP number x, the δ_x such that there is no FP number between x and the FP number $x + \delta_x$ is called *ulp*, for "unit in the last place". More precisely, if $x = m\beta^e$ is the canonical representation of an FP number, then we have

$$\text{ulp}(x) = \beta^e.$$

This definition only covers FP numbers, but the ulp may be defined for any real number, given an FP format (see section 3.1.3.2 for a formal definition). An equivalent definition for this chapter is that $\text{ulp}(x)$ is the ulp of the floating-point number bracketing x with the smallest magnitude (that is to say the rounding to zero of x, see section 1.1.3).

The successor x^+ of an FP number x is the smallest FP number larger than x. For a positive x, it is $x + \text{ulp}(x)$, except in case of overflow. The predecessor x^- can be defined accordingly as the largest FP number smaller than x. These values are represented in Figure 1.2.

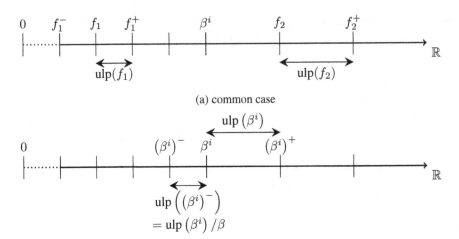

(a) common case

(b) when the FP number is a power of the radix

Figure 1.2. *Unit in the last place (ulp), successor, and predecessor.*

For a positive FP number x, the successor of x is always $x + \text{ulp}(x)$, but the predecessor of x may or may not be $x - \text{ulp}(x)$. The most common case is given in Figure 1.2a, where the predecessor of x is $x - \text{ulp}(x)$. But when x is equal to β^i with $i \geq e_{\min} + \varrho$, then its predecessor lies in the previous binade and therefore has a smaller ulp. This is the case of Figure 1.2b: $x = \beta^i$ and $\text{ulp}(x) = \beta^{i-\varrho+1}$, but the predecessor of x is $x^- = \beta^i - \beta^{i-\varrho}$ and $\text{ulp}(x^-) = \beta^{i-\varrho} = \text{ulp}(x)/\beta$. This specific case also happens when considering the successor of a negative power of the radix. Note that for all positive x, we have $x^- + \text{ulp}(x^-) = x$.

More properties (and the associated Coq definitions) can be found in section 3.1.4.

1.1.3. *FP rounding*

Even if some FP computations may be exact (in particular in Chapter 5), most are not. When a real result is between two FP numbers, one of the bracketing FP numbers should be returned instead. The choice of the FP results depends on the *rounding mode*. A generic unknown rounding mode is denoted by $\square$. Note that, for any rounding mode, when x is an FP number, then $\square(x) = x$.

Several rounding modes exist. The IEEE-754 standard defines five of them [IEE 08, section 4.3]; some of them are depicted in Figure 1.3. The rounding of x to $+\infty$, denoted by $\triangle(x)$, is the smallest FP number greater or equal to x. The rounding of x to $-\infty$, denoted by $\triangledown(x)$, is the largest FP number smaller or equal to x. The rounding of x to 0 is its rounding to $-\infty$ when x is nonnegative and its rounding to $+\infty$ otherwise. These three rounding modes are the standard directed rounding modes.

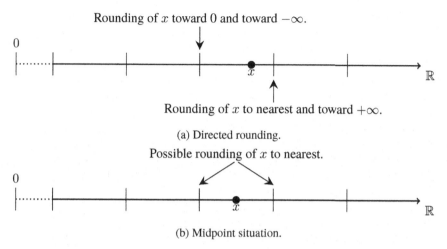

(a) Directed rounding.

(b) Midpoint situation.

Figure 1.3. *FP rounding.*

Another rounding mode is *to nearest*, meaning the result is expected to be the FP number nearest to x. When the real to be rounded is exactly the midpoint of two consecutive FP numbers as in Figure 1.3b, there is a choice to be made. This choice is called a tie-breaking rule and is denoted by τ in this book. The rounding of x to nearest is denoted by $\circ^\tau(x)$. The IEEE-754 standard defines two tie-breaking rules: tie breaking to even and tie breaking away from zero. For binary formats, the default rounding mode of processors is usually rounding to nearest, tie breaking to even, denoted by $\circ$. When there is a tie, the chosen FP number is the one with the even integer significand. The other standard tie-breaking rule is tie breaking away from zero: when there is a tie, the chosen FP number is the one with the larger magnitude.

When various formats are used, the rounding notation will indicate the precision of the destination format, for instance $\circ_{53}(x)$ for the rounding of x to nearest, tie breaking to even in a format whose precision is 53, typically *binary64*. The minimal exponent will not be indicated as it can be deduced from the context.

Another notation used throughout this book is the following one. For an arbitrary arithmetic expression $\mathcal{E}$, we denote by $\circ[\mathcal{E}]$ the FP expression where all the operations are FP operations (rounded to nearest, tie-breaking to even) and all the literals are rounded if needed. For instance, let a, b, c, and d be FP numbers, $\circ[a \times b + (c + d)]$ means $\circ(\circ(a \times b) + \circ(c + d))$ and $\circ[0.1 \times a + b]$ means $\circ(\circ(\circ(0.1) \times a) + b)$. We have similar notations for other rounding modes, for instance $\square[\cdots]$ or $\circ^\tau[\cdots]$.

1.1.4. *FP operations*

We have seen in the preceding section how to round a real number. Let us see how FP operations are defined. The ideal case is called *correct rounding*: the FP result is the rounding of the mathematical result. Correct rounding means that the operation is performed as if it was an exact operation followed by a rounding. In practice, the mathematical value does not have to be computed with full accuracy, which would be impossible for many values, such as $\sqrt{2}$ for instance.

Correct rounding brings several benefits. The first one is portability: as there is only one possible result for each operation, the result does not depend on the processor or the programming language or the operating system. This assertion should be mitigated and we refer the reader to sections 1.4.2 and 8.4 and to Chapter 7. The second benefit is accuracy: the error between the exact value and the computed value is usually guaranteed to be small relative to these values (see section 1.2). The third benefit is preservation of monotony: as rounding is increasing, the correct rounding of a monotone function is also monotone. For instance, a correctly-rounded exponential is increasing.

In practice, correct rounding is achieved for basic operations: addition, subtraction, multiplication, division, and square root. Hardware implementations usually rely on three additional bits called *guard*, *rounding*, and *sticky* bits for that purpose [ERC 04]. Correct rounding is also achieved for the *fused-multiply-add* (FMA) operator that computes $\Box(\pm a \times b \pm c)$, that is without a rounding after the product. Correct rounding is also mandatory for the *remainder* operator that computes $x - n \cdot y$ for two FP numbers x and y and for an integer n near x/y (see section 5.3). The correct rounding of these operations is mandated by the IEEE-754 standard [IEE 08, section 4.3].

All supported signed and unsigned integers may be converted into any FP format, with possibly a rounding for large values. All FP numbers may be converted into any integer format, with possibly a rounding and with an undefined result in case of overflow. Conversions to and from decimal character sequences, with possibly a rounding, are also required by the IEEE-754 standard. All these conversions are required to be correctly rounded.

For other functions, e.g. exp, sin, and pow, correct rounding is only recommended by the standard. The reason is that correct rounding is much harder to ensure due to the *table maker's dilemma* [GOL 91, MUL 16]: the mathematical result may be very near the midpoint between two FP numbers for rounding to nearest and very near an FP number for directed rounding. In these cases, one needs a very accurate intermediate computation to be able to decide which FP number is the closest. Consider this example taken from [MUL 16]. For some specific *binary64* FP number $x \approx 2^{678}$, we have

$$\log(x) = \overbrace{111010110.0100011110011110101 \cdots 110001}^{53 \text{ bits}}$$

$$\underbrace{00000000000000000 \cdots 000000000000000}_{65 \text{ zeros}} 1110 \cdots$$

This has been proved to be a worst case for directed rounding for computing the logarithm in *binary64*, meaning an FP number whose logarithm is the nearest to an FP number. Such worst cases tell which accuracy is sufficient to correctly round the result of an elementary function and help make correctly-rounded libraries more efficient [LEF 01, MUL 16, DIN 07].

1.2. Error analysis

As explained above, correct rounding is the best we might expect from an FP computation: it is the most accurate possible result. Unfortunately, it is not usually exact and we may want to bound the round-off error. There are several kinds of error that are described and bounded in this section. For a real number x and its approximation $\tilde{x}$, the *absolute error* is $\tilde{x} - x$ and the *relative error* is $\frac{\tilde{x}-x}{x}$ (provided that $x \neq 0$).

1.2.1. *Absolute and relative errors of a single operation*

Let us consider a single rounding $\square$. For a given real x, its absolute error is the value $\square(x) - x$ and its relative error is $\frac{\square(x)-x}{x}$ (provided that $x \neq 0$).

Let us begin with several bounds on the absolute error. First, $\bigtriangledown(x) \leq x \leq \bigtriangleup(x)$. Second, the rounding of x is the rounding either to $+\infty$ or to $-\infty$ of x, that is either $\square(x) = \bigtriangledown(x)$ or $\square(x) = \bigtriangleup(x)$, therefore $|\square(x) - x| \leq \bigtriangleup(x) - \bigtriangledown(x)$. When x is an FP number, $x = \square(x) = \bigtriangledown(x) = \bigtriangleup(x)$. When x is not an FP number, the distance between $\bigtriangledown(x)$ and $\bigtriangleup(x)$ is $\mathrm{ulp}(x)$. In any case, we deduce (see lemma 3.37 of section 3.2.2.2) that

$$|\square(x) - x| < \mathrm{ulp}(x).$$

Similarly (see lemma 3.39), when rounding to nearest (whatever the tie-breaking rule), we have

$$|\circ^{\tau}(x) - x| \leq \mathrm{ulp}(x)/2.$$

In an FP format with precision ϱ and minimal exponent $e_{\min}$, we may get rid of the ulp function in order to have formulas depending only on $|x|$ and on the format:

$$|\square(x) - x| < \max\left(|x| \cdot \beta^{1-\varrho}, \beta^{e_{\min}}\right);$$

$$|\circ^\tau(x) - x| \leq \max\left(|x| \cdot \beta^{1-\varrho}/2, \beta^{e_{\min}}/2\right).$$

Note that $\beta^{1-\varrho}/2$ is quite small, for instance it is equal to 2^{-53} in *binary64*. The max may be turned into an addition for the sake of simplicity. Note also that these bounds may be slightly improved [JEA 17].

The second argument of the max is due to subnormal numbers. It may be removed when underflow is known not to occur, which then gives the following error bound:

$$\left|\frac{\circ^\tau(x) - x}{x}\right| \leq \beta^{1-\varrho}/2.$$

For instance, let f_1 and f_2 be FP numbers with $f_1 + f_2 \neq 0$, then

$$\left|\frac{\circ^\tau(f_1 + f_2) - (f_1 + f_2)}{f_1 + f_2}\right| \leq \beta^{1-\varrho}/2.$$

This always holds because, when the result of an addition is subnormal, it is exact (see section 5.1.2.2). Another case is the square root. Let f be a positive FP number, then

$$\left|\frac{\circ^\tau\left(\sqrt{f}\right) - \sqrt{f}}{\sqrt{f}}\right| \leq \beta^{1-\varrho}/2.$$

This always holds as the result of a square root cannot be a subnormal number.

1.2.2. *Direct and inverse errors*

More often than not, numerical programs use more than one FP computation, therefore the errors may accumulate. The way they accumulate heavily depends on the operations at hand.

If we consider the FP addition, then the absolute error on the output is the sum of the absolute errors on the inputs plus the absolute rounding error of the addition. If we consider the FP multiplication, then the relative error of the output is the sum of the relative errors on the inputs plus the relative rounding error of the multiplication (we must ensure that there is no underflow for this error to be small). For the multiplication, this rule is only a first-order approximation of the errors.

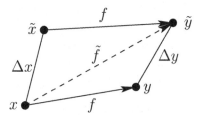

Figure 1.4. *Direct error Δy and inverse error Δx.*

All the previously bounded errors are called direct errors (or sometimes forward errors). They bound the difference between a computed value $\tilde{y}$ and the value y that would have been obtained without rounding. When y is the result of a function f applied to an input x, we have another error called the inverse error (or backward error). It is the value Δx such that $f(x + \Delta x) = \tilde{y}$. Said otherwise, it is the perturbation on the input that would have produced $\tilde{y}$ if there were no rounding, as explained in Figure 1.4. This is linked to the condition number [HIG 02].

1.3. Exceptional values

Viewing FP numbers as real numbers is sometimes not enough. For instance, overflow and division by zero may happen. To avoid triggering traps and to allow recovery in subsequent operations, the IEEE-754 standard defines several exceptional values corresponding to results of exceptional computations [IEE 08, section 3.3].

The full set of FP data is therefore:

– normal FP numbers: $m\beta^e$ with $\beta^{\varrho-1} \leq |m| < \beta^\varrho$ and $e_{\min} \leq e \leq e_{\max}$,

– subnormal nonzero FP numbers: $m\beta^{e_{\min}}$ with $0 < |m| < \beta^{\varrho-1}$,

– infinities: two values $+\infty$ and $-\infty$,

– zeros: two values $+0$ and -0,

– NaNs (for Not-a-Number).

An FP number is a finite or infinite number that is representable in an FP format, while an FP datum is either an FP number or a NaN.

The NaNs are obtained when there is no reasonable number that may be the result of the operation, for instance when computing $\sqrt{-1}$, $+\infty + (-\infty)$, or $0/0$. There are two kinds of NaNs: signaling and quiet.

It is always possible to compare two FP data and all the comparison results are fully specified in the IEEE-754 standard. For non-exceptional values, it is the comparison of the corresponding real values. FP infinities behave as mathematical infinities, for instance $1 < +\infty$. Zeros are seen as equal for FP comparisons: $+0 = -0$. The strangest fact about FP comparisons is when it operates on one (or several) NaN. In this case, the comparison always returns *false*. In particular, $0 < $ NaN is *false* whatever the NaN and NaN ≤ 0 is *false* too. A way to test whether an FP number x is a NaN is to test whether $x = x$. This test is *false* if and only if x is a NaN.

This section is organized as follows: section 1.3.1 presents the IEEE-754 levels, that is to say several abstraction levels useful to consider FP data. Section 1.3.2 presents the binary encoding of the FP data. Section 1.3.3 presents how the FP operations behave on all the values.

1.3.1. *IEEE-754 levels*

The IEEE-754 standard offers four specification levels describing how to represent an FP number [IEE 08, section 3.2]:

1) an extended real (either a real number or a signed infinity);

2) either a rational number or a signed zero or a signed infinity or a NaN;

3) either a triple (sign, significand, and exponent) or a signed infinity or a quiet NaN or a signaling NaN;

4) a string of bits.

At level 1 (the most abstract level for IEEE-754), numbers are just regarded as real numbers or infinities. Note that level 1 is independent of the FP format and allows numbers that might not be representable in a given format.

At level 2, some constraints appear. Only a subset of real numbers can be represented by FP numbers, according to the FP format (see section 1.1.1). Some new exceptional values appear too: zero gets a sign and there is now a Not-a-Number for representing the result of operations such as $\infty - \infty$.

At level 3, the representable numbers are now regarded as triples. A single rational number might be represented by several triples, if the format allows some redundancy. This is the case with decimal formats, as the number 1 might be represented by the cohort $(+, 1, 0)$, $(+, 10, -1)$, $(+, 100, -2)$, and so on. Another change is that there are now two NaNs, a quiet one and a signaling one.

Finally, level 4 tells how these triples are encoded as bit strings (the standard does not say anything about endianness though). There is again some redundancy, as an exceptional value might be encoded in several different ways. For instance, NaNs carry a payload, while the zeros of decimal formats have several representations.

So, from level 4 to level 2, implementation details are progressively erased and similar values are conflated. From level 2 to level 1, FP numbers are embedded in the set of extended real numbers.

1.3.2. Binary encoding

Let us now present in more details about level 4: i.e. FP data as strings of bits. The way they are encoded is very precise. Before going into the gory details of the exceptional values, let us consider a normal FP number in the *binary32* format. In order to give it a meaning, we split it as follows:

$$\boxed{1}\quad \boxed{11000110}\quad \boxed{10010011110000111000000}$$
$$s\qquad\quad E\qquad\qquad\qquad f$$

The most significant bit is the sign s. The next 8 bits are the biased exponent E (seen as an integer) and the remaining 23 bits are the fraction f. The values 8 and 23 are set by the *binary32* format. The value of this normal FP number is

$$\boxed{1}\qquad\qquad \boxed{11000110}\qquad\qquad \boxed{10010011110000111000000}$$
$$s\qquad\qquad\quad E\qquad\qquad\qquad\qquad f$$
$$\downarrow\qquad\qquad\qquad\downarrow\qquad\qquad\qquad\qquad\downarrow$$
$$(-1)^s\quad \times\quad 2^{E-B}\quad \times\qquad\quad 1 \bullet f$$

$$(-1)^1\quad \times\quad 2^{198-127}\quad \times\quad 1.10010011110000111000000_2$$

$$-2^{54} \times 206727 \approx -3.7 \times 10^{21}$$

with B being the bias, also set by the *binary32* format.

This is the simplest case. Let us detail what happens for all the possible FP values. Let us consider a radix-2 FP format with w bits for the exponent and $\varrho - 1$ for the fraction. Let the bias be $B = 2^{w-1} - 1$ and let $E_{\min} = 2 - 2^{w-1}$. The value of the FP number with the string of bits (s, E, f) is then fully defined:

– If $1 \le E \le 2^w - 2$, the FP number is normal and its value is $(-1)^s \times 2^{E-B} \times 1 \bullet f$ (as in the example above).

– If $E = 0$ and $f \ne 0$, the FP number is nonzero subnormal and its value is $(-1)^s \times 2^{E_{\min}} \times 0 \bullet f$.

– If $E = 0$ and $f = 0$, the FP number is a zero and its value is $(-1)^s \cdot 0$.

– If $E = 2^w - 1$ and $f = 0$, the FP number is an infinity and its value is $(-1)^s \cdot \infty$.

– If $E = 2^w - 1$ and $f \neq 0$, the FP number is a NaN. The fraction of a NaN is called the payload. It distinguishes between quiet and signaling NaNs and may be used for debugging purpose as seen fit by the processor maker.

For radix 10, the encoding is rather similar, only more complex. We refer the interested reader to the IEEE-754 standard [IEE 08, section 3.5] for the details.

The ordering (s, E, f) of the bit string allows for an easy comparison. If you ignore exceptional values and deal with the sign, then the comparison of FP values is the same as the comparison of their binary encoding (seen as integers).

1.3.3. *Rounding and exceptional values*

We have seen what the exceptional values are and how to encode them. We have left to see how FP operations behave on them. We have several sub-cases depending on the kind and values of the inputs:

– When the inputs are finite, we compute an ideal result as in section 1.1.4 (as the exact operation followed by the rounding). If the result is representable with an exponent smaller or equal to $e_{\max}$, it is returned. If the result is too large, then either the largest FP number or an infinity is returned. For instance, consider the largest FP number Ω, then the rounding to zero of $2 \times \Omega$ is Ω and its rounding to nearest is $+\infty$.

– The previous sub-case is not specified enough when the result is zero. For instance computing $1 - 1$, the result is $+0$ in all rounding modes except in rounding to $-\infty$ where it is -0.

– Except for invalid operations, when an input is infinite, the value is the limiting case of real arithmetic with an operand of arbitrarily large magnitude. When no such limit exists, a NaN is returned. For instance, $2 - (+\infty)$ is $-\infty$, $\sqrt{+\infty}$ is $+\infty$, and $+\infty - (+\infty)$ is NaN.

– When an input is a NaN, it is propagated. When two inputs are NaNs, then one of them is propagated, meaning the payload of the result is that of one of the inputs. The choice of which payload is not specified.

– The case of zero inputs is the more complicated, and therefore not given exhaustively here. An example is that $\sqrt{-0}$ is -0. For multiplications, $0 \times x$ is a zero for a finite x, its sign being the exclusive OR of the operand signs. Also, $0 \times \infty$ produces a NaN, $x/0$ produces an infinity for a finite nonzero x, and $0/0$ produces a NaN.

There are some more invalid operations that return a NaN. See the standard for an exhaustive list [IEE 08, section 7.2].

This inventory seems arbitrary, but it has nice properties. For instance, provided that neither the inputs nor the output are a NaN (they may be zeros or infinities), the sign of a product or quotient is the exclusive OR of the signs of the operands. Another example is that $x + x$ and $x - (-x)$ are equal and they have the same sign as x, even when x is a zero.

Another part of the standard is about exceptions [IEE 08, section 7] such as invalid, division by zero, or inexact. This is out of the scope of this book.

1.4. Additional definitions and properties

After this general introduction about FP arithmetic, we need some additional definitions or results to be used later in this book. Section 1.4.1 is about faithful rounding, which is a way to characterize a computation less accurate than the correct rounding defined above. Section 1.4.2 is about double rounding: due to some hardware specificities, some computations may be done first in an extended format, possibly followed by a second rounding toward the working format. If the extended format is not precise enough, we may get a less accurate result than what is obtained with a direct rounding toward the working format. Finally, section 1.4.3 gives a way to ensure that double rounding is innocuous. Using a specific rounding for the intermediate computation leads to a final correct rounding (as if it was done directly in the working format) as soon as the extended format has a slightly larger precision than the working format.

1.4.1. *Faithful rounding*

Correct rounding has been described in section 1.1.4: the FP operation is performed as if it was an exact operation followed by a rounding. Unfortunately, it is not always possible or practical to get such a correct rounding for all operations. A weaker rounding is called *faithful rounding*. The faithful rounding of a real number x is either its rounding to $+\infty$ or its rounding to $-\infty$ as shown in Figure 1.5.

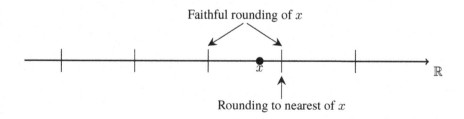

Figure 1.5. *Faithful rounding and rounding to nearest.*

Faithful rounding does not ensure monotony or portability. It nevertheless ensures a good accuracy (as good as a directed rounding mode). It will be used in section 6.3.

1.4.2. *Double rounding*

The IEEE-754 standard defines FP formats, including *binary64*. But x86-compatible processors provide an extended format with 15 bits for the exponent and 64 bits for the integer significand (not the fraction as in section 1.3.2 since the first bit is explicit). We denote this 80-bit format by *binary80*.

This seems a great idea as hardware operations and registers support this extended format, therefore bringing more accuracy. The main drawback is that double rounding may occur: the FP operation is first performed in extended precision, then the value is stored in memory as a *binary64* FP number.

In some cases as in Figure 1.6a, this is innocuous as the result is the same as if a direct rounding in the *binary64* format did occur. In other cases as in Figure 1.6b, double rounding (first in extended, then in *binary64*) may yield a result different from a direct rounding.

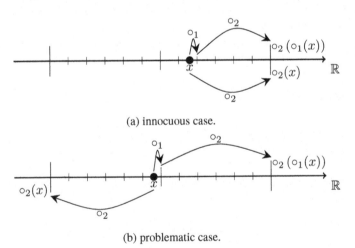

(a) innocuous case.

(b) problematic case.

Figure 1.6. *Double rounding. The small ticks are the FP numbers in the extended format. The large ticks are the FP numbers in the working format. The medium tick is the midpoint between two FP numbers in the working format (it also belongs to the extended format).*

Even in the problematic case of Figure 1.6b, the accuracy is not so bad. In particular, double rounding provides a result that is a faithful rounding of the mathematical operation. But the error bound is even better than directed rounding.

Let us denote by $\circ_{53}$ the rounding to nearest in the *binary64* format and $\circ_{64}$ in the *binary80* format.

Theorem 1.1. For a real number x, let f be either $\circ_{53}(x)$ or $\circ_{64}(x)$ or the double rounding $\circ_{53}(\circ_{64}(x))$. Then, we have the following implications.

$$\text{If } |x| \geq 2^{-1022}, \text{ then } \left| \frac{f - x}{x} \right| \leq 2050 \times 2^{-64}.$$

$$\text{If } |x| \leq 2^{-1022}, \text{ then } |f - x| \leq 2049 \times 2^{-1086}.$$

These bounds are a little larger than the *binary64* bounds of section 1.2.1 (2050 and 2049 instead of 2048). The various subcases to be investigated are sketched in Figure 1.7: the various cases for f and intervals for x are successively studied and the corresponding errors are bounded using the formulas of section 1.2.1.

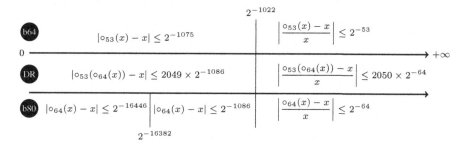

Figure 1.7. *Bound on the rounding error of x in* binary64 *(b64),* binary80 *(b80), and in double rounding (DR) depending on the value of x. The latter corresponds to theorem 1.1.*

This theorem will be used in section 8.4.1 when verifying the correctness of programs whatever the compilation choices.

There is another case of double rounding: instead of directly computing an operation to *binary32*, it is first computed in *binary64* and the result is then rounded to *binary32*. Surprisingly, this final result is correctly rounded as this double rounding is innocuous [FIG 95, ROU 14].

1.4.3. *Rounding to odd*

As seen above, when the intermediate precision is not large enough, double rounding is not innocuous. It can be made innocuous by introducing a new rounding mode, called *rounding to odd*, and using it for the first rounding. This was used by

Goldberg when converting binary FP numbers to decimal representations [GOL 91] and formally studied later, notably to emulate the FMA operator [BOL 08].

The informal definition of this rounding is the following one: when a real number is not representable, it is rounded to the adjacent FP number with an odd integer significand. Assume an even radix β and two different formats: a working format and an extended format, both FP formats with gradual underflow. Assume that the extended format has at least two more digits, both for the precision and the minimal exponent. If $\square_{ext}^{odd}$ denotes the rounding to odd in the extended format, and if $\circ^\tau$ denotes a rounding to nearest with tie-breaking rule τ in the working format (see theorem 3.47 of section 3.2.2.3), we have

$$\forall x \in \mathbb{R}, \quad \circ^\tau \left(\square_{ext}^{odd}(x) \right) = \circ^\tau(x).$$

This property will be used in section 7.3 to convert between integers and FP numbers.

2

The Coq System

The Coq software is a proof assistant, that is, its primary purpose is to let users state theorems, to help them write the proofs of these theorems, and to finally check that these proofs are correct with respect to some logical rules. This chapter gives only the most basic concepts about Coq. It is not meant to replace the reference manual [COQ 86] or some actual books on the topic [BER 04, CHL 13, PIE 17].

Section 2.1 describes the logical formalism of Coq: how to define types and functions, and how to state theorems. Formal proofs of these theorems can be built by parts and section 2.2 presents some basic tools, named *tactics*, for expressing basic proof steps (more advanced tactics will be presented in Chapter 4). Finally, section 2.3 presents some parts of the standard library shipped with Coq; of particular interest are the theories on integer and real arithmetics.

2.1. A brief overview of the system

Let us describe briefly the logic that underlies Coq, as it impacts the way theorems are stated and proved. It is based on the correspondence between function types and logical propositions, which is described in section 2.1.1. These function types are slightly more expressive than in most mainstream programming languages, since Coq supports dependently-typed programming (section 2.1.2). Moreover, types themselves can be built using functions (section 2.1.3). One of the features of Coq is the ubiquitous use of inductive types, which are described in section 2.1.4. Since Coq is both a programming language and a proof assistant, it makes it possible to turn a formal development into an actual program, through the process of extraction (section 2.1.5). In fact, functions do not even have to be extracted to be executed, they can also be evaluated within Coq and thus can even serve the purpose of proving theorems (section 2.1.6).

2.1.1. *Propositions as types*

At its core, Coq is based on *type theory*. Such a theory associates with each *term* a *type*, and operations on terms are permitted only if they satisfy some typing rules. For example, when x is of type A, (f x) (which means applying f to x) is of type B if and only if f is of type $A \rightarrow B$. In this case, the term (f x) is said to be well typed. Coq ultimately rejects any term that is not well typed, as would a compiler for a statically-typed programming language. This notion of typing extends to theorem proofs: a Coq proof is correct if and only if it is a well-typed term. This term is then seen as the proof of a mathematical proposition given by the type of the term. In other words, a user wanting to formally prove a proposition T has to exhibit a term for which Coq can check the type is T. Note that the actual proof term does not matter much here; as long as the user shows that T is an *inhabited* type, proposition T holds.

For example, an implication $A \Rightarrow B$ can be understood as representing the type of functions building a proof of B from a proof of A. Therefore, if the user finds a proper way of transforming any proof of A into a proof of B, then A implies B. Thus the -> symbol interchangeably represents both the function type and the implication in Coq syntax. Here are two simple examples. First, the identity function of type $A \rightarrow A$ is a type-theoretic proof of the implication $A \Rightarrow A$ (for an arbitrary proposition A), since it takes a proof of A as argument and returns it. Second, a proof of $A \Rightarrow (B \Rightarrow A)$ is a function that takes a proof of A as argument, stores it, and returns a function that takes a proof of B as argument, ignores it, and returns the previously stored proof of A. Both lemmas and their proofs could be stated as follows in Coq:

```
Lemma L1 : A -> A.
Proof (fun (a:A) => a).
```

```
Lemma L2 : A -> (B -> A).
Proof (fun (a:A) => (fun (_:B) => a)).
```

The (fun (a:A) => e) construct expresses an anonymous function that takes an argument of type A, binds it to the name a, evaluates an expression e in which a might occur, and returns the result of this evaluation. Mathematically, it could be denoted by $a \in A \mapsto e$. When a name is _, the argument passed to the function is just ignored.

Note that the arrow/implication is right-associative (so parentheses are not needed around $B \Rightarrow A$ in the statement of L2), the functions are curried (so several arguments can follow the fun keyword in the proof of L2), and Coq tries to infer types (so the types A and B do not have to be explicitly written in the proof of L2). Thus the second lemma could also be stated and proved as follows.

```
Lemma L2 : A -> B -> A.
Proof (fun a _ => a).
```

Writing proofs as functions is fine when they are as tiny as in these two examples, but it soon becomes cumbersome for larger proofs. So Coq supports an interactive way of devising such proofs that will be presented in section 2.2. For now, we will keep talking about functions.

Note that the proposition $A \Rightarrow B \Rightarrow A$ above is logically equivalent to $A \wedge B \Rightarrow A$ where $\wedge$ denotes the conjunction. The former proposition using only implications is favored in Coq as it is better suited to the way one can write proof terms. Similarly, rather than defining functions of type $(A \times B) \to C$ taking a pair argument, users tend to define functions of type $A \to B \to C$ instead.

2.1.2. *Dependent types*

An important point about functions in Coq is that their types can be much more intricate than in most other programming languages. Indeed, Coq allows for *dependent types*: the return type of a function might depend on the values of its arguments. So for a function $a \in A \mapsto e$ with e of type B, the argument a can actually occur in the type B. In Coq, the type of such a function is denoted (forall a:A, B). As the syntax suggests, a function of this type is a proof of the proposition $\forall a \in A, B$. For instance, a proof of $\forall n \in \mathbb{N}, n + 0 = n$ is a function that takes a natural number n as argument and returns a proof of the equality between $n + 0$ and n.

Dependent types are useful not only for expressing the types of the proofs of universally quantified theorems, but also for plain functions. For instance, in section 3.3.1.3, a function Fmult will be defined; it takes two pairs of integers (m_1, e_1) and (m_2, e_2), and it returns a pair (m_3, e_3) such that $m_3 \cdot \beta^{e_3} = (m_1 \cdot \beta^{e_1}) \times (m_2 \cdot \beta^{e_2})$, for some β passed as the first argument. The result of this function is meaningful only if both inputs are using the same radix β. Moreover, using the returned value with a different radix is meaningless and thus an error. We can use dependent types to prevent such an incorrect usage. Rather than having Fmult take pairs of integers of types Z * Z, we can define it as taking arguments of type (float beta) with float a family of types (whose values are just pairs of integers) indexed by radices. This gives the following type for Fmult:

```
Fmult : forall beta:radix,
        float beta -> float beta -> float beta.
```

When given a term (Fmult radix10 x (Fmult radix2 y z)), Coq complains that (Fmult radix2 y z) is of type (float radix2) while it is expecting an argument of type (float radix10) as the last argument of the outer Fmult.

To conclude on dependent types, note that the function type (A -> B) is just syntactic sugar for the dependent type (forall a:A, B) when a does not occur in B. Similarly, the implication $A \Rightarrow B$ is no different from $\forall a \in A, B$ when a does not occur in B as a *free variable*.

2.1.3. *Propositions and types, predicates and sets*

It should be understood that types are not different from the other terms in Coq. In particular, they can be the result of evaluating a function on some arguments. For instance, the type Z * Z, which is syntactic sugar for (prod Z Z), is the result of applying the function prod to the two arguments Z and Z. This prod function is thus of type (Type -> Type -> Type), where Type denotes the type of all the types in Coq. Similarly, the float family of types mentioned above is just a function of type (radix -> Type). Proposition types are of type Prop, a subset of Type. For instance, the theorem statement $\forall n \in \mathbb{N}, n + 0 = 0$ is of type Prop (and of type Type too).

Finally, let us focus on the kind of functions of type A -> Prop, for some type A. They are useful for denoting some subsets of A. Indeed, for any subset $P \subseteq A$, we can associate a function (P : A -> Prop) such that $x \in P$ holds if and only if (P x) holds. For instance, the following definition characterizes the subset of positive real numbers:

Definition positive_real (x : R) : Prop := 0 < x.

In the following chapters, we will not always make explicit the distinction between types and subsets. For instance, in Chapter 3, a format F is defined as a subset of real numbers, hence as a function of type (R -> Prop). Thus whenever a mathematical proposition states $\forall x \in F, Q$, it should be understood as the Coq type (forall x:R, F x -> Q).

Assume now that we want to declare the natural logarithm function on positive real numbers. Its type could be as follows:

ln: forall x:R, positive_real x -> R.

This is a binary function: the first argument is a real number while the second one is a proof that the input is positive. For example, ln 2 would be (ln 2 Rlt_0_2) with Rlt_0_2 a proof that 2 is a positive real number. In practice, using such partial functions is cumbersome since one has to devise a proof that the argument is in the domain before writing the application. So the standard library of Coq has chosen to define the logarithm as a total function of type (R -> R) with a default result for non-positive inputs (see section 2.3.2).

2.1.4. *Inductive types*

There are two kinds of types the user can define in Coq: function types, described in sections 2.1.1 and 2.1.2, and inductive types. The latter are a generalization of the *algebraic datatypes* proposed by programming languages such as OCaml or Haskell. Note that the various inductive types presented below hardly deserve to be qualified as inductive, since their definitions are not even *recursive*. An example of an inductive type with a recursive definition is the *list* datatype, since a list can be defined as being either empty or the pair of an element (the head of the list) and a list (its tail). Similarly, a natural number is either zero or the successor of a natural number. Such recursive types will not be detailed in this chapter, since the formalizations presented in this book make little use of them. In particular, the Z type of integers is defined on top of a recursive inductive type (akin to a list of bits) but the proofs will not care about this formalization peculiarity.

One of the simplest kinds of inductive type is the *record* type. The float family above is such a type and it is defined as follows (see also section 3.1.1).

```
Record float (beta : radix) : Type :=
  Float { Fnum : Z ; Fexp : Z }.
```

This command declares that a value of type (float beta) (for some radix beta) is a record with two fields, both of them of type Z. Moreover, this command defines three functions: Float, Fnum, and Fexp. The Float function is a *constructor*, that is, its purpose is to produce a value whose type is in the float family. Note that, if the name Float (or any other identifier) had not been indicated by the user, Coq would have named the constructor Build_float by default. The constructor takes a radix and two integers and it returns a record whose fields are initialized with those integers. It has the following type:

```
Float : forall beta:radix, Z -> Z -> float beta.
```

The two functions Fnum and Fexp make it possible to access the field values of a float record. They have the same type:

```
Fnum, Fexp : forall beta:radix, float beta -> Z.
```

For instance, the following Coq term extracts the exponent part from a record of type (float radix2) representing $12 \cdot 2^{23}$; it evaluates to 23:

```
Fexp radix2 (Float radix2 12 23).
```

Note that, contrarily to the float family, a record type will usually have some fields whose types depend on the parameters (here beta). In fact, field types can depend not only on the family parameters but also on the value of the earlier fields in a record, in a way similar to dependent types. The radix type is an example that presents such a dependency, as the type of its second field depends on the value of the first field. Here is a simplified definition of this type (the actual definition is given in section 3.1.1).

```
Record radix := {
  radix_val : Z ;
  radix_prop : 2 <= radix_val
}.
```

This leads to the following type for the accessor to the second field:

```
radix_prop : forall r:radix, 2 <= radix_val r.
```

Notice how the type of radix_prop is actually a theorem statement (in Prop). It means that, for any value r of type radix, the value v of its first field is such that the logical proposition $2 \leq v$ holds. Mathematically, this means that a value of type radix is an integer larger or equal to 2. This also implies that, to create a value of type radix, the user has to pass a proof that the integer is larger than or equal to 2 to the Build_radix constructor.

Note that this way of defining a subset of the integers plays a role similar to the definition of positive real numbers in section 2.1.3. Which style of definition to choose should be evaluated on a case-by-case basis. For instance, since a radix is usually set once and for all, it is a bit more convenient to pack both the integer and its associated proof into a single value. That way, instead of having theorem statements that start with (forall beta:Z, 2 <= beta -> ...), they just start with (forall beta:radix, ...).

Inductive types are not just (dependent) records. They can present several alternatives. The simplest kind of such types is an enumeration. For instance, the comparison type is defined as follows in the standard library of Coq.

```
Inductive comparison : Type := Eq | Lt | Gt.
```

This definition causes Coq to create a type comparison, as well as three constant constructors Eq, Lt, and Gt, of this type. This definition also implies that these three constants are different and that any value of type comparison is necessarily equal to one of these three constants. This type is used as the return type of some comparison functions such as Zcompare, which takes two real integers as arguments and returns how these two numbers are ordered (equal, less than, or greater than).

More complicated inductive types combine features from enumerations and record types; the various cases are not just constant but they can carry data. An example is the type used to represent an IEEE-754 format in section 3.4.1. This type should tell that an FP number can be either a signed zero or a signed infinity or some other case. This leads to the following definition binary_float for binary formats.

```
Inductive binary_float (prec : Z) (emax : Z) : Type :=
  | B754_zero : bool -> binary_float prec emax
  | B754_infinity (sign : bool)
  | ... (* NaN and finite cases *)
```

This definition causes Coq to create several functions. First, it declares the binary_float function, which takes two integers denoting the precision and the maximal exponent of a format and returns the type used to represent the values of this format. Second, Coq declares several constructors for generating values of type (binary_float prec emax), one per case of the inductive type. Both B754_zero and B754_infinity have the same type (presented with two different syntaxes); these functions take two integers (the precision and the maximal exponent) and a Boolean value (the sign), and they return the value describing the corresponding signed zero or infinity (depending on the function):

```
B754_zero, B754_infinity : forall (prec emax : Z),
  bool -> binary_float prec emax.
```

When an inductive type presents several cases, Coq does not generate accessor functions, since those only make sense when there is a single case (as in records). To find which case a value falls in and what the associated data are, one has to use *pattern matching*. For instance, if x is a value of type (binary_float ...), the following Coq term executes the branch corresponding to the case of the value. Note that pattern matching can be nested, so the example term branches not only on x but also on its associated Boolean data in the case of a signed zero.

```
match x with
| B754_zero true => ... (* evaluated when x is -0 *)
| B754_zero false => ... (* evaluated when x is +0 *)
| B754_infinity b => ... (* evaluated when x is infinite,
                            with b made to hold its sign *)
| ... (* the other cases *)
end.
```

Note that the Record command is just some syntactic sugar for defining an inductive type along some accessor functions. For instance, the float type family and its Fnum accessor could have been defined as follows. (The other accessor is similar.)

```
Inductive float (beta : radix) : Type :=
  | Float : Z -> Z -> float beta.
```

```
Definition Fnum (beta : radix) (x : float beta) : Z :=
  match x with Float m e => m end.
```

2.1.5. Extraction

The Fmult function mentioned in section 2.1.2 can be defined as follows.

```
Definition Fmult (beta : radix)
                 (x y : float beta) : float beta :=
  Float beta (Fnum x * Fnum y) (Fexp x + Fexp y).
```

The correctness of this definition can be guaranteed by proving the following simple lemma, assuming F2R is a function that takes a value of type (float beta) and returns the real number it represents.

```
Lemma F2R_Fmult :
  forall (beta : radix) (x y : float beta),
  F2R (Fmult x y) = (F2R x) * (F2R y).
Proof ...
```

So one can first define functions using Coq, then state which specification they should comply with, and finally prove that they actually do. But it is possible to go further due to the *extraction* mechanism [LET 04]. Coq is indeed able to turn a formal development into an OCaml or Haskell program. In the case of OCaml, the function corresponding to Fmult is extracted as follows. Notice that, since OCaml does not support dependent types, they have been erased from the generated code.

```
(** val fmult : radix -> float -> float -> float **)
let fmult _ x y =
  { fnum = (Z.mul x.fnum y.fnum);
    fexp = (Z.add x.fexp y.fexp) }
```

Since Fmult has a concrete definition and so do all the functions its body depends on, the resulting OCaml function can be executed and it behaves the same way as the original Coq function. Note that Coq does not extract functions whose type lies in Prop, so F2R_mult is not part of the generated program. Nonetheless, since the original Fmult function was proved correct, the generated function is expected to be correct too. For instance, this extraction mechanism is used to turn the CompCert formal development into an actual C compiler [LER 09] (see section 7.2).

2.1.6. Computation

The Fmult function above can be executed once extracted to an external language but it can also be evaluated within Coq. This evaluation mostly amounts to unfolding definitions, binding values to variables, and performing pattern matching; these are *reduction* steps. Two terms that reduce to the same value are said to be *convertible*. An important point is that the logic of Coq does not distinguish between convertible terms, so rather than proving a given statement, the user can choose to prove any other statement it can be converted to.

To illustrate this mechanism, assume that one wants to prove (4 < 5 + 2)%Z, where %Z means that the enclosed literals and operators should work on integers (for example, rather than real numbers). Section 2.3.1 will give a bit more detail on the definition of these operators and literals but for now what matters is that they have concrete definitions. In particular, (5 + 2)%Z reduces to 7%Z and (4 < 7)%Z reduces to (Lt = Lt). As a consequence, since reflexivity of equality is a suitable proof of (Lt = Lt), it is also a suitable proof of (4 < 5 + 2)%Z.

Tactics of Chapter 4 make heavy use of reduction as a fast and lightweight mechanism to automate proofs. Indeed, most of them turn the statement to be proved into an equality $e = true$ with e an expression involving only concrete definitions. Thus, if e reduces to *true*, the transformed statement has a trivial proof. This is the basis for performing proofs by *reflection* [BOU 97].

2.2. Tactics

In the formalism of Coq, proving a theorem is the same as exhibiting some well-typed term whose type matches the statement of the theorem. As we have seen, the user can directly exhibit terms for the simplest theorems. For the more complicated ones, this approach is no longer practical. So Coq proposes a way to interactively build terms by using *tactics*. As far as this book is concerned, whether terms are being built underneath when proving a theorem is of no consequences. The only thing that matters is that user proofs have been formally verified by the system.

At any point of the proof process, the user is presented with several goals which represent the missing pieces of a proof. A goal is split into a *context* containing several variables and hypotheses, and a conclusion that has to be proved given these variables and hypotheses (from the perspective of building functions, these variables and hypotheses are the arguments of a function and the conclusion is the return type of the function). Initially, there is only one goal, which represents the theorem to be proved. Successfully applying a tactic replaces the current goal by some new goals (possibly none). Once there are no goals left, the theorem has been successfully proved.

In Coq, the proof process is mostly *backward reasoning*. Let us illustrate this expression with an example. Assume that we have a hypothesis H of type A in the context and that we want to prove that the conclusion B holds. We can denote this goal by $H : A \vdash B$. Let us also suppose that we have proved some lemma L of type

$A \Rightarrow B$. There are two ways to proceed. We can apply the lemma either to the hypothesis H (section 2.2.5) or to the conclusion (section 2.2.2). If we apply it to the hypothesis, we now have to prove the following goal: $H : A, H' : B \vdash B$. This is called *forward reasoning* and it corresponds to the way most pen-and-paper proofs are written. If we apply the lemma to the conclusion, we obtain instead the following goal: $H : A \vdash A$. This is called backward reasoning. Note that, in both cases, the conclusion of the goal now matches one of the hypotheses, so we can conclude and get rid of this goal (section 2.2.1). Both kinds of reasoning are possible in Coq, but proof scripts tend to be more verbose when done forward, due to the way standard tactics are designed.

Let us present some of the tactics commonly encountered when performing proofs. Sections 2.2.1–2.2.4 present tactics for acting on the conclusion of the current goal (backward reasoning). Sections 2.2.5 and 2.2.6 present tactics for acting on the hypotheses of the current goal (forward reasoning). All these tactics are provided by the default distribution of Coq, but one can also use tactics provided by external packages (section 2.2.7). In this chapter, we focus on the most basic tactics. Chapter 4 will later present some tactics for automatically proving more complicated goals that might occur when proving statements specific to computer arithmetic.

2.2.1. *Discharging a goal*

As mentioned above, a proof is complete once there are no goals left. In order to remove a goal, its conclusion has to be immediately deducible from its hypotheses. Telling the tool how to perform this deduction is called *discharging* the goal.

The simplest tactic for discharging a goal is the easy tactic. It will successfully discharge the current goal, for example, when its conclusion is convertible to a hypothesis, or when two hypotheses are trivially contradictory (one is the negation of another), or when the goal is convertible to an equality between syntactically equal terms, or when a hypothesis is convertible to an equality between two terms that are different.

For example, as mentioned in section 2.1.6, the proposition (4 < 5 + 2)%Z reduces to the equality (Lt = Lt). The easy tactic would thus succeed in discharging a goal whose conclusion is (4 < 5 + 2)%Z. The tactic can also discharge any goal containing a hypothesis of type (4 * 2 < 3 + 5)%Z. Indeed, the latter reduces to (Eq = Lt), which is contradictory because by definition of inductive types, these two constructors of the comparison type are distinct.

2.2.2. *Applying a theorem*

The apply tactic can be used to reason by sufficient conditions, that is to apply a previously proven theorem to the current conclusion. For instance, assume that there is a lemma L of type $\forall x,\; A_1 \Rightarrow A_2 \Rightarrow B$. Assume also that the current goal is a specialization of B, where some term t occurs in place of x. Then applying L will produce two new goals. The conclusion of the first one is A_1 where the occurrences of the variable x have been replaced by t. Similarly, the conclusion of the second one is A_2 where the occurrences of the variable x have been replaced by t.

Let us consider some simple examples. Lemma Rmult_le_compat_l states the compatibility of the multiplication with the ordering of real numbers:

$$\forall r, r_1, r_2 \in \mathbb{R}, \quad 0 \le r \Rightarrow r_1 \le r_2 \Rightarrow r \cdot r_1 \le r \cdot r_2.$$

If the conclusion of the current goal is

$$(x + 1) \cdot (y \cdot z) \le (x + 1) \cdot 2,$$

one can execute the following tactic: apply Rmult_le_compat_l.

Coq finds out that, by instantiating r with $x + 1$, r_1 with $y \cdot z$, and r_2 with 2, the lemma can be applied to the conclusion. So it replaces the current goal with two new goals. The conclusions of these goals are $0 \le x + 1$ and $y \cdot z \le 2$.

In the example, all the universally-quantified variables of the lemma appear in its conclusion, so Coq can find suitable values in the conclusion of the current goal (assuming the lemma can be applied). This is not always the case, since some variables might occur only in the hypotheses of the lemma. For instance, Rle_trans states the transitivity of the ordering of real numbers:

$$\forall r_1, r_2, r_3 \in \mathbb{R}, \quad r_1 \le r_2 \Rightarrow r_2 \le r_3 \Rightarrow r_1 \le r_3.$$

When applying this lemma, Coq can find suitable values for instantiating r_1 and r_3 from the conclusion of the goal, but r_2 is not constrained in any way. So the user has to manually provide the value of r_2. For instance, if the conclusion of the goal is $0 \le x+1$ and if it is possible to prove that x is nonnegative, then one could execute the following tactic: apply Rle_trans with x.

Coq would then produce two goals with the following conclusions: $0 \le x$ and $x \le x+1$. It might happen that some hypotheses of the lemma to be applied are already present in the context. In this case, the user can indicate the name of the hypothesis to the apply tactic. So, if the goal were to contain a hypothesis x_nonneg of type $0 \le x$, the user could have used the following tactic instead:

apply Rle_trans with (1 := x_nonneg).

This time, Coq generates only one new goal of conclusion $x \leq x + 1$, since the other one was discharged using x_nonneg. The integer 1 on the left of := means that x_nonneg has to be used as the first hypothesis of the lemma. Note that the user no longer has to specify that the value of r_2 is x, since Coq can now infer this term from the type of x_nonneg.

2.2.3. *Dealing with equalities*

We have seen how Rmult_le_compat_l can be applied to the following conclusion:

$$(x + 1) \cdot (y \cdot z) \leq (x + 1) \cdot 2. \qquad [2.1]$$

But, if the conclusion had instead been

$$(1 + x) \cdot (y \cdot z) \leq (x + 1) \cdot 2,$$

Coq would have complained that it cannot find a value for r, since it would have to be both $1 + x$ and $x + 1$. So, before applying the lemma, we have to transform $1 + x$ into $x + 1$. The replace tactic can be used for that purpose:

replace (1 + x) with (x + 1).

In addition to the current goal where $1 + x$ has been replaced by $x + 1$, there is now a goal with $x + 1 = 1 + x$ as a conclusion. This second goal can be discharged by executing apply Rplus_comm, since the latter lemma states the commutativity of addition on real numbers:

$$\forall r_1, r_2 \in \mathbb{R}, \ r_1 + r_2 = r_2 + r_1.$$

The rewrite tactic makes it possible to perform both replace and apply steps at once: rewrite Rplus_comm. Note that $1 + x$ and $x + 1$ are not mentioned any longer. Indeed, when rewriting using Rplus_comm, Coq notices that the operation on the left-hand side of the equality is an addition applied to two universally quantified variables, so it looks for an addition in the conclusion of the current goal. The first occurrence it finds is $1 + x$, so the tactic instantiates r_1 with 1 and r_2 with x. Then, it replaces all the occurrences of $1 + x$ in the conclusion by $x + 1$. Note that some variants of the replace and rewrite tactics make it possible to replace only some selected occurrences of a term.

2.2.4. *Handling logical connectives and quantifiers in the conclusion*

The most primitive construct of Coq's logic is the universal quantifier. Instead of proving a goal whose conclusion is $\forall x \in T, \ P$, it is equivalent to prove a goal with a new variable x in the context and with P as a conclusion. This transformation is performed by executing intros x. The argument of the intros tactic is the name given to the variable introduced in the context; it has to be a fresh name. For instance, if x had already been part of the context, the user could have typed intros y instead, assuming that y was not. Coq would have substituted all the free occurrences of x in the proposition P by y when setting it as the conclusion of the goal.

Implication is handled in a similar way. If the conclusion is $P \Rightarrow Q$, then the goal can be changed into a goal with P in the context and Q as a conclusion. The argument passed to the intros tactic serves to name the hypothesis whose type is P. If the conclusion contains several universal quantifiers and/or implications, they can all be introduced at once by passing several names to intros.

The logical connectives $\wedge$ and $\vee$ and the existential quantifier in the conclusion can be handled by using only the apply tactic, but some tactics avoid the need to remember the name of the lemmas. When the conclusion is a conjunction $P \wedge Q$, the split tactic creates two goals with P and Q as their respective conclusions, so that the user can prove each proposition separately. When the conclusion is a disjunction $P \vee Q$, the user can choose which of P or Q is meant to be proved. This is done by executing one of the left or right tactics. As for a conclusion $\exists x$, P, it can be discharged by proving that P holds when substituting all its free occurrences of x by some specific *witness*. The value of this witness is passed as an argument to the exists tactic.

Let us consider the proposition

$$\forall x \in \mathbb{R},\ 0 \leq x \Rightarrow \exists y \in \mathbb{R},\ 0 \leq y \wedge y \cdot y = x.$$

The following tactics would be a sensible way to start the Coq proof.

```
intros x H.    (* introduce x and (0 <= x) *)
exists (sqrt x). (* substitute y with (sqrt x) *)
split.         (* create two goals *)
```

At this point, there are two goals. They have the same context: a variable x of type $\mathbb{R}$ and a hypothesis H of type $0 \leq x$. The conclusions of the two goals are $0 \leq \sqrt{x}$ and $\sqrt{x} \cdot \sqrt{x} = x$, respectively.

2.2.5. *Handling logical connectives and quantifiers in the context*

When a hypothesis exhibits an existential quantifier or a logical connective $\wedge$ or $\vee$, the destruct tactic can be used to decompose this hypothesis into new ones. Assume that there is a hypothesis H that we want to decompose. The following variants of the tactic will deal with the various connectives:

```
destruct H as [K L]. (* H is a conjunction "P /\ Q" *)
destruct H as [y K]. (* H is "exists x, P" *)
destruct H as [K|K]. (* H is a disjunction "P \/ Q" *)
```

When H is a conjunction or an existential property, the current goal is replaced by a new goal with the same conclusion but with a context in which H is replaced by some new hypotheses. For the conjunction $P \wedge Q$, two new hypotheses of types P and Q are added, and the pattern [K L] causes them to be named K and L, respectively. For the existential property $\exists x \in T$, P, a variable of type T is added and it is named y according to the pattern [y K]. A hypothesis K is also added; its type is P except that all the free occurrences of x have been replaced by y.

When H is a disjunction $P \vee Q$, two new goals are created. As before, the conclusion is left unchanged. The context of the first goal now contains a hypothesis K of type P, while the second goal contains a hypothesis K of type Q (the names of the hypotheses could have been chosen so that they are different).

Note that the intros tactic supports the same bracketed patterns as destruct, so both tactics can be combined to shorten some proofs. For instance, the following two scripts have the same effect on the goal but the second one avoids the naming of intermediate objects.

```
(* script 1 *)
intros x H1 H2.
destruct H1 as [K L].
destruct H2 as [M|N].

(* script 2 *)
intros x [K L] [M|N].
```

2.2.6. *Forward reasoning*

Except for destruct, all the tactics presented so far act on the conclusion of the goal. Even destruct does not do much to the hypotheses since it only decomposes them into simpler hypotheses. That is why reasoning in Coq is mostly done in a backward way: conclusions are strengthened until they can be discharged.

There are some ways though to perform forward reasoning; one of them is the assert tactic. On a goal $H_1 : A_1, \ldots, H_n : A_n \vdash C$, executing assert P with P a proposition produces two goals. The first goal has the same set of hypotheses $H_1 : A_1, \ldots, H_n : A_n$, but its conclusion is now P. The second goal has the old conclusion C but its hypotheses have been extended to $H_1 : A_1, \ldots, H_n : A_n, H : P$. It is a generalization of the *modus ponens*: in order to prove $A_1 \wedge \cdots \wedge A_n \Rightarrow C$, one can instead prove both $A_1 \wedge \cdots \wedge A_n \Rightarrow P$ and $A_1 \wedge \cdots \wedge A_n \wedge P \Rightarrow C$.

The `assert` tactic can be used to prove once and for all a proposition that will be the conclusion of several goals later on. Defining an auxiliary lemma beforehand would have the same effect, but at the expense of a greater verbosity.

Another way to perform a step of forward reasoning is the `apply in` tactic. Given a lemma L of type $P \Rightarrow Q$ and a goal $H : P \vdash C$, executing `apply L in H` produces a goal $H : Q \vdash C$.

2.2.7. *The SSReflect tactic language*

The tactics presented above are only a small subset of what the Coq tactic language provides for proving theorems in an incremental way. Moreover, it is not the only set of tactics that can be used, since several Coq developments come with their own language meant to complement or to replace the vanilla language. One of the most popular ones is SSReflect [GON 16], which provides a smaller but slightly more expressive set of tactics than vanilla Coq. It has been used for parts of two developments presented in this book: CoqInterval (see section 4.2) and Coquelicot (see section 9.2).

2.3. Standard library

Coq does not just ship with a proof/type checker and a tactic language, it also contains a standard library that provides numerous definitions and lemmas, some of which have already been mentioned. For instance, Coq's standard library defines a `bool` inductive type (with two constructors `true` and `false`), which models Boolean values, as well as a few functions on it, such as `andb` which behaves like the Boolean conjunction. It also comes with a few datatypes, such as the `list` inductive type family of homogeneous lists. We will focus only on the parts of the standard library that have an immediate interest for formalizing FP arithmetic and verifying FP algorithms, that is, the theory of integer arithmetic (section 2.3.1) and the theory of real arithmetic (section 2.3.2).

Note that the parsing engine of Coq interprets literals and operators with respect to a *scope*. For example, in the scope of integers, `*` is interpreted as the integer product, while in the scope of types, it is interpreted as the Cartesian product of two types. The user can set an ambient scope which is used by default to interpret notations. For all the snippets of this book, the ambient scope will be that of real numbers. For instance, the following statement is a proposition on real numbers:

```
Goal forall x, 0 <= x -> 0 < x + 1.
```

Other scopes can be enabled explicitly by using the `(...)%...` syntax. For instance, the following statement now talks about integer constants and operators. In particular, it means that x is now of an integer (of type Z) since it is an argument of the integer comparison and addition.

Goal forall x, (0 <= x -> 0 < x + 1)%Z.

Finally, scopes might also be enabled implicitly for function arguments. For instance, if a function has an argument of type Z then literals and operators parsed in an application of this function will be parsed in the scope of Z. Thus, in the expression (Zlt_bool 0 (x + 1)) the constants and operators are interpreted in the scope of Z, since the Zlt_bool function takes two integer arguments.

2.3.1. *Integers*

The standard library of Coq provides a type Z for representing (unbounded) integers. It also provides the usual operations on this type: opposite (Zopp), absolute value (Zabs), addition (Zplus), subtraction (Zminus), and multiplication (Zult). When in the %Z scope, common notations can also be used: +, -, *. Some other available functions are Euclidean division, power function, square root, and so on. All these functions have concrete definitions and can thus be used for performing computations within Coq.

Some comparison relations are also provided: <= (Zle) and < (Zlt). There is no specific relation for the integer equality; the standard polymorphic equality = (eq) can be used instead. Zle and Zlt are built upon the Zcompare function, which takes two integers as arguments and returns a value of type comparison:

Definition Zle (x y : Z) : Prop :=
  Zcompare x y <> Gt.
Definition Zlt (x y : Z) : Prop :=
  Zcompare x y = Lt.

These relations having a return type Prop, they cannot be used by functions meant for computing. Instead, one should directly use the Zcompare function or one of its Boolean instances: Zle_bool, Zlt_bool, and Zeq_bool. As an illustration, one could define the absolute value on integers as follows (note that Zabs is defined in a slightly different way in the standard library).

Definition Zabs (x : Z) : Z :=
  if Zle_bool 0 x then x else (- x)%Z.

Since (Zle_bool 0 (-5)) reduces to false, the term (Zabs (-5)) reduces to (Zopp (-5)), which reduces to 5%Z.

Due to these arithmetic operators and comparison functions, integer computations can be performed within Coq. One should, however, not expect performances to be on par with programs relying on a library such as GMP [GRA 15b]. Indeed, the representation of the Z integers, the algorithms chosen when defining the integer functions, and the inherent cost of the reduction engine of Coq, make verified computations several orders of magnitude slower. To alleviate this issue, Coq provides a second standard library of integers [GRÉ 06]. Combined with a modified reduction engine [ARM 10], these BigZ integers make it possible to tackle much larger computations within Coq (though still much slower than GMP). This will be useful for the tactics of section 4.2.

2.3.2. Real numbers

The standard library of Coq also comes with a type R of real numbers and a formalization of arithmetic and analysis [MAY 01]. The naming of operators is similar to that of integers: Ropp, Rabs, Rplus, Rminus, Rmult, Rdiv, and Rinv (for the multiplicative inverse). As for the comparison relations, they are denoted Rle and Rlt. There is also a scope in which some of these operators can be denoted using the common symbols: +, -, *, /, <=, <. This is the ambient scope for the examples of this book.

There is a major difference with integer arithmetic though: R is an abstract type and basic arithmetic operators are axiomatized. As a result, terms on real numbers do not reduce to anything useful. For example, (5 + 2)%R cannot be reduced further. In particular, it is not convertible to 7%R. So, contrarily to the proposition (4 < 5 + 2)%Z, there is no trivial proof of (4 < 5 + 2)%R using computations. One should instead use the fact that the conversion of integers to real numbers is a morphism for addition and comparison to turn the statement into a proposition about integers which can then be discharged using computations. This peculiarity of R also means that real operators cannot be extracted to useful functions.

Note that it is possible to give a concrete definition of the real numbers and of some of the operators using Cauchy sequences, as was done in external libraries [CRU 04, SPI 11, KRE 13]. Unfortunately, since the equality of real numbers is undecidable, a function like Rle_bool, which compares two real numbers and returns a Boolean value, would still be out of reach of a definition suitable for performing computations. A different solution would be to consider only a subset of the real numbers on which equality is decidable, e.g. real algebraic numbers [COH 12]. But that would preclude the definition of functions such as the trigonometric ones, which are provided by the standard library.

As mentioned in section 2.1.3, some partial functions have been made total in Coq. For example, the multiplicative inverse of 0 is a real number by axiom but no theorem tells which one it is. Note, however, that (0 / 0)%R is provably equal to zero, since it expands to the product of 0 by some real number. Functions such as square root and logarithm, however, are explicitly defined as returning zero outside their definition domain.

Regarding analysis, the standard library provides some notions of limit, differentiability, and Riemann integrability for real-valued functions. Unfortunately, the corresponding operators are quite unwieldy in practice since they are partial functions. For example, differentiability of a function at a given point is provided by the following predicate:

```
derivable_pt : (R -> R) -> R -> Type.
```

Note that the return type is Type rather than Prop for historical reasons and, while it impacts the formalization, it can be considered to be Prop from a user point of view. The derivative then has the following dependent type:

```
derive_pt : forall (f : R -> R) (x : R),
            derivable_pt f x -> R.
```

In other words, this operator takes a function f, a real number x, and a proof that f is differentiable at point x, and it returns the value of the derivative of f at x. This has several consequences. First, any statement about the derivative of a function has to embed a previously proved lemma about the differentiability of that function. Second, rewriting an expression such as $(f + g)'(x)$ into $f'(x) + g'(x)$, while straightforward in mathematics, is tedious in Coq, as the proof terms have to be manually built by the user beforehand. Third, these proof terms can quickly blow up, and thus become next to impossible for the user to forge by hand, e.g. in the case of iterated derivatives or differentiation under an integral.

The external library Coquelicot, described in section 9.2, alleviates these issues by providing total functions for limits, series, power series, derivatives, integrals, and so on, while being a conservative extension of the standard library of Coq [BOL 15d].

3

Formalization of Formats
and Basic Operators

Chapter 1 has given an informal description of FP arithmetic and Chapter 2 has given a short overview of the Coq proof assistant. Let us now explain how this arithmetic can be formalized in Coq. More precisely, this chapter describes how it is formalized in the Flocq library [BOL 11b].[1] Such a formalization has two purposes. First, it should properly model the various arithmetics we are interested in. In particular, it should be a good match for any arithmetic compliant with the IEEE-754 standard [IEE 08], but it should also be able to cover more exotic formats. For instance, numerous lemmas can be applied just as well to fixed-point formats as to floating-point formats. Second, the formalization should be suitable for verifying theorems and algorithms relying on such an arithmetic.

Let us clarify what we mean by suitable. At the highest level, FP numbers are just a way to approximate real numbers, so regarding them as real numbers makes it simpler to reason about them. Unfortunately, this approach does not capture all their expressiveness (e.g. signed zeros) and some properties are not available at this level of abstraction. On the other side of the spectrum, at the hardware level, an FP number is nothing more than a string of bits. This is a faithful representation of FP numbers, but the relation to real numbers is so tenuous that it makes reasoning more cumbersome. As such, it is important to let the user choose an abstraction level that makes it simple to state and prove properties about FP algorithms, programs, and circuits.

To this end, the IEEE-754 standard provides several levels of abstraction to characterize the various ways to represent and manipulate FP numbers. The Flocq library follows a similar approach, though there will not be a one-to-one

---

1 http://flocq.gforge.inria.fr/

correspondence between the various ways to represent FP formats in Coq and the IEEE-754 levels. For instance, level 1 in the IEEE-754 standard includes infinities as possible values of FP numbers. Yet most proof assistants, including Coq, favor a pure theory of real numbers, if only because they form a field. So the Flocq library supports only real numbers as the most abstract representation of FP numbers. Infinities may be either taken into account, as done in section 3.4, or prevented at the program level, as done in section 8.1.

We will start with this abstract representation as, in most cases, verifying programs is made simpler by considering that an FP format such as *binary64* is just a set of real numbers. Section 3.1 explains how formats are just predicates of the Coq type R -> Prop, which must satisfy some additional properties. Section 3.1.2 describes the formalization of the formats commonly encountered. Section 3.1.3 then explains how all these formats can be expressed in a common setting that encompasses both fixed-point and floating-point arithmetics. Section 3.1.4 finally presents some facts that hold for most formats. Table 3.2 compares some FP formats of Flocq, with their definitions and main features.

Once the formats have been formalized, one can turn to basic arithmetic operations on FP numbers. The IEEE-754 standard describes these FP operations by relating them to the *infinitely-precise* operations, that is, the exact operations on real numbers. This approach is especially suited for verifying programs that regard FP numbers as an approximation of real numbers. In the IEEE-754 standard, the relation between exact and FP operations is described through the use of *rounding* operators. The same approach is followed here. Section 3.2 shows how rounding is defined and the kind of properties we can expect about it. Rounding operators can be seen either as relations between real numbers and their rounded values, as shown in section 3.2.1, but they can also be seen as functions taking real numbers as arguments and returning their rounded values, as shown in section 3.2.2.

While seeing FP numbers as a subset of real numbers is sufficient for most proofs, there might be situations where one wants to actually perform a computation inside a proof. Chapters 4 and 7 (and to a lesser extent section 5.5) show why this feature might be useful. A Coq formalization of FP arithmetic purely based on real numbers would make it hard to perform such computations. So section 3.3 shows how the Flocq library also provides some way to effectively compute with formalized FP numbers inside Coq. More precisely, section 3.3.1 shows how to implement rounding operators and prove they are correct. Once they are available, it is easy to obtain FP addition and FP multiplication. For division and square root, the situation is slightly more complicated and section 3.3.2 explains how to implement them. Note that most details of section 3.3 are not strictly needed for understanding the remainder of this book and can be skipped at first reading.

Finally, we can look at how an IEEE-754-compliant arithmetic is formalized in Flocq. Such an arithmetic will be most useful when specifying and proving programs that manipulate infinities and NaNs, and/or perform bit manipulations on the representation of FP numbers. The details of how IEEE-754 numbers are represented is given in section 3.4. In particular, section 3.4.1 shows how a Coq inductive type is used to model FP numbers, in a way reminiscent of level 3 of the IEEE-754 standard, while section 3.4.2 shows how to represent FP numbers as strings of bits, which is what level 4 of the standard is about. Then, section 3.4.3 shows how Flocq implements FP operators that support not only finite FP numbers but also infinities, signed zeros, and NaNs, so that one can recover an effective arithmetic compliant with the IEEE-754 standard. Again, section 3.4 is quite intricate and most of its details can be skipped at first reading.

3.1. FP formats and their properties

As mentioned above, while the IEEE-754 levels are meant to abstract away from the low-level details of FP formats, they cannot be reused as is in a formal system. At the highest level of abstraction (which corresponds to levels 1 and 2 of IEEE-754), we prefer to ignore infinities; we simply represent FP numbers by real numbers of R (see section 2.3.2) and FP formats by predicates of type R -> Prop. There is another important difference when it comes to formats. We assume them to be unbounded, that is, they contain arbitrarily large FP numbers.

For instance, the IEEE-754 *binary32* format can represent any real number of the form $m \cdot 2^e$ with m and e two integers such that $|m| < 2^{24}$ and $-149 \le e \le 104$. The formalism of Flocq lifts the restriction on the upper bound of the exponent. So the corresponding Coq format contains all the real numbers such that $|m| < 2^{24}$ and $-149 \le e$. That way, we do not have to introduce infinities yet, since any real number can be rounded up or down to a number of that format. This makes proofs simpler as overflow can be ignored at that level. Underflow, however, is taken into account.

First, section 3.1.1 describes how FP numbers are represented as pairs significand-exponent and how they are embedded into real numbers. Then, section 3.1.2 shows some predicates that characterize commonly encountered formats. For instance, the predicate FLT_format : R -> Prop is satisfied by all the real numbers that can be represented in a given IEEE-754 format. Remember that the abstract formats defined by such predicates are not bounded and thus also contain larger values (but not more precise ones). All these formats share some common properties, so the notion of format is being generalized in section 3.1.3. That way, all the common formats can be expressed using the generic_format predicate. Finally section 3.1.4 states some properties satisfied by these formats.

3.1.1. *Real numbers and FP numbers*

Formats will be represented as predicates over real numbers, but they mostly deal with the subset of rational numbers that can be represented as $m \cdot \beta^e$ with β a fixed radix and m and e two integers.

Let us start with the radix, which is a nonnegative integer β. In order to simplify theorem statements, we constrain its value, so as to avoid degenerate cases such as $\beta = 0$ or $\beta = 1$. So the radix type is a dependent record containing an integer radix_val and a proof that this integer is at least 2.

```
Record radix := {
  radix_val :> Z ;
  radix_prop : Zle_bool 2 radix_val = true
}.
```

While the usual radices are 2 and 10, the formalization does not preclude more exotic radices such as 3 [BRU 11]. Next comes the bpow function, which performs an exponentiation by a signed integer exponent and returns a real number, so that we get the real number β^e.

```
Definition bpow : radix -> Z -> R.
```

When defining generic formats in section 3.1.3, we will need a way to somehow perform the converse operation: given a real number x, the mag_β function tells us in which range $[\beta^{e-1}; \beta^e)$ lies $|x|$. In the following, we call this range the *slice* of x and e is its *magnitude*. The integer $e = \text{mag}_\beta(x)$ is given by

$$e = \lfloor \log |x| / \log \beta \rfloor + 1,$$

but the actual way it is obtained does not matter much. So the mag function is defined as taking a real x and returning a dependent record containing an integer e and a proof that $|x| \in [\beta^{e-1}; \beta^e)$ holds when $x \neq 0$.

```
Record mag_prop (beta : radix) (x : R) := {
  mag_val :> Z ;
  _ : x <> 0 -> bpow beta (mag_val - 1) <= Rabs x < bpow beta mag_val
}.
```

```
Definition mag (beta : radix) (x : R) :
  mag_prop beta x.
```

As explained, $\text{mag}(x)$ is the integer such that $|x| \in [\beta^{\text{mag}(x)-1}; \beta^{\text{mag}(x)})$. Another possibility would have been to choose the integer e such that $x \in [\beta^e; \beta^{e+1})$. It would have made some properties related to the IEEE-754 standard simpler to state, at the expense of making some proofs a bit more verbose, especially when it comes to formatting functions (see section 3.1.3).

Note that the mag function is partial, in the sense that the integer it returns is specified only when x is nonzero. Since the radix β is usually fixed, the integer returned by (mag beta x) will be denoted on mag(x) in mathematical formulas.

Both records radix and mag_prop define their first field using the :>Z coercion. It means that a value of type radix or mag_prop can be used in any place where Coq expects an integer; a projection is then implicitly inserted so that the first field is used in place of the record value. In particular, (mag beta x) can be handled as if it was an integer. It makes for lighter definitions and statements.

We can now define a type that represents numbers of the form $m \cdot \beta^e$. This is the purpose of the float type. It is a record type containing two fields: an integer significand and an exponent. It is parameterized by the radix of the format. This radix is used by the F2R function when converting an element of the float type to a real number.

```
Record float (beta : radix) :=
  Float { Fnum : Z ; Fexp : Z }.
```

```
Definition F2R {beta : radix} (f : float beta) :=
  Z2R (Fnum f) * bpow beta (Fexp f).
```

As its name implies, the Z2R function maps an integer to the corresponding real number. Note that, in Coq, the fields Fnum and Fexp also act as accessor functions of type (float beta -> Z). That is the way they will be used from now on.

3.1.2. *Main families of FP formats*

Now that we have a type to represent numbers of the form $m \cdot \beta^e$, we can move on to defining formats. The main ones are floating-point formats and fixed-point formats. The two most important families of FP formats are FLT, which models formats with gradual underflow (section 3.1.2.1), and FLX, which extends the range of normal numbers to arbitrary small values (section 3.1.2.2). In other words, an FLT format models an IEEE-754 format when one assumes that no overflow can occur, while an FLX format can be used when one also assumes that there is no underflow. Section 3.1.2.3 then presents the family of predicates for describing fixed-point formats. Finally section 3.1.2.4 shows some other predicates representing less common formats, such as those that support abrupt underflow.

3.1.2.1. *The FLT formats*

Let us first present the format family that best models IEEE-754 arithmetic: the FLT formats. They are parameterized by a radix, a minimal exponent, and a precision.

Variable beta : radix.
Variable emin prec : Z.

A real number is part of an FLT format defined by these parameters if

1) it can be represented by a record of type float beta;

2) the significand of that record is compatible with the precision;

3) the exponent of that record is compatible with the minimal exponent.

The corresponding predicate is named FLT_format. Its definition makes use of the %Z specifier to indicate that the enclosed comparison operators should be interpreted in the scope of integer arithmetic.

```
Inductive FLT_format (x : R) : Prop :=
  FLT_spec (f : float beta) :
    x = F2R f -> (Zabs (Fnum f) < Zpower beta prec)%Z ->
    (emin <= Fexp f)%Z -> FLT_format x.
```

The minimal exponent e_{min} is to be understood as the logarithm in radix β of the smallest positive subnormal number that fits into the format. Note that this differs slightly from the definition of the IEEE-754 standard, for which the minimal exponent corresponds to the smallest positive normal number. Also, the exponent (Fexp f), which we usually denote by e, corresponds to the q notation of the IEEE-754 standard. The parameters for the IEEE-754 formats are described in Table 3.1.

Format	β	ϱ	e_{min}
binary32	2	24	-149
binary64	2	53	-1074
binary128	2	113	-16494
decimal32	10	7	-101
decimal64	10	16	-398
decimal128	10	34	-6176

Table 3.1. *Parameters for the IEEE-754 formats.*

As explained above, there is no upper bound on the exponents, so arbitrarily large numbers can fit in these formats. As a result, if a property holds for an algorithm when proved using such a format, it does not mean that this property still holds in practice. For instance, consider two representable numbers $\beta^e \geq 1$ and x. Given an algorithm that first performs an FP multiplication of x and β^e, and then divides the product by β^e, we can prove that the final result is x using the formalization above. Yet, during the actual computation, the multiplication could overflow and the final result would no longer be equal to x.

Thus, some extra care has to be taken. Indeed, proved properties do not translate to the actual execution of an algorithm in general, but they do when no overflow occurs. This absence of overflow can be ensured in several ways. First, one could devise the algorithm in a way such that an alternate execution path is taken whenever an overflow occurs, as in section 6.5. Second, one could just prove that no overflow can occur given the preconditions on the inputs of the algorithm, as in section 8.1.

3.1.2.2. The FLX formats

The FLT family of formats does not put an upper bound on the exponents and thus simplifies the proofs at that level of formalization. Yet this family supports underflow, so it might still make some proofs a bit harder than needed. For instance, the relative error of a multiplication might not be bounded when underflows occur, which might complicate proofs based on error analysis.

As with overflows, such exceptional situations might have been proven impossible given the inputs of the algorithm, or the algorithm might have an alternate execution path to handle them. In both cases, it is worth using a simplified formalization of formats, so as to ease the verification effort. Thus, the FLX_format predicate is defined similarly to FLT_format, but without a lower bound on the exponents; as such, it is parameterized by some radix β and some precision ϱ only.

```
Inductive FLX_format (x : R) : Prop :=
  FLX_spec (f : float beta) :
    x = F2R f -> (Zabs (Fnum f) < Zpower beta prec)%Z ->
    FLX_format x.
```

Any real number that is part of an FLT format is also part of an FLX format with the same radix and the same precision. Also, any real number that is part of an FLX format is part of all the FLX formats with a larger precision.

3.1.2.3. The FIX formats

The third predicate represents the family of fixed-point formats. This time, there is no concept of precision, but the minimal exponent e_{min} is back. A real number x satisfies the predicate FIX_format if there is a number of type float beta equal to x with an exponent equal to e_{min}.

```
Inductive FIX_format (x : R) : Prop :=
  FIX_spec (f : float beta) :
    x = F2R f -> Fexp f = emin -> FIX_format x.
```

Any real number part of an FLT format is also part of a FIX format with the same radix and the same minimal exponent. Also, any real number part of a FIX format is part of all the FIX formats with a smaller minimal exponent.

3.1.2.4. *Some other formats*

There are a few other formats that might be useful in some specific cases. Let us first consider the FTZ_format predicate. It represents FP formats that have a minimal exponent but do not provide subnormal numbers. A nonzero real number x part of an FTZ format is represented by a normalized FP number f: the integer significand of f has exactly ϱ digits.

```
Inductive FTZ_format (x : R) : Prop :=
  FTZ_spec (f : float beta) :
    x = F2R f ->
    (x <> 0 ->
      (Zpower beta (prec - 1) <= Zabs (Fnum f)
                          < Zpower beta prec)%Z) ->
    (emin <= Fexp f)%Z -> FTZ_format x.
```

Any real number part of an FTZ format is also part of the FLT format with the same parameters β, ϱ, and e_{min}. The converse is true only for zero and for numbers in the range of normal numbers.

Another interesting family of formats is the FLXN_format variant of FLX_format. An FLXN format is parameterized by the same arguments β and ϱ as an FLX format and it describes the exact same subset of real numbers. The main difference lies in the constraints on the FP number f representing a real number x part of an FLXN format. If x is nonzero, then f is normalized: the integer significand of f has exactly ϱ digits.

```
Inductive FLXN_format (x : R) : Prop :=
  FLXN_spec (f : float beta) :
    x = F2R f ->
    (x <> 0 ->
      (Zpower beta (prec - 1) <= Zabs (Fnum f)
                          < Zpower beta prec)%Z) ->
    FLXN_format x.
```

This format is mostly meant as a proof helper. If a theorem assumes that some real number x is part of an FLX format, then by first changing this hypothesis into x being part of an FLXN format, one gets that x is representable as a normal number.

3.1.3. *Generic formats*

All the formats presented so far can be seen as specializations of a more generic family of formats. Such formats are specified by a radix β and an integer function *fexp* (denoted by φ in mathematical formulas).

Variable beta : radix.
Variable fexp : Z -> Z.

Section 3.1.3.1 shows how the generic_format predicate is built using β and φ. Section 3.1.3.2 gives a few basic properties of this predicate. Section 3.1.3.3 then states some requirements φ should satisfy in order for the resulting format to be more useful in practice. Finally section 3.1.3.4 shows how φ has to be defined to recover the formats from the previous section (FLT, FLX, *etc*).

3.1.3.1. *Definition*

First, let us define the *canonical exponent* of a real number. It is given by computing the magnitude of a number and then applying φ to the resulting exponent.

Definition cexp (x : R) : Z :=
  fexp (mag beta x).

If the real number x is part of the format specified by φ, then it should be representable as an FP number in radix β with an exponent equal to its canonical exponent. In other words, we have

$$\exists m \in \mathbb{Z},\ x = m \cdot \beta^{\mathrm{cexp}(x)}. \tag{3.1}$$

We call the integer m the *scaled mantissa* of x. This value is defined as follows:

Definition scaled_mantissa (x : R) : R :=
  x * bpow beta (- cexp x).

The scaled mantissa of an arbitrary real number x is not an integer in general. When it is, the number is part of the format defined by φ. Another way to state this property is that x is part of the format if it is equal to the FP number that has a significand equal to the truncated scaled mantissa of x and an exponent equal to the canonical exponent of x. That is how the generic_format predicate is defined.

Definition generic_format (x : R) :=
  x = F2R (Float beta (Ztrunc (scaled_mantissa x)) (cexp x)).

This definition of the format is equivalent to the one given by equation [3.1]. It has two benefits, though. First, it gives an explicit representation of x as a number of type `float beta`. Second, it can be used as a rewriting rule to recover this representation in a simple way.

Another equivalent way to express format membership would be to say that x is equal to an FP number f that is canonical according to the following predicate:

```
Definition canonical (f : float beta) :=
  Fexp f = cexp (F2R f).
```

The following two lemmas make it possible to go from one characterization to the other.

```
Lemma canonical_generic_format :
  forall x : R, generic_format x ->
  exists f : float beta, x = F2R f /\ canonical f.
```

```
Lemma generic_format_canonical :
  forall f : float beta, canonical f ->
  generic_format (F2R f).
```

3.1.3.2. Basic properties

Let us detail some basic properties of the formats characterized by the `generic_format` family of predicates. Let us suppose we have such a format F described by a function φ. First, this format contains zero and it is stable when taking the opposite. We can also characterize the powers of the radix it contains.

Lemma 3.1 (`generic_format_0`).
$0 \in F$.

Lemma 3.2 (`generic_format_opp`).
$\forall x \in \mathbb{R},\ x \in F \Rightarrow -x \in F$.

Lemma 3.3 (`generic_format_bpow` and `generic_format_bpow_inv'`).
$\forall e \in \mathbb{Z},\ \varphi(e+1) \leq e \Leftrightarrow \beta^e \in F$.

Proof. $\beta^e \in F$ means that there exists $m \in \mathbb{Z}$ such that $\beta^e = m \cdot \beta^{\varphi(e+1)}$ by equation [3.1]. So it means that $\beta^{e-\varphi(e+1)}$ is an integer, which happens if and only if $\varphi(e+1) \leq e$ holds. ∎

Later, once we have put some mild constraints on φ, we will present a variant of the previous lemma. Instead of having $\varphi(e+1) \leq e$ as a left-hand side of the equivalence, lemma 3.5 will be using $\varphi(e) \leq e$, which makes some proofs simpler.

In the meantime, let us have a closer look at the representable numbers of the interval $[\beta^{e-1}; \beta^e)$ for some integer e. If we have $e \leq \varphi(e)$, then no point of the interval is part of the format, as they cannot be multiples of $\beta^{\varphi(e)}$. So let us assume that $\varphi(e) < e$ holds. The representable numbers of the interval $[\beta^{e-1}; \beta^e)$ are then of the form $\beta^{e-1} + k \cdot \beta^{\varphi(e)}$ with k a nonnegative integer less than $(\beta - 1) \cdot \beta^{e-1-\varphi(e)}$. The factor $\beta^{\varphi(e)}$ is the *unit in the last place* for the numbers in the interval $[\beta^{e-1}; \beta^e)$ (see section 1.1.2).

Definition ulp (x : R) : R :=
    bpow beta (cexp x).

Note that the above definition of ulp has been simplified for readability. The actual definition is slightly more complicated so as to return a meaningful result even when x is zero. For example, $\mathrm{ulp}_{\mathrm{FLX}}(0) = 0$ and $\mathrm{ulp}_{\mathrm{FLT}}(0) = \mathrm{ulp}_{\mathrm{FIX}}(0) = \beta^{e_{\min}}$.

Finally, let us characterize a bit more how representable numbers can be expressed with respect to their canonical exponent.

Lemma 3.4 (generic_format_discrete). Let x be a real number and let e_x be its canonical exponent in the format.

$$\forall m \in \mathbb{Z}, \ m \cdot \beta^{e_x} < x < (m + 1) \cdot \beta^{e_x} \Rightarrow x \notin F.$$

Proof. If x was representable, there would be an integer m_x such that $x = m_x \cdot \beta^{e_x}$. Thus we would have $m \cdot \beta^{e_x} < m_x \cdot \beta^{e_x} < (m + 1) \cdot \beta^{e_x}$, which is impossible. ∎

3.1.3.3. *Requirements for having a useful exponent function*

Given a representable number $x = m \cdot \beta^{\varphi(e)} \in [\beta^{e-1}; \beta^e)$, if the number $x^+ = x + \mathrm{ulp}(x) = (m + 1) \cdot \beta^{\varphi(e)}$ is part of the format too, then lemma 3.4 tells us that x^+ is the *successor* of x. Having x^+ be the successor of x seems like a sensible property, but there is unfortunately no reason for x^+ to be representable. So this section exhibits some constraints on φ that ensure that it is, as well as a few other properties. These constraints are embedded into the valid_exp predicate, which we now construct.

First, let us see when x^+ might not be the successor of x. From $\varphi(e) < e$ and $m \cdot \beta^{\varphi(e)} < \beta^e$, we deduce $(m + 1) \cdot \beta^{\varphi(e)} \le \beta^e$, since m and $\beta^{e-\varphi(e)}$ are integers. When $x^+ = (m+1) \cdot \beta^{\varphi(e)}$ is strictly less than β^e, it is already written in canonical form so x^+ is representable. Thus, x^+ cannot be the successor of x only if $x^+ = \beta^e$. So far, nothing prevents one from building a function φ such that β^{e-1} is part of the format, but β^e is not. By adding the following constraint, we ensure that β^e is representable when β^{e-1} is.

$$\forall e \in \mathbb{Z}, \ \varphi(e) < e \Rightarrow \varphi(e + 1) \le e. \qquad [3.2]$$

Note that, as soon as a power of β is representable, all the larger ones are representable too, by induction on constraint [3.2]. As a result, if β^e is part of the format, then for any $x \geq \beta^e$ that is not representable, $\mathrm{ulp}(x)$ is the distance between the two surrounding FP numbers. Moreover, when $x \geq \beta^e$ is representable, then $\mathrm{ulp}(x)$ is the distance to its successor. Thus, if there are representable powers of β arbitrarily close to zero, then we do not need any other constraint to get a sensible format.

Unfortunately, if some power of β is not representable in the format, the behavior of the ulp function might still not be that useful for inputs close to zero. In particular, we would like ulp on these inputs to return the distance between zero and the closest nonzero representable number. For now, let us ignore the situation where the format contains only zero. In other words, we assume that there exists a smallest representable power of β, denoted by $\eta = \beta^{e_\eta}$, which is also the smallest positive representable number. The property we would like is that, for any number $x \in (0; \eta)$, we have $\mathrm{ulp}(x) = \eta$. Since we want

$$\beta^{e_\eta} = \eta = \mathrm{ulp}(x) = \beta^{\varphi(\mathrm{mag}(x))}$$

for any $x \in [\beta^{e-1}; \beta^e) \subseteq (0; \beta^{e_\eta})$, we need

$$e \leq e_\eta \Rightarrow \varphi(e) = e_\eta.$$

We prefer that e_η does not appear explicitly, so that a format is described by φ only. To do so, we make e_η implicit in the previous formula. We set $e_\eta = \varphi(e)$ for any e such that $e \leq \varphi(e)$. This gives the following two constraints. The first one expresses that $\eta = \beta^{\varphi(e)}$ is representable; the second one ensures that all the elements of $(0; \eta)$ have the same canonical exponent $e_\eta = \varphi(e)$.

$$\forall e \in \mathbb{Z}, \ e \leq \varphi(e) \Rightarrow \varphi(\varphi(e) + 1) \leq \varphi(e); \qquad\qquad [3.3]$$

$$\forall e, e' \in \mathbb{Z}, \ e \leq \varphi(e) \wedge e' \leq \varphi(e) \Rightarrow \varphi(e') = \varphi(e). \qquad\qquad [3.4]$$

Note that, for formats for which all the powers of β are representable, the previous two propositions are vacuously true, so they do not constrain these formats in any way. Note also that the conjunction of constraint [3.3] and lemma 3.3 excludes the degenerate format where 0 is the only representable number.

The conjunction of the three constraints about φ, [3.2], [3.3], and [3.4], is expressed by the valid_exp predicate:

```
Definition valid_exp (fexp : Z -> Z) :=
  forall e : Z,
  ( fexp e < e -> fexp (e + 1) <= e )%Z /\
```

```
( e <= fexp e ->
  fexp (fexp e + 1) <= fexp e /\
  forall e' : Z, e' <= fexp e -> fexp e' = fexp e )%Z.
```

From now on, whenever we mention a function φ, we assume that it is valid according to the predicate above. Let us now provide another lemma for characterizing the powers of the radix that are part of the format.

Lemma 3.5 (generic_format_bpow' and generic_format_bpow_inv).
$\forall e \in \mathbb{Z}, \varphi(e) \leq e \Leftrightarrow \beta^e \in F$.

Proof. By lemma 3.3, it suffices to prove the equivalence $\varphi(e) > e \Leftrightarrow \varphi(e+1) > e$. To do so, we prove that either side of the equivalence implies $\varphi(e) = \varphi(e+1)$, which is a consequence of constraint [3.4]. ■

3.1.3.4. *Usual FP formats*

Let us now see how to express the main families of FP formats as exponent functions. The simpler ones are for the FIX and FLX formats. For FIX, all the representable numbers have the same canonical exponent: e_{min}. This gives the following exponent function:

Definition FIX_exp (e : Z) := emin.

For an FLX format, the scaled mantissa of any representable nonzero numbers is an integer with exactly ϱ digits. This gives the following exponent function:

Definition FLX_exp (e : Z) := (e - prec)%Z.

An FLT format is a combination of both previous characterizations. The scaled mantissa should have ϱ digits, unless it causes the canonical exponent to drop below e_{min}. In that latter case, the canonical exponent is e_{min}. This gives the following function:

Definition FLT_exp (e : Z) := Zmax (e - prec) emin.

The situation is slightly more complicated for an FTZ format. Similarly to an FLT format, the scaled mantissa of a representable number has ϱ digits as long as it is in the range of normal numbers. On the range of subnormal numbers, the canonical exponent is no longer e_{min} however. It is now the power of the smallest representable positive number, that is, $e_{min} + \varrho - 1$. This gives the following function:

Definition FTZ_exp (e : Z) :=
  if Zlt_bool (e - prec) emin
  then (emin + prec - 1)%Z
  else (e - prec)%Z.

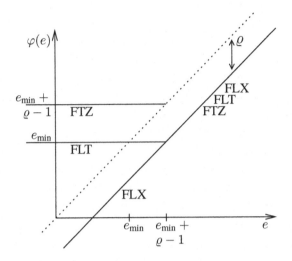

Figure 3.1. *Behavior of φ depending on the format.*

Lemmas `FIX_exp_valid`, `FLX_exp_valid`, and so on, ensure that all these four exponent functions satisfy the `valid_exp` predicate. But for these functions to be useful, we also have to make sure that they define the same families of formats as the predicates from section 3.1.2. For instance, the following two theorems ensure that any real satisfying the `FLT_format` predicate is part of the format described by the `FLT_exp` function, and conversely.

Theorem `generic_format_FLT` :
  `forall x : R, FLT_format x ->`
  `generic_format FLT_exp x.`

Theorem `FLT_format_generic` :
  `forall x : R, generic_format FLT_exp x ->`
  `FLT_format x.`

These theorems do not just ensure the consistency of the formalism. They also make it possible to juggle between the various representations of formats. For instance, it is simpler to show that a number is representable using the `FLT_format` predicate, while talking about ulp is simpler using `generic_format`.

Let us now illustrate visually some similarities and some differences between the various exponent functions for FP formats: `FLX_exp`, `FLT_exp`, and `FTZ_exp`. Figure 3.1 plots the value of the canonical exponent of a representable number depending on its magnitude. Note that the figure extends infinitely to the left, for magnitudes of numbers closer and closer to zero. The dotted line represents the limit at which there are no representable numbers anymore. When an exponent function lies above it, the resulting canonical exponent is too large to accommodate any number of such a magnitude.

If the number x is in the range of normal numbers, that is, $|x| \geq \beta^{e_{\min}+\varrho-1}$, all three functions behave the same: the canonical exponent is ϱ units below the dotted line, so that the scaled mantissa has ϱ digits. The behavior of the exponent functions differs for numbers whose magnitude is less than $e_{\min} + \varrho - 1$. For an FLX format, there is no change of behavior; its plot is still ϱ units below the dotted line. The other exponent functions, however, become constant on the range of subnormal numbers. For an FLT format, the plot looks continuous: the underflow is gradual. For an FTZ format, however, there is a gap between the canonical exponents of normal numbers and of (what would be) subnormal numbers: the underflow is abrupt.

Let us give some clues about the way the plot would look for more exotic formats. First, if the exponent function is above the dotted line at some point, then it has to be constant until it reaches the dotted line (case of the FTZ_exp function). This is a consequence of constraints [3.3] and [3.4]. Once the plot is below the dotted line, however, constraint [3.2] tells us that it can vary as wildly as it wants, as long as it does not cross the dotted line further to the right. In particular, it does not have to be monotone increasing, contrarily to what the usual formats might lead us to believe.

The most-used formats in this book are the FLT, FLX, and FIX formats. Table 3.2 summarizes their definitions and main features.

Format F	FIX	FLX	FLT
parameters	$e_{\min}$	ϱ	$e_{\min}$ and ϱ
$x \in F$ iff	$\exists m \in \mathbb{Z},$ $x = m \cdot \beta^{e_{\min}}$	$\exists m, e \in \mathbb{Z},$ $x = m \cdot \beta^e \wedge |m| < \beta^\varrho$	$\exists m, e \in \mathbb{Z},$ $x = m \cdot \beta^e \wedge$ $|m| < \beta^\varrho \wedge e \geq e_{\min}$
$\varphi(e)$	$e_{\min}$	$e - \varrho$	$\max(e - \varrho, e_{\min})$
ulp(0)	$\beta^{e_{\min}}$	0	$\beta^{e_{\min}}$
ulp(β^e)	$\beta^{e_{\min}}$	$\beta^{e-\varrho+1}$	if $e \geq e_{\min} + \varrho - 1$ then $\beta^{e-\varrho+1}$ else $\beta^{e_{\min}}$
smallest positive normal number			$\beta^{e_{\min}+\varrho-1}$

Table 3.2. *Characteristics of the three main families of formats*

3.1.4. *Main properties*

Now that we have a unified formalism to talk about the usual formats, let us describe a few properties that any generic format satisfies, especially those related to the *ulp* function.

3.1.4.1. *Unit in the last place*

As mentioned in section 3.1.3.2, the definition of *ulp* is chosen so that the following property holds:

Lemma 3.6 (ulp_neq_0).
$$\forall x \in \mathbb{R},\ x \neq 0 \Rightarrow \mathrm{ulp}(x) = \beta^{\mathrm{cexp}(x)}.$$

More interestingly, we can relate x and $\mathrm{ulp}(x)$ as follows. Notice the hypothesis $x \in F$. Otherwise $|x|$ might lie in a range $[\beta^{e-1}; \beta^e)$ containing no representable numbers, in which case the *ulp* would be larger than β^e.

Lemma 3.7 (ulp_le_abs).
$$\forall x \in F,\ x \neq 0 \Rightarrow \mathrm{ulp}(x) \leq |x|.$$

When the format is part of the FLX family, we can use the precision of the format to better characterize the relation between $\mathrm{ulp}(x)$ and x. We also no longer need to suppose that x is a representable number.

Lemma 3.8 (ulp_FLX_le and ulp_FLX_ge).
$$\forall x \in \mathbb{R},\ |x| \cdot \beta^{-\varrho} \leq \mathrm{ulp}_{\mathrm{FLX}}(x) \leq |x| \cdot \beta^{1-\varrho}.$$

Note that the left-most inequality is not tight (except in the case $x = 0$).

Since any FLX and FLT formats with the same precision coincide on the range of normal numbers, the upper bound above is also valid for any FLT format on that range.

Lemma 3.9 (ulp_FLT_le).
$$\forall x \in \mathbb{R},\ \beta^{e_{\min}+\varrho-1} \leq |x| \Rightarrow \mathrm{ulp}_{\mathrm{FLT}}(x) \leq |x| \cdot \beta^{1-\varrho}.$$

For an FLT format, we can also precisely characterize the *ulp* for subnormal numbers (and for the first slice of normal numbers).

Lemma 3.10 (ulp_FLT_small).
$$\forall x \in \mathbb{R},\ |x| < \beta^{e_{\min}+\varrho} \Rightarrow \mathrm{ulp}_{\mathrm{FLT}}(x) = \beta^{e_{\min}}.$$

Finally, the *ulp* function is constant for FIX formats.

Lemma 3.11 (ulp_FIX).
$$\forall x \in \mathbb{R},\ \mathrm{ulp}_{\mathrm{FIX}}(x) = \beta^{e_{\min}}.$$

Formats with gradual underflow are slightly better behaved than those with abrupt underflow. For instance, if the rounded sum of two FP numbers is zero, then the exact sum is zero, too (see section 5.1.2.1). The following lemma characterizes such formats without abrupt underflow:

Lemma 3.12 (generic_format_ulp, not_FTZ_generic_format_ulp, ulp_ge_ulp_0, not_FTZ_ulp_ge_ulp_0). The following three propositions are equivalent:

- $\forall e \in \mathbb{Z}, \varphi(\varphi(e) + 1) \leq \varphi(e);$

- $\forall x \in \mathbb{R}, \text{ulp}(x) \in F;$

- $\forall x \in \mathbb{R}, \text{ulp}(0) \leq \text{ulp}(x).$

3.1.4.2. Predecessor and successor

We can now use the ulp function to define the predecessor and successor of a representable number. Given a positive real number $x \in F$, we know that there is no representable number between x and $x + \text{ulp}(x)$, which makes the latter a suitable candidate for the successor of x. This gives us part of the definition of succ:

```
Definition succ (x : R) : R :=
  if Rle_bool 0 x then x + ulp x
  else - pred_pos (-x).
```

For $x = 0$, the returned value is $\text{ulp}(0)$, which is the successor of 0 for any format with a minimal exponent. For a format without minimal exponent, 0 has no successor and the succ function above will just return 0. As for negative inputs, the successor of x is just defined as the opposite of the predecessor of $-x$. Note that, while the successor is defined for any real number, it is meaningful only for representable numbers.

To complete the definition of the successor, we thus have to characterize the predecessor of a positive representable number. The situation is slightly more complicated than for the successor: $x - \text{ulp}(x)$ is almost always the predecessor of x except generally for the powers of the radix. Indeed the predecessor of β^e is not in the same slice as β^e, so $\text{pred}(\beta^e) = \beta^e - \beta^{\varphi(e)}$ is possibly different from $\beta^e - \text{ulp}(\beta^e)$. This gives the following definition for the predecessor of positive numbers:

```
Definition pred_pos (x : R) : R :=
  if Req_bool x (bpow (mag beta x - 1))
  then x - bpow (fexp (mag beta x - 1))
  else x - ulp x.
```

Now that the successor is fully defined, we can just define the predecessor by taking the opposite:

```
Definition pred (x : R) : R := - succ (-x).
```

The advantage of this definition is that it gives us for free the following two identities.

Lemma 3.13 (pred_opp, succ_opp). For any $x \in \mathbb{R}$,

$$
\begin{aligned}
\operatorname{pred}(-x) &= -\operatorname{succ}(x), \\
\operatorname{succ}(-x) &= -\operatorname{pred}(x).
\end{aligned}
$$

Among the basic properties of the predecessor and successor of a representable number, we have that they are representable.

Lemma 3.14 (generic_format_pred, generic_format_succ).
$\forall x \in F$, $\operatorname{pred}(x) \in F \wedge \operatorname{succ}(x) \in F$.

We also know how $\operatorname{pred}(x)$ and $\operatorname{succ}(x)$ are located with respect to x. Note that the hypothesis $x \neq 0$ comes from the fact that there exist FP numbers arbitrarily close to zero in formats such as FLX. Note also that x does not have to be representable for the propositions to hold.

Lemma 3.15 (pred_le_id, succ_ge_id, pred_lt_id, succ_gt_id).
For any $x \in \mathbb{R}$,

$$
\operatorname{pred}(x) \leq x \leq \operatorname{succ}(x),
$$
$$
x \neq 0 \quad \Rightarrow \quad \operatorname{pred}(x) < x < \operatorname{succ}(x).
$$

Transitivity gives us the next lemma.

Lemma 3.16 (pred_lt_le, lt_succ_le). For any $x, y \in \mathbb{R}$,

$$
\begin{aligned}
x \neq 0 \wedge x \leq y &\quad \Rightarrow \quad \operatorname{pred}(x) < y, \\
y \neq 0 \wedge x \leq y &\quad \Rightarrow \quad x < \operatorname{succ}(y).
\end{aligned}
$$

More interestingly, when x and y are part of the format, we get the following relations.

Lemma 3.17 (pred_lt_le, lt_succ_le). For any $x, y \in F$,

$$
\begin{aligned}
x < y &\quad \Rightarrow \quad x \leq \operatorname{pred}(y), \\
x < y &\quad \Rightarrow \quad \operatorname{succ}(x) \leq y.
\end{aligned}
$$

We also have that pred and succ are inverse from each other when considering representable numbers.

Lemma 3.18 (pred_succ, succ_pred). For any $x \in F$,

$$\begin{aligned} \mathrm{pred}(\mathrm{succ}(x)) &= x, \\ \mathrm{succ}(\mathrm{pred}(x)) &= x. \end{aligned}$$

Finally, these two functions are strictly increasing for representable numbers.

Lemma 3.19 (pred_lt, succ_lt). For any $x, y \in F$,

$$\begin{aligned} x < y &\Rightarrow \mathrm{pred}(x) < \mathrm{pred}(y), \\ x < y &\Rightarrow \mathrm{succ}(x) < \mathrm{succ}(y). \end{aligned}$$

This concludes the section on FP formats. We have seen how the usual formats can be described by predicates and how they could be unified into a single family of formats using φ functions. We have also seen how to define the unit in the last place, the predecessor, and the successor, for all these formats, as well as some of their properties.

3.2. Rounding operators and their properties

An informal definition of rounding and the standard rounding modes have been given in section 1.1.3. Formally, the rounding property can be expressed either implicitly or explicitly; which way is best depends on the context. The implicit representation is a relation that tells whether some FP number f is the rounded value of some real number x. The explicit representation expresses f as a function of x. In both cases, the rounded value depends on the format and on the rounding direction.

Section 3.2.1 explains how rounding can be expressed as a relation. Section 3.2.2 explains how to express it as a function over the real numbers in the specific case of the generic formats described in section 3.1.3. Note that, as a function over the real numbers, this explicit form is hardly useful for performing actual computations, unless its inputs are such that one can perform effective arithmetic operations on them, e.g. rational numbers. Section 3.3.1 will later give a more suitable way to perform effective arithmetic on FP operands.

3.2.1. Rounding relations

In the implicit setting, rounding is simply defined as a relation $\mathcal{R}$ between real numbers: $x \mathrel{\mathcal{R}} f$ means that f is the rounded value of x. The relation has to satisfy two properties so that it can be useful in practice. First, one should be able to round any real number. This is expressed by the round_pred_total predicate which states:

$$\forall x \in \mathbb{R}, \exists f \in \mathbb{R}, x \, \mathcal{R} \, f.$$

From a set-theoretic point of view, $\mathcal{R}$ is the graph of the rounding operator and this predicate ensures that it is a total (multi-)function.

Second, the rounding operator should be monotone increasing. The round_pred_monotone predicate states this property as follows:

$$\forall x_1, x_2, f_1, f_2 \in \mathbb{R}, x_1 \, \mathcal{R} \, f_1 \Rightarrow x_2 \, \mathcal{R} \, f_2 \Rightarrow x_1 \leq x_2 \Rightarrow f_1 \leq f_2.$$

The following lemma expresses that any relation satisfying the monotony property is the graph of a (partial) function: for any $x \in \mathbb{R}$, there is at most one $f \in \mathbb{R}$ such that $x \, \mathcal{R} \, f$.

Lemma 3.20 (round_unique). If $\mathcal{R}$ is monotone,

$$\forall x, f_1, f_2 \in \mathbb{R}, x \, \mathcal{R} \, f_1 \Rightarrow x \, \mathcal{R} \, f_2 \Rightarrow f_1 = f_2.$$

Proof. This is easily proved by verifying that both $f_1 \leq f_2$ and $f_1 \geq f_2$ hold as a consequence of the monotony property applied to $x_1 = x_2 = x$. ∎

Therefore, if $\mathcal{R}$ is a relation satisfying both the totality and monotony properties, then it is the graph of a total function. So, in a sense, once we have an implicit representation of a rounding operator, we also have an explicit representation. But this explicit representation is useless in practice (except as syntactic sugar), so section 3.2.2 will later come with a different representation.

3.2.1.1. *Directed rounding*

Let us now focus on the usual rounding modes, starting with directed rounding. A rounding mode is defined with respect to a subset $F \subseteq \mathbb{R}$. Given a real x, the real f obtained by rounding x toward $-\infty$ is defined as the largest element of F that is less than or equal to x. The following Coq relation expresses it:

```
Definition Rnd_DN_pt (F : R -> Prop) (x f : R) :=
  F f /\ f <= x /\
  forall g : R, F g -> g <= x -> g <= f.
```

Note that Rnd_DN_pt does not necessarily satisfy the totality property. For instance, it might happen that, for some x, there is no $f \in F$ such that $f \leq x$. Even the absence of a lower bound for F is not sufficient to ensure that the relation is total. Indeed, F also has to be closed from above. For example, with $F = (-\infty; 0)$, nonnegative numbers cannot be rounded down, because the accumulation point 0 is not part of F. With $F = (-\infty; 0]$ however, rounding toward $-\infty$ is a total function: nonpositive numbers are rounded to themselves, while positive numbers are rounded to 0.

Let us denote $\downarrow$ the relation Rnd_DN_pt. While it is not necessarily total, it is nonetheless monotone increasing:

Lemma 3.21 (Rnd_DN_pt_monotone). Let x_1, x_2, f_1, and f_2 be some real numbers such that $x_1 \downarrow f_1$ and $x_2 \downarrow f_2$. If $x_1 \leq x_2$, then $f_1 \leq f_2$.

Proof. Since $x_1 \downarrow f_1$, we have $f_1 \leq x_1$, so $f_1 \leq x_2$ by transitivity. Since $x_2 \downarrow f_2$, f_2 is the largest element of F that is less than or equal to x_2, so f_1 cannot be larger than f_2, which concludes the proof. ∎

A consequence of lemmas 3.20 and 3.21 is Rnd_DN_pt_unique which states that Rnd_DN_pt defines a (partial) function.

There is another important property of rounding operators: idempotence. In terms of relation, if we have $x \mathcal{R} f$, then we also have $f \mathcal{R} f$. This property does not hold in general, even if the relation is both monotone and total. But it does for the usual rounding modes. There are several equivalent ways of stating the idempotence property and Flocq provides two of them for Rnd_DN_pt. As with monotony, the following two lemmas are straightforward to prove.

Lemma 3.22 (Rnd_DN_pt_refl).
$\forall x \in \mathbb{R}, \ x \in F \Rightarrow x \downarrow x.$

Lemma 3.23 (Rnd_DN_pt_idempotent).
$\forall x, f \in \mathbb{R}, \ x \downarrow f \Rightarrow x \in F \Rightarrow x = f.$

Rounding toward $+\infty$ is defined in a symmetric way and denoted by $\uparrow$. It satisfies the same properties as rounding toward $-\infty$. Once we can round toward both infinities, we can state and prove that there are no element of F between x rounded down and x rounded up.

Lemma 3.24 (Only_DN_or_UP). Let x be a real number x. Let f_- and f_+ be its rounded values such that $x \downarrow f_-$ and $x \uparrow f_+$. For any number $f \in F$, if $f_- \leq f \leq f_+$, then f is equal to f_- or f_+.

Proof. Let us assume that $f \leq x$ holds first. Since f_- is the largest element of F less than or equal to x, we have $f \leq f_-$. Thus, $f = f_-$, since $f_- \leq f$ holds by hypothesis. Similarly, if $f \geq x$ holds instead, we have $f = f_+$. ∎

Rounding toward zero is defined as rounding toward $-\infty$ for nonnegative numbers and toward $+\infty$ for nonpositive numbers:

```
Definition Rnd_ZR_pt (F : R -> Prop) (x f : R) :=
  (0 <= x -> Rnd_DN_pt F x f) /\
  (x <= 0 -> Rnd_UP_pt F x f).
```

This rounding relation is not necessarily monotone. Let us suppose for instance that 1 and -1 are both part of the format, but no real numbers between them are. As a result, numbers in $(-1; 0]$ are rounded toward 1, while numbers in $[0; 1)$ are rounded toward -1. So, we have to make an assumption that F: 0 is representable.

Lemma 3.25 (Rnd_ZR_pt_monotone). If $0 \in F$, then rounding toward zero is a monotone relation.

3.2.1.2. Rounding to nearest

Let us focus now on rounding to nearest. A real number x rounds to $f \in F$ if f is closer to x than any other element of F. The Coq definition is as follows:

```
Definition Rnd_N_pt (F : R -> Prop) (x f : R) :=
  F f /\
  forall g : R, F g -> Rabs (f - x) <= Rabs (g - x).
```

Let this relation be denoted by $\mathcal{N}$. Since the midpoint of two consecutive elements of F is in relation with both of them, $\mathcal{N}$ cannot be a monotone relation. But it does satisfy the following weaker property, in which the ordering hypothesis of monotony $x_1 \leq x_2$ has been changed to $x_1 < x_2$:

Lemma 3.26 (Rnd_N_pt_monotone). Given some real numbers x_1, x_2, f_1, and f_2 such that $x_1 \mathcal{N} f_1$ and $x_2 \mathcal{N} f_2$, if $x_1 < x_2$, then $f_1 \leq f_2$.

Proof. We deduce from the hypotheses that $|f_1 - x_1| \leq |f_2 - x_1|$ and $|f_2 - x_2| \leq |f_1 - x_2|$. By studying the various ways the four numbers can be ordered (so as to remove the absolute values in the inequalities), we can show that they all contradict $f_2 < f_1$. ∎

The idempotence of $\mathcal{N}$, however, does not need to be weakened:

Lemma 3.27 (Rnd_N_pt_idempotence).
$\forall x, f \in \mathbb{R}, \; x \mathcal{N} f \Rightarrow x \in F \Rightarrow x = f.$

Various other lemmas can be stated about $\mathcal{N}$, which will be useful when proving properties about the two rounding modes to nearest we are interested in. The difference between any two rounding modes to nearest lies in the way it rounds a real number that is halfway between two consecutive representable FP numbers. Let us start with the definition of rounding to nearest with tie breaking away from zero:

```
Definition Rnd_NA_pt (F : R -> Prop) (x f : R) :=
```

```
Rnd_N_pt F x f /\
forall g : R, Rnd_N_pt F x g -> Rabs g <= Rabs f.
```

This relation $\mathcal{N}_A$ can be expressed as follows. In order for x $\mathcal{N}_A$ f to hold, we should have x $\mathcal{N}$ f and any g such that x $\mathcal{N}$ g should be no farther away from zero than f. As with rounding toward zero, there is an issue if 0 is not representable, since there might then be two ways to round zero that are equally far from zero.

Lemma 3.28 (Rnd_NA_pt_unique). Given some real numbers x, f_1, and f_2, if $0 \in F$ and x $\mathcal{N}_A$ f_1 and x $\mathcal{N}_A$ f_2, then $f_1 = f_2$.

Proof. If x is equal to 0, it is representable, so lemma 3.27 makes it possible to conclude. Let us assume that x is nonzero. By definition of $\mathcal{N}_A$, we have both $|f_2| \leq |f_1|$ and $|f_1| \leq |f_2|$, so either f_1 and f_2 are equal or they are opposite and nonzero. The latter case is impossible since one of f_1 and f_2 would be strictly closer to x, thus contradicting the fact that both are the closest to x among the elements of F. ∎

The next lemma is a direct consequence of lemmas 3.26 and 3.28.

Lemma 3.29 (Rnd_NA_pt_monotone). If $0 \in F$, then $\mathcal{N}_A$ is a monotone relation.

3.2.1.3. *Rounding to nearest, tie breaking to even*

The case of rounding to nearest with tie breaking to even is slightly more complicated, since it involves a notion of integer significand and thus of radix. Moreover, this significand has to be somehow bounded, otherwise any number $x = m \cdot \beta^e$ can be represented with an even scaled mantissa if β is even. Due to these requirements, we can no longer use an arbitrary format F. We use instead the generic formats defined in section 3.1.3. The definition of the relation $\mathcal{N}_E$ is otherwise similar to the one for $\mathcal{N}_A$.

Definition Rnd_NE_pt (fexp : Z -> Z) (x f : R) :=
  Rnd_N_pt (generic_format fexp) x f /\
  ((exists g : float beta, f = F2R g /\
    canonical fexp g /\ Zeven (Fnum g) = true) \/
  (forall h : R,
    Rnd_N_pt (generic_format fexp) x h -> h = f).

As with $\mathcal{N}_A$, we have to show that, even when x is the midpoint of two consecutive floating-point numbers, only one of them is in relation with x. The main difficulty lies in proving that for any two consecutive representable numbers, one and only one of them has a canonical representation with an even scaled mantissa. If both numbers have the same canonical exponents, their scaled mantissas are consecutive integers so the property holds. So let us consider the case of some representable power β^e and its predecessor $(\beta^e)^-$. Since β^e is representable, we have both $e \geq \varphi(e+1)$ and $e \geq \varphi(e)$ by lemmas 3.3 and 3.5. The two numbers have the following canonical representations:

$$(\beta^e)^- = (\beta^{e-\varphi(e)} - 1) \cdot \beta^{\varphi(e)} \quad \text{and} \quad \beta^e = \beta^{e-\varphi(e+1)} \cdot \beta^{\varphi(e+1)}.$$

Thus, in order for $\mathcal{N}_E$ to be a rounding relation, we need both integers $\beta^{e-\varphi(e)}$ and $\beta^{e-\varphi(e+1)}$ to have the same parity. This is obviously the case if β is odd. If β is even, we need either both exponents to be positive or both to be zero. This can be stated as follows:

$$(\varphi(e) < e \Rightarrow \varphi(e+1) < e) \wedge (e = \varphi(e) \Rightarrow e = \varphi(e+1)).$$

This formula can be modified to make it look closer to the definition of valid_exp. It can also be generalized to ignore the requirement on β^e to be representable. Therefore, a generic format makes it possible to break ties to even if and only the radix is odd or we have for any $e \in \mathbb{Z}$

$$(\varphi(e) < e \Rightarrow \varphi(e+1) < e) \wedge (e \leq \varphi(e) \Rightarrow \varphi(\varphi(e) + 1) = \varphi(e)).$$

For a FIX format, the φ function is constant, so the property trivially holds. For an FLX format, $\varphi(e) = e - \varrho$, so tie breaking to even can be performed if and only if the precision is at least 2 when the radix is even. This is the same for an FLT format.

An FTZ format, however, does not always make it possible to break ties to even when the radix is even. Indeed, the successor of zero is $\beta^{\varrho-1} \cdot \beta^{e_{\min}}$, which has an even integer significand for $\varrho \geq 2$. So the midpoint between zero and its successor can be rounded to both of them and one of them has to be chosen arbitrarily, e.g. zero.

3.2.2. Rounding functions

Section 3.1.3 explained how the usual formats can be expressed as integer functions that take a magnitude and return the canonical exponent of all the real numbers in the corresponding slice. This section explains how, given such an integer function and a real number, one can express the closest FP number to this real number in a given direction.

3.2.2.1. Definition and basic properties

Let us first give an informal description of the process. Let x be a positive real number and e an integer such that $\beta^{e-1} \leq x < \beta^e$. For now, let us assume that the closest FP number can be written as $m \cdot \beta^{\varphi(e)}$ with m an integer. Rounding x is therefore just a matter of finding m. Let us denote m_- and m_+ the following integers:

$$m_- = \lfloor x \cdot \beta^{-\varphi(e)} \rfloor \quad \text{and} \quad m_+ = \lceil x \cdot \beta^{-\varphi(e)} \rceil.$$

Since we have the following inequalities

$$\lfloor x \cdot \beta^{-\varphi(e)} \rfloor \le x \cdot \beta^{-\varphi(e)} \le \lceil x \cdot \beta^{-\varphi(e)} \rceil,$$

we deduce

$$m_- \cdot \beta^{\varphi(e)} \le x \le m_+ \cdot \beta^{\varphi(e)}.$$

Therefore, m will be either m_- or m_+, depending on the rounding direction. Note that, if x is representable in the format described by φ, we have $m = m_- = m_+$. Otherwise, we have $m = m_-$ when rounding toward zero or toward $-\infty$, and $m = m_+$ when rounding away from zero or toward $+\infty$.

In the general case, we want to compute m from the value of the scaled mantissa $x \cdot \beta^{-\varphi(e)}$. The way to do so is given by the *rnd* parameter which turns a scaled mantissa into an integer. This leads to the following definition for a generic rounding function:

Definition round (fexp : Z -> Z) (rnd : R -> Z) (x : R) :=
  F2R (Float beta (rnd (scaled_mantissa fexp x))
              (cexp fexp x)).

In other words, we have

$$\text{round}(x) = rnd \left(x \cdot \beta^{-\text{cexp}(x)} \right) \cdot \beta^{\text{cexp}(x)}. \tag{3.5}$$

Note that the final result $m \cdot \beta^{\varphi(e)}$ is not necessarily in canonical form, as the rounded value of x might be equal to β^e, whose canonical representation is $\beta^{e-\varphi(e+1)} \cdot \beta^{\varphi(e+1)}$.

Let us detail the constraints the *rnd* parameter should satisfy for round to behave as a proper rounding function. In particular, we expect two properties from a rounding function: identity on the elements of the format, and monotony. Let us consider the simple case of a FIX format described by $\varphi(e) = e_{\min}$. The function is the identity on the elements of this FIX format if and only if the following equality holds for any $m \cdot \beta^{e_{\min}}$ with $m \in \mathbb{Z}$:

$$rnd((m \cdot \beta^{e_{\min}}) \cdot \beta^{-e_{\min}}) \cdot \beta^{e_{\min}} = m \cdot \beta^{e_{\min}}.$$

In other words, the *rnd* function has to be the identity on integers. Regarding monotony, let us consider two real numbers $x_1 \le x_2$. We need the following property to hold:

$$rnd(x_1 \cdot \beta^{-e_{\min}}) \cdot \beta^{e_{\min}} \le rnd(x_2 \cdot \beta^{-e_{\min}}) \cdot \beta^{e_{\min}}.$$

Thus *rnd* also has to be monotone increasing.

Let us now prove that these two properties are sufficient not just for FIX formats but for any format F described by a valid exponent function φ. In the following lemmas, the *rnd* function is assumed to be the identity over integers and to be monotone increasing. The unary operator $\square : \mathbb{R} \to \mathbb{R}$ designates round applied to φ and *rnd*.

The first lemma states that round is the identity over elements of F. The second one is specialized for the case 0.

Lemma 3.30 (round_generic).
$$\forall x \in \mathbb{R}, \; x \in F \Rightarrow \square(x) = x.$$

Proof. By defining a generic format with equation [3.1], we know that the scaled mantissa $x \cdot \beta^{-\text{cexp}(e)}$ is an integer. Thus, it is left unchanged by the *rnd* function, which concludes the proof. ∎

Lemma 3.31 (round_0).
$$\square(0) = 0.$$

We have just proved that round is the identity over F, but we also need to prove that the image of round is F. The following two lemmas will be useful to do so, as they give some clues about the slice of $\square(x)$ depending on the slice of x.

Lemma 3.32 (round_bounded_large_pos). If $\beta^{e-1} \leq x < \beta^e$,

$$\varphi(e) < e \Rightarrow \beta^{e-1} \leq \square(x) \leq \beta^e.$$

Proof. By monotony of *rnd* and the fact that it is the identity function over the integers, we know that $\square(x) = m \cdot \beta^{\varphi(e)}$ with m as an integer such that

$$\lfloor \beta^{e-1-\varphi(e)} \rfloor \leq m \leq \lceil \beta^{e-\varphi(e)} \rceil.$$

Using the hypothesis $\varphi(e) < e$, we can remove the integer parts from the previous inequality, which makes it possible to conclude. ∎

Lemma 3.33 (round_bounded_small_pos). If $\beta^{e-1} \leq x < \beta^e$,

$$e \leq \varphi(e) \Rightarrow \square(x) = 0 \vee \square(x) = \beta^{\varphi(e)}.$$

Proof. The proof starts as in lemma 3.32, but the hypothesis $e \leq \varphi(e)$ causes m to be equal to either 0 or 1. ∎

We can now prove that F is the image of round. Note that the following proofs will only detail the case of positive inputs. The negative case can then be handled by considering a function $rnd' : x \mapsto -rnd(-x)$ as it is both monotone increasing and the identity function over the integers when rnd is.

Lemma 3.34 (generic_format_round).

$\forall x, \Box(x) \in F.$

Proof. Without loss of generality, we assume $\beta^{e-1} \leq x < \beta^e$. Let us first suppose that $e > \varphi(e)$, so that the hypotheses of lemma 3.32 hold. We thus have $\beta^{e-1} \leq \Box(x) \leq \beta^e$. If $\Box(x)$ is not equal to β^e, then it has the same canonical exponent as x and is thus representable in the format. Otherwise, $\Box(x)$ is equal to β^e, which is representable according to lemma 3.5.

Let us now suppose that $e \leq \varphi(e)$. So lemma 3.33 tells us that $\Box(x)$ is equal to either 0 or $\beta^{\varphi(e)}$. Both these numbers are part of the format, so $\Box(x)$ is as well. ∎

Lemma 3.35 (round_le).

$\forall x, y \in \mathbb{R}, \; x \leq y \Rightarrow \Box(x) \leq \Box(y).$

Proof. Without loss of generality, we assume that both x and y are positive and that $\beta^{e_x-1} \leq x < \beta^{e_x}$ and $\beta^{e_y-1} \leq y < \beta^{e_y}$. By monotony of rnd, we can show that $\Box(x) \leq \Box(y)$ holds if $\varphi(e_x) = \varphi(e_y)$. From $x \leq y$, we deduce $e_x \leq e_y$. If they are equal, then $\varphi(e_x) = \varphi(e_y)$ and we conclude. So let us assume instead that $e_x < e_y$.

Let us first suppose that $e_y \leq \varphi(e_y)$. Constraint [3.4] ensures that $\varphi(e_x) = \varphi(e_y)$, so $\Box(x) \leq \Box(y)$ holds. So we can assume $\varphi(e_y) < e_y$ and thus $\beta^{e_y-1} \leq \Box(y)$ by lemma 3.32. As a consequence, it suffices to prove that $\Box(x) \leq \beta^{e_y-1}$ holds.

Let us suppose that $e_x \leq \varphi(e_x)$. By lemma 3.33, we have $\Box(x) \leq \beta^{\varphi(e_x)}$, so it is sufficient to prove that $\varphi(e_x) \leq e_y - 1$ holds. By absurd, if $e_y \leq \varphi(e_x)$ holds, constraint [3.4] induces $\varphi(e_x) = \varphi(e_y)$, which contradicts $\varphi(e_y) < e_y$.

The last case to consider is $e_x > \varphi(e_x)$. By lemma 3.32, we get $\Box(x) \leq \beta^{e_x}$. Thus, $\Box(x) \leq \beta^{e_y-1}$ since $e_x < e_y$. ∎

3.2.2.2. Rounding errors

Now that rounding has been defined, we can give some bounds on the rounding error. First, let us relate the values of the rounding to $-\infty$ (denoted by $\bigtriangledown(x)$) and the rounding to $+\infty$ (denoted by $\triangle(x)$) for a real number x. Note that the φ function describing the format F is implicit in the rounding operators and in the ulp function.

Lemma 3.36 (round_UP_DN_ulp).

$x \notin F \Rightarrow \triangle(x) = \bigtriangledown(x) + \text{ulp}(x).$

Proof. By definition of ulp and by equation [3.5] defining the rounding function, the statement is equivalent to

$$\lceil x \cdot \beta^{-\text{cexp}(x)} \rceil = \lfloor x \cdot \beta^{-\text{cexp}(x)} \rfloor + 1,$$

which holds, since x is not representable.   ■

As a corollary, we get the following bound on the absolute error $\Box(x) - x$.

Lemma 3.37 (error_lt_ulp).
   $x \neq 0 \Rightarrow |\Box(x) - x| < \text{ulp}(x).$

Proof. The proof of the inequality depends on whether x is representable. If it is, we have $\Box(x) = x$. The assumption $x \neq 0$ ensures that $\text{ulp}(x)$ is not zero, which concludes this case.

Let us assume x is not representable. lemma 3.24 tells us that $\Box(x)$ is either $\triangledown(x)$ or $\triangle(x)$. By definition of $\triangledown$ and $\triangle$, we also know that $\triangledown(x) < x < \triangle(x)$. As a result, $|\Box(x) - x| < \triangle(x) - \triangledown(x) = \text{ulp}(x).$ ■

The following corollary avoids the hypothesis $x \neq 0$ by relaxing the inequality.

Lemma 3.38 (error_le_ulp).
   $|\Box(x) - x| \leq \text{ulp}(x).$

When rounding is to nearest, we get a sharper bound on the absolute error.

Lemma 3.39 (error_le_half_ulp).
   $|\circ^\tau(x) - x| \leq \frac{1}{2}\text{ulp}(x).$

Proof. The proof is similar to that of lemma 3.37. Indeed, $\circ^\tau(x)$ is still either $\triangledown(x)$ or $\triangle(x)$, but now it is which one is the nearest to x, by definition of $\circ^\tau$.   ■

Depending on the situation, it might be better to have a bound on the absolute error that is expressed using $\text{ulp}(\Box(x))$. This bound is given by the following two lemmas, whose proofs are much more complicated.

Lemma 3.40 (error_lt_ulp_round).
   $x \neq 0 \Rightarrow |\Box(x) - x| < \text{ulp}(\Box(x)).$

Lemma 3.41 (error_le_half_ulp_round).
   $|\circ^\tau(x) - x| \leq \frac{1}{2}\text{ulp}(\circ^\tau(x)).$

Let us now see the case of the relative error in the case of an FLT format of precision ϱ and of minimal exponent $e_{\min}$.

Lemma 3.42 (relative_error_FLT).
$$\beta^{e_{\min}+\varrho-1} \leq |x| \Rightarrow |\square(x) - x| < \beta^{-\varrho+1} \cdot |x|.$$

Proof. This is a corollary of lemma 3.37 using the fact that $\text{ulp}(x) \leq \beta^{-\varrho+1} \cdot |x|$ when $|x| \geq \beta^{e_{\min}+\varrho-1}$. ∎

Lemma 3.43 (relative_error_N_FLT).
$$\beta^{e_{\min}+\varrho-1} \leq |x| \Rightarrow |\circ^\tau(x) - x| \leq \frac{1}{2}\beta^{-\varrho+1} \cdot |x|.$$

Lemma 3.44 (relative_error_N_FLT_round).
$$\beta^{e_{\min}+\varrho-1} \leq |x| \Rightarrow |\circ^\tau(x) - x| \leq \frac{1}{2}\beta^{-\varrho+1} \cdot |\circ^\tau(x)|.$$

3.2.2.3. *Double rounding*

Let us now see what happens when one rounds a number representable in a format F_{ext} (given by φ_{ext}) to a format F (given by φ), both with the same radix. First, the result, while rounded, is still representable in F_{ext}.

Theorem 3.45 (generic_round_generic).
$$x \in F_{ext} \Rightarrow \square(x) \in F_{ext}.$$

This theorem holds, whatever the two formats F and F_{ext} (with the same radix), even if one does not extend the other. Consider for instance a *binary64* number x. When this number is truncated to an integer (fixed-point format with minimal exponent 0), the result $\lfloor x \rfloor$ is still a *binary64* number, though it is not necessarily equal to x. Conversely, the result of rounding an integer to *binary64* is still an integer, though possibly different.

As in section 1.4.2, let us now consider the case of an original value obtained by rounding to F_{ext} a basic arithmetic operation on values of F, e.g. $\square_{ext}(x \cdot y)$ for $x, y \in F$. Under some hypotheses, we can still relate the final result $\square(\square_{ext}(x \cdot y))$ to the result $x \cdot y$ of the infinitely-precise arithmetic operation [FIG 95]. For instance, the following theorem states that the first rounding $\square_{ext}$ is innocuous in the case of a product [ROU 14].

Theorem 3.46 (double_round_mult). Assume that $\square_{ext}$ and $\square$ are the same rounding mode (though to different formats). If the formats satisfy the following inequalities

$$\forall u, v \in \mathbb{Z}, \quad \varphi_{ext}(u+v) \leq \varphi(u) + \varphi(v) \quad \wedge \quad \varphi_{ext}(u+v-1) \leq \varphi(u) + \varphi(v),$$

then

$$\forall x, y \in F, \quad \square(\square_{ext}(x \cdot y)) = \square(x \cdot y).$$

In the case of two FLX formats, the requirement on the formats simply means that F_{ext} offers at least twice the precision of F. The interested reader should refer to [ROU 14] for the proof of this theorem and its variants for the other basic operations: addition, division, and square root.

Finally, we can also take a look at what happens when the first rounding is the rounding to odd presented in section 1.4.3. Note that the definition of rounding to odd necessitates the same format constraints as for rounding to nearest, tie breaking to even, given in section 3.2.1.3. Indeed, one and only one of two consecutive representable FP numbers will have an odd integer significand. The following theorem assumes that both formats satisfy these constraints and that the radix β is even.

Theorem 3.47 (round_odd_prop). If the formats satisfy the following inequality,

$$\forall e \in \mathbb{Z}, \quad \varphi_{ext}(e) \le \varphi(e) - 2,$$

then

$$\forall x \in \mathbb{R}, \quad \circ^\tau(x) = \circ^\tau\left(\Box_{ext}^{odd}(x)\right).$$

The need for an even radix could be removed in some cases, depending on the parity of the smallest positive FP number. This would have greatly complicated the proof though, and for few possible uses.

3.3. How to perform basic FP operations

The *as-if* definition (see section 1.1.4) of FP arithmetic is suitable for most proofs, since it relates FP operations to real operations. For instance, one usually does not need to know which FP number is the result of an operation, only that it is the closest (for some rounding direction) of the infinitely-precise result. There are, however, a few situations where one might need to have effective FP algorithms described using a formal system.

First, these algorithms can be used as reference implementations. For instance, by running these algorithms (either inside the system, or outside as an extracted program), one can generate guaranteed test cases against which a program or a circuit can be tested. Similarly, they will be used in Chapter 7 as a way to perform constant propagation at compilation time.

Second, these algorithms can be used inside the system in order to perform proofs by computation. Indeed, since FP arithmetic approximates real arithmetic, one can use it to deduce properties on real numbers. For instance, from the monotony of rounding operators, one can deduce the following implication:

$$\circ(14/10) < \circ(\sqrt{2}) \Rightarrow 14/10 < \sqrt{2}$$

Thus, if the left-hand side can actually be evaluated by performing the FP operations inside the system, the right-hand side becomes proved and it is as guaranteed as if a more algebraic reasoning had been used, e.g. by squaring both sides of the inequality and comparing them by means of rational computations. This FP-based reasoning will be presented in Chapter 4, where interval arithmetic is combined with effective FP operators to make it possible to automatically prove some properties related to real or FP arithmetic.

This section shows how to implement such effective yet verified operators. While the presented implementation is in no way competitive with hardware FP units for fixed precision or with a library such as MPFR for multi-precision [FOU 07], it is still useful as a reference implementation. Section 3.3.1 shows how to round a real number described by its position with respect to a close enough FP number. One can then easily deduce operators for addition and multiplication. Finally, section 3.3.2 shows how to perform FP divisions or square roots.

3.3.1. *Rounding, addition, and multiplication*

Since operations such as $+$, $\times$, $\div$, $\sqrt{\cdot}$ on the Z type of Coq are defined rather than axiomatized, this type is suitable for performing integer computations inside the system. As a result, the float type being a pair of Z integers, it can be used as a support for effective FP computations. It can represent arbitrary rational numbers of the form $m \cdot \beta^e$, so the first step is to devise an operator able to effectively round such a rational number so that it fits into a target format.

3.3.1.1. *Locating real numbers and rounding them*

Our approach to rounding a number is closely related to the concepts of *rounding* and *sticky* bits [ERC 04]. We define a ternary relation between a real number x, a float number, that is a pair of integers (m, e) representing the real number $m \cdot \beta^e$, and those two bits. Rather than manipulating two bits, we use an inductive type that reflects our intended semantics better. There are two constructors; the first one indicates that x is equal to $m \cdot \beta^e$, while the second one indicates how x is located with respect to the midpoint between $m \cdot \beta^e$ and $(m + 1) \cdot \beta^e$. This location is represented by a value of type comparison (section 2.1.4).

```
Inductive location :=
  | loc_Exact
  | loc_Inexact (pos : comparison).
```

Values of type location are related to the rounding (r) and sticky (s) bits as follows. Let us assume that x is positive and that we have two consecutive representable FP numbers that enclose it: $m \cdot \beta^e \leq x < (m + 1) \cdot \beta^e$. A location loc_Exact means that $x = m \cdot \beta^e$. In that case, the rounding and sticky bits for x are $r = 0$ and $s = 0$; the rounded value of x is x itself, whatever the rounding mode.

A location (loc_Inexact Eq) means that x is the midpoint between $m \cdot \beta^e$ and $(m + 1) \cdot \beta^e$, that is, $r = 1$ and $s = 0$. A location (loc_Inexact Lt) means that x is smaller than the midpoint: $r = 0$ and $s = 1$. A location (loc_Inexact Gt) means that x is larger than the midpoint: $r = 1$ and $s = 1$. Table 3.3 summarizes these three cases and shows which significand m or $m + 1$ we have to choose for the rounded value of x, depending on the rounding mode. Directed rounding modes do not care about the location with respect to the midpoint, and neither does rounding to odd (see section 1.4.3). For rounding modes to nearest, the difference lies in the handling of the (loc_Inexact Eq) case.

loc_Inexact	Lt	Eq	Gt
rounding bits	$r = 0 \quad s = 1$	$r = 1 \quad s = 0$	$r = 1 \quad s = 1$
toward zero	m	m	m
toward infinity	$m + 1$	$m + 1$	$m + 1$
to nearest even	m	$m + odd(m)$	$m + 1$
to nearest away	m	$m + 1$	$m + 1$
to odd	$m + even(m)$	$m + even(m)$	$m + even(m)$

Table 3.3. *Rounding and sticky bits depending on the location of the value to round assuming $m \geq 0$; resulting significand according to the rounding mode ($even(m)$, resp. $odd(m)$, is 1 if m is even, resp. odd, and 0 otherwise).*

The notion of location does not have to be restricted to characterizing how a real number x is located with respect to two enclosing FP numbers $m \cdot \beta^e$ and $(m+1) \cdot \beta^e$. More generally, we can compare x to any two real numbers u and v such that $x \in [u; v)$, using the following relation.

```
Inductive inbetween (u v x : R) : location -> Prop :=
  | inbetween_Exact : x = u -> inbetween u v x loc_Exact
  | inbetween_Inexact (l : comparison) : u < x < v ->
    Rcompare x ((u + v) / 2) = l ->
    inbetween u v x (loc_Inexact l).
```

In that definition, the Rcompare function compares two real numbers and returns a value of type comparison: Lt if the first argument is smaller, Eq if both arguments are equal, Gt if the first one is larger. Since this function is defined on real numbers, it is not computable and will be used only for specifications and proofs.

We can then specialize the inbetween relation above to the case where $u = m \cdot \beta^e$ and $v = (m + 1) \cdot \beta^e$.

Definition inbetween_float (m e : Z) (x : R) (l : location) :=
  inbetween (F2R (Float m e)) (F2R (Float (m + 1) e)) x l.

Notice that (inbetween_float m e x l) does not put many requirements on m and e. The property only forces $m \cdot \beta^e \leq x < (m + 1) \cdot \beta^e$. In particular, $m \cdot \beta^e$ and $(m + 1) \cdot \beta^e$ might not be representable in a given format or, if they are, they might not be consecutive. As a result, the location ℓ is not immediately usable for rounding x, unless we know additional properties about m and e.

Let us now assume that e is the canonical exponent of x for the format described by φ. If the location ℓ is loc_Exact, then we have $x = m \cdot \beta^e$. By definition of e, this means that x is representable in the format. Note that, as before, this does not -tell us anything about $(m + 1) \cdot \beta^e$; for instance, if x is negative and its absolute value is a power of β, then $(m + 1) \cdot \beta^e$ might not be representable, or, if it is, it might not be the successor of $m \cdot \beta^e = x$. This does not matter. Indeed, since x is representable in the format, whatever the rounding mode, the rounded value of x is x itself, as stated by lemma 3.30.

On the contrary, if the location ℓ is loc_Inexact, then $m \cdot \beta^e$ and $(m + 1) \cdot \beta^e$ are two consecutive numbers of the format, regardless of the sign of x, as stated by lemma 3.4.

Thus, when rounding toward $-\infty$ or $+\infty$, we have enough information for selecting the rounded value of x. More precisely, when rounding toward $-\infty$, inbetween_float_DN states that the rounded value $\bigtriangledown(x)$ is always $m \cdot \beta^e$. Note that the format is an implicit argument of the cexp and round functions in the statement below.

Theorem inbetween_float_DN :
  forall (x : R) (m : Z) (l : location),
  let e := cexp x in
  inbetween_float m e x l ->
  round Zfloor x = F2R (Float m e).

When rounding toward $+\infty$, the rounded value $\triangle(x)$ is $(m + 1) \cdot \beta^e$ when the location is not exact, $m \cdot \beta^e$ when otherwise. For other rounding modes too, the result will be either $m \cdot \beta^e$ or $(m + 1) \cdot \beta^e$. So, for each rounding mode, we define a function that decides which of the two possible results the rounded value is. When rounding toward $+\infty$, the function is defined as follows:

Definition round_UP (l : location) :=
  match l with
  | loc_Exact => false
  | _ => true
  end.

We then introduce the `cond_incr` function, which takes an integer and a Boolean, and returns either the integer or its successor, depending on the value of the Boolean input. It is used for stating the following `inbetween_float_UP` theorem.

Theorem `inbetween_float_UP` :
```
  forall (x : R) (m : Z) (l : location),
  let e := cexp x in
  inbetween_float m e x l ->
  let m' := cond_incr (round_UP l) m in
  round Zceil x = F2R (Float m' e).
```

When rounding to nearest, the argument of `loc_Inexact` tells us how to round x, as shown in Table 3.3. If ℓ is (`loc_Inexact Lt`), we choose $\circ^\tau(x) = m \cdot \beta^e$. If ℓ is (`loc_Inexact Gt`), we choose $\circ^\tau(x) = (m+1) \cdot \beta^e$. When x is at equal distance of $m \cdot \beta^e$ and $(m+1) \cdot \beta^e$, the end result depends on the tie breaking rule τ. So, to account for that last case, the `round_N` function is parameterized by the Boolean result expected in the case of a tie.

Definition `round_N (p : bool) (l : location) :=`
```
  match l with
  | loc_Exact => false
  | loc_Inexact Lt => false
  | loc_Inexact Eq => p
  | loc_Inexact Gt => true
  end.
```

The theorem below characterizes rounding to nearest, with ties broken to even significands. As can be seen from the statement, the Boolean argument to `round_n` tells that m should not be increased when it is even and the location is (`loc_Inexact Eq`).

Theorem `inbetween_float_NE` :
```
  forall (x : R) (m : Z) (l : location),
  let e := cexp x in
  inbetween_float m e x l ->
  let m' := cond_incr (round_N (negb (Zeven m)) l) m in
  round ZnearestE x = F2R (Float m' e).
```

In some situations, it might be easier to compute the location of $|x|$ rather than of x. In other words, the property we have at hand is (`inbetween_float m e (Rabs x) l`). Corollaries of the theorems above can be devised to handle this kind of hypothesis. For instance, the following function tells how to compute $\bigtriangledown(x)$ given the sign of x (*true* if negative) and the location of $|x|$:

```
Definition round_sign_DN (s : bool) (l : location) :=
  match l with
  | loc_Exact => false
  | _ => s
  end.
```

```
Theorem inbetween_float_DN_sign :
  forall (x : R) (m : Z) (l : location),
  let e := cexp x in
  inbetween_float m e (Rabs x) l ->
  let sx := Rlt_bool x 0 in
  let m' := cond_incr (round_sign_DN sx l) m in
  round Zfloor x = F2R (Float (cond_Zopp sx m') e).
```

In the above statement, sx designates the sign of x; it is *true* when x is negative. The round_sign_DN function tells which of $m \cdot \beta^e$ and $(m+1) \cdot \beta^e$ to choose for representing $|\triangledown(x)|$. As with theorems inbetween_float_UP and inbetween_float_NE, we use cond_incr to compute the significand accordingly. Note that this significand m' is the significand of $|\triangledown(x)|$. So we use cond_Zopp to conditionally negate m', depending on sx, in order to recover the significand of $\triangledown(x)$.

3.3.1.2. *Effective rounding*

As mentioned before, in order to round a nonnegative real number x, we look for two integers m and e and some location ℓ such that e is the canonical exponent of x and (inbetween_float m e x l) holds. We could use cexp(x) for e and (Ztrunc (scaled_mantissa x)) for m. As for the location, comparing x with $m \cdot \beta^e$ and with $(m + \frac{1}{2}) \cdot \beta^e$ would give enough information. Unfortunately, none of these operations can effectively compute. So we need a bit more information about x than just its value as a real number.

We focus on rounding a positive number x that satisfies (inbetween_float m0 e0 x l0) with e_0 being an exponent no larger than the canonical exponent of x.[2] To obtain the canonical exponent of x, we start by counting how many radix-β digits the significand m_0 contains using the Zdigits functions. This function computes an integer d such that $\beta^{d-1} \leq |m_0| < \beta^d$. Note that the powering function used in that formula is Zpower from the standard library, which means that β^{-1} returns 0. As a result, this couple of inequalities holds even for $m_0 = 0$, since Zdigits returns 0 in that case.

---

2 If e_0 is larger than the canonical exponent of x, we do not have enough information to characterize the rounded value unless ℓ_0 is loc_Exact. In that latter case, x is its own rounded value.

We deduce that $\text{mag}(x) = d + e_0$. If φ is computable (which is the case for functions such as FLT_exp), we can thus effectively compute the canonical exponent of x: $e = \varphi(d + e_0)$. Remember that we supposed that e_0 is no larger than the canonical exponent of x, so $e_0 \leq e$. When they are equal, we simply choose $m = m_0$ and $\ell = \ell_0$. Let us now consider the case $e_0 < e$. We look for the integer significand m such that

$$m \cdot \beta^e \leq x < (m+1) \cdot \beta^e.$$

Since we have $e_0 < e$, the integer m has to satisfy the property

$$m \cdot \beta^e \leq m_0 \cdot \beta^{e_0} \leq x < (m+1) \cdot \beta^e,$$

which can be rewritten as $m \leq m_0 \cdot \beta^{e_0 - e} < m + 1$. This gives the computable expression $m = \lfloor m_0 / \beta^{e - e_0} \rfloor$.

Finally, we have to effectively compute ℓ. Let us consider a slightly more general problem first. Instead of having the property (inbetween_float m0 e0 x l0), we suppose that (inbetween u0 v0 x l0) holds with u as a real, δ as a positive real, k as an integer, $u_0 = u + k \cdot \delta$, and $v_0 = u_0 + \delta$. We also suppose that we have an integer $n \geq 2$ and that $0 \leq k < n$. We want to find the location ℓ such that (inbetween u v x l) holds, with $v = u + n \cdot \delta$. Figure 3.2 illustrates the problem.

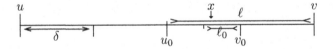

Figure 3.2. *From* (inbetween u0 v0 x l0) *to* (inbetween u v x l).

The new location ℓ is loc_Exact if and only if $x = u$, that is, if and only if the original location ℓ_0 is loc_Exact and $k = 0$. Let us now assume that either k is nonzero or ℓ_0 is Inexact. Since the midpoint between u and v is $u + n/2 \cdot \delta$, the choice of the final location is better explained separately, depending on whether n is even or odd. Let us start with n even. We get the following cases:

$$\ell = \text{loc\_Inexact} \begin{cases} \text{Lt} & \text{if } 2 \cdot k < n, \\ \text{Eq} & \text{if } 2 \cdot k = n \text{ and } \ell_0 = \text{loc\_Exact}, \\ \text{Gt} & \text{if } 2 \cdot k = n \text{ and } \ell_0 \neq \text{loc\_Exact}, \\ \text{Gt} & \text{if } 2 \cdot k > n. \end{cases}$$

When n is odd, the midpoint can be rewritten as $u + (n-1)/2 \cdot \delta + \delta/2$, so the case study is as follows:

$$\ell = \texttt{loc\_Inexact} \begin{cases} \texttt{Lt} & \text{if } 2 \cdot k < n-1, \\ \texttt{Lt} & \text{if } 2 \cdot k = n-1 \text{ and } \ell_0 = \texttt{loc\_Exact}, \\ \ell_0' & \text{if } 2 \cdot k = n-1 \text{ and } \ell_0 = \texttt{loc\_Inexact } \ell_0', \\ \texttt{Gt} & \text{if } 2 \cdot k > n-1. \end{cases}$$

The new_location function implements the above formulas. Given k, n, and ℓ_0, it returns ℓ. Figure 3.3 summarizes both even and odd cases.

Let us go back to the original problem. We have $u_0 = m_0 \cdot \beta^{e_0}$, $v_0 = (m_0+1) \cdot \beta^{e_0}$, $u = m \cdot \beta^e$, and $v = (m+1) \cdot \beta^e$. From $v_0 = u_0 + \delta$, we deduce $\delta = \beta^{e_0}$. From $v = u + n \cdot \delta$, we deduce $n = \beta^{e-e_0}$. Finally, from $u_0 = u + k \cdot \delta$, we deduce

$$k = m_0 - \lfloor m_0/\beta^{e-e_0} \rfloor \cdot \beta^{e-e_0},$$

which means that k is simply the remainder of the integer division that led to m. The value of ℓ can then be deduced from the values of k, n, and ℓ_0, using the formulas above. Note that the parity of n is the parity of β when $e > e_0$, so the formulas for computing ℓ only depend on radix β. Moreover, when m_0 is represented using radix-β digits, the division can be replaced by a simple shift and ℓ can be obtained by not actually computing k but by looking at digits that are truncated by the shift.

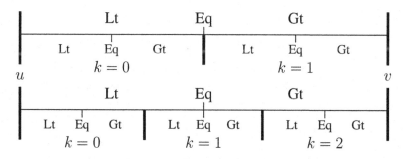

Figure 3.3. *Choice of locations for $n = 2$ and $n = 3$. Locations above the lines are the new locations. Only inexact locations are indicated; exact locations are on the bold ticks on the left of* Lt *locations.*

Let us turn the above mathematical description into a Coq function. The truncate function takes a triple as argument and returns a triple. These two triples are comprised of a significand, an exponent, and the location of the exact value with respect to this FP number.

```
Definition truncate (t : Z * Z * location) :=
  let '(m0, e0, l0) := t in
  let k := (fexp (Zdigits beta m0 + e0) - e0)%Z in
  if Zlt_bool 0 k then
    let p := Zpower beta k in
    (Zdiv m0 p, (e0 + k)%Z, new_location p (Zmod m0 p) l0).
  else t.
```

First the function computes the amount k of radix-β digits that have to be truncated, since the expression (Zdigits beta m0 + e0) is the magnitude of $m_0 \cdot \beta^{e_0}$. Then the function performs the truncation of the significand and adjusts the exponent accordingly. Finally it computes the location of x with respect to the truncated FP number. So not only does the returned triple locate x, but it also has the property that the exponent is guaranteed to be the canonical exponent of x, assuming that the input triple was precise enough. The following theorem formalizes what precise enough means.

Theorem 3.48 (truncate_correct). Let x be a nonnegative real, m_0 and e_0 be two integers, and ℓ_0 be a location such that (inbetween_float m0 e0 x l0) holds. If either of the following two properties hold,

– we have $e_0 \leq \varphi(s_0)$ with s_0 the magnitude of $m_0 \cdot \beta^{e_0}$, or

– ℓ is loc_Exact,

then the triple $(m, e, \ell) = \text{truncate}(m_0, e_0, \ell_0)$ satisfies

– inbetween_float m e x l, and

– e is the canonical exponent of x, or ℓ is loc_Exact and $x \in F$.

Using the truncate function, we can now compute the rounding of x. Indeed, section 3.3.1.1 explains how to build, for some rounding modes, a function *choice* such that

$$\square(x) = choice(m, \ell) \cdot \beta^e$$

holds when e is the canonical exponent of x and (inbetween_float m e x l) holds. This leads to the following theorem.

Theorem 3.49 (round_trunc_any_correct). Let x be a nonnegative real, m_0 and e_0 be two integers, and ℓ_0 be a location such that (inbetween_float m0 e0 x l0) holds. If either of the following two properties hold,

– we have $e_0 \leq \varphi(s_0)$ with s_0 the magnitude of $m_0 \cdot \beta^{e_0}$, or

– ℓ is loc_Exact,

then the triple $(m, e, \ell) = \text{truncate}(m_0, e_0, \ell_0)$ is such that

$$\square(x) = choice(m, \ell) \cdot \beta^e.$$

The theorems above take (inbetween_float m0 e0 x 10) and $x \geq 0$ as hypotheses. But, as in section 3.3.1.1, there might be situations where x is possibly negative and it is the location of $|x|$ we know: (inbetween_float m0 e0 (Rabs x) 10). In that case, the round_trunc_sign_any_correct corollary can be of use.

We now have enough tools to perform some rounded computations: addition, multiplication, division, and square root.

3.3.1.3. *Addition and multiplication*

Now that we are able to round any number of the form Float m0 e0, we can implement simple functions for computing the FP sum and FP product. Indeed, an FP operation should behave as if it were first computing the result with infinite precision and then rounding it to the target format (see section 1.1.4). For addition and multiplication, we can actually compute this intermediate result, since the precision needed is effectively finite.

Let us consider the case of multiplication first. The exact operation can be defined as follows. Given two FP numbers as inputs, it computes the product of their significands and the sum of their exponents.

```
Definition Fmult (f1 f2 : float beta) : float beta :=
  let '(Float m1 e1) := f1 in
  let '(Float m2 e2) := f2 in
  Float (m1 * m2) (e1 + e2).
```

```
Theorem F2R_mult :
  forall f1 f2 : float beta,
  F2R (Fmult f1 f2) = F2R f1 * F2R f2.
```

For addition, the implementation of the exact operation is slightly more complicated, since we need to align the two input FP numbers on the same exponent, so that we can add their significand. This is the purpose of the Falign function. Given two FP numbers, it returns the minimum of the two exponents and the significands of the inputs with respect to this exponent. The definition and correctness of the addition are then stated as follows.

```
Definition Fplus (f1 f2 : float beta) : float beta :=
  let '(m1, m2, e) := Falign f1 f2 in
  Float (m1 + m2) e.
```

Theorem F2R_plus :
  forall f1 f2 : float beta,
  F2R (Fplus f1 f2) = F2R f1 + F2R f2.

Since the algorithms above compute the exact sum or product of two FP inputs, the truncate function can be applied to this exact result with a loc_Exact location. Theorem 3.49 then shows how to obtain the rounded sum or product from the result of truncate (assuming the exact result was nonnegative).

Note that, while these algorithms effectively compute an FP result in Coq, they are not the most efficient ones to do so. In particular, when adding a negligible number, the algorithm performs two large shifts (one for the input alignment and one for the truncation during rounding), while the correct result could have been guessed from the signs and magnitudes of the inputs.

Flocq does not provide any FMA operator, but the same approach as above could be used. First, we use a combination of Fmult and Fplus to compute the infinitely-precise result, then we apply truncate to obtain the rounded result.

3.3.2. *Division and square root*

For addition and multiplication, we are able to simply compute the infinitely-precise result and then round it to the target format. While possibly inefficient, it gives a simple yet correct and effective implementation of these operators. For division and square root, the situation is different, since we can no longer represent the intermediate infinitely-precise result z as an FP number. So we use instead a truncated FP number $\tilde{z}$ and a location ℓ to represent this intermediate result. We can then reuse the mechanism described in section 3.3.1.2 to get the final rounded result $\square(z)$ from $\tilde{z}$ and ℓ.

For this final stage to succeed in computing $\square(z)$, the exponent of the intermediate result $\tilde{z}$ must be less than or equal to the canonical exponent of the infinitely-precise result z (either that or the location ℓ is loc_Exact). We should strive to get the exponent of $\tilde{z}$ as close to the canonical exponent as possible, so that $\tilde{z}$ is not computed slower than needed due to some extra accuracy.

3.3.2.1. *Division*

Given two FP numbers $x = m_x \cdot \beta^{e_x}$ and $y = m_y \cdot \beta^{e_y}$, we want to compute the rounded value $\square(z)$ of the exact quotient $z = x/y$. The case $x = 0$ is immediate and the case $y = 0$ is undefined, so let us consider nonzero inputs. Since the theorems of section 3.3.1 make it possible to round z given the location of $|z| = |x|/|y|$, we can assume without any loss of generality that m_x and m_y are positive integers. Let us denote d_x and d_y as the numbers of radix-β digits of m_x and m_y. We thus have the following enclosure:

$$\beta^{d_x+e_x-d_y-e_y-1} < z < \beta^{d_x+e_x-d_y-e_y+1}.$$

As a result, there are only two possible canonical exponents for z: $\varphi(d_x+e_x-d_y-e_y)$ or $\varphi(d_x+e_x-d_y-e_y+1)$. Let e be the smallest of these two exponents. Note that, for the FIX, FLX, and FLT formats, the first exponent is either equal to the second one, or one less. So, for these common formats, there will not be much performance overhead in always choosing $e = \varphi(d_x+e_x-d_y-e_y)$.

Now that we have a suitable exponent e, we need to compute a significand m and a location ℓ such that (inbetween_float m e (x / y) l) holds. This means that the significand has to satisfy the following property:

$$m \cdot \beta^e \leq z < (m+1) \cdot \beta^e,$$

which is equivalent to

$$m \leq (m_x/m_y) \cdot \beta^{e_x-e_y-e} < m+1.$$

In other words, m is the Euclidean quotient of two integers u and v. If $e_x - e_y \geq e$, then these two integers are $u = m_x \cdot \beta^{e_x-e_y-e}$ and $v = m_y$. Otherwise, they are $u = m_x$ and $v = m_y \cdot \beta^{e-e_x+e_y}$. Note that the case $e_x - e_y \geq e$ is the most common one since it occurs whenever the result is in the normal range and m_x does not have an excessive amount of digits. Thus, one could further tweak the heuristic choice of e by choosing a sufficiently small exponent, so that only the common case remains, for reasons of simplicity.

As in section 3.3.1.2, the location ℓ can be obtained by looking at the remainder w of the Euclidean division $u \div v$. The computation of ℓ can be simplified though, since we do not have to deal with any imprecision on the inputs. In particular, we no longer need to make any distinction between the cases v even and v odd. This leads to the following formula when the division is inexact ($w \neq 0$):

$$\ell = \text{loc\_Inexact} \begin{cases} \text{Lt} & \text{if } 0 < 2 \cdot w < v, \\ \text{Eq} & \text{if } 2 \cdot w = v, \\ \text{Gt} & \text{if } 2 \cdot w > v. \end{cases}$$

Looking at the FLXN format (section 3.1.2.4) gives some intuition on the values at hand. Since this format only allows normalized numbers, we have $d_x = d_y = \varrho$, so the chosen exponent is

$$e = \varphi(d_x + e_x - d_y - e_y) = e_x - e_y - \varrho.$$

As a result, $e_x - e_y - e = \varrho$. So the FP operator performs an integer division between a 2ϱ-digit integer and a ϱ-digit integer. The resulting significand m has at most one extra digit, which has to be truncated during the final rounding.

3.3.2.2. Square root

The case of square root is quite similar to that of division. Given a positive FP number $x = m_x \cdot \beta^{e_x}$, we want to compute the rounded value $\square(z)$ of the exact square root $z = \sqrt{x}$. If the integer significand m_x has d_x digits, we have the following enclosure:

$$\beta^{(d_x + e_x - 1)/2} \leq z < \beta^{(d_x + e_x)/2}.$$

This time, the canonical exponent of z is known exactly: $\varphi(\lceil (d_x + e_x)/2 \rceil)$. As for the significand m, it has to satisfy the following property, assuming that e is any exponent less than or equal to the canonical one:

$$m \cdot \beta^e \leq \sqrt{m_x \cdot \beta^{e_x}} < (m+1) \cdot \beta^e.$$

This is equivalent to

$$m \leq \sqrt{m_x \cdot \beta^{e_x - 2e}} < m + 1.$$

To use an integer square root to compute m, we need $m_x \cdot \beta^{e_x - 2e}$ to be an integer; choosing e so that $e_x \geq 2e$ ensures that. The largest exponent satisfying both constraints is

$$e = \min(\varphi(\lceil (d_x + e_x)/2 \rceil), \lfloor e_x/2 \rfloor).$$

The significand m is the integer square root of $v = m_x \cdot \beta^{e_x - 2e}$, that is, $m = \lfloor \sqrt{v} \rfloor$. As with division, we get the location ℓ by looking at the remainder $w = v - m^2$. In the inexact case, we have to compare $\sqrt{v}$ with $m + 1/2$, which is the same as comparing v with $m^2 + m + 1/4$, which is the same as comparing w with $m + 1/4$. Since w and m are integers, this gives the following formula when the square root is inexact:

$$\ell = \texttt{loc\_Inexact} \begin{cases} \texttt{Lt} & \text{if } 0 < w \leq m, \\ \texttt{Gt} & \text{if } w > m. \end{cases}$$

Notice that the case loc_Inexact Eq cannot occur. This does not mean that $\sqrt{x}$ can never be a midpoint, though. Indeed, since e is potentially smaller than the canonical exponent, the subsequent normalization could change the location to loc_Inexact Eq when the radix is even. For odd radices, however, $\sqrt{x}$ is never a midpoint, whatever the format.

Let us look at an FLXN format to see how the operator behaves. Since $d_x = \varrho$ in this case, we have

$$
\begin{aligned}
e &= \min(\varphi(\lceil (d_x + e_x)/2 \rceil), \lfloor e_x/2 \rfloor) \\
&= \min(\lfloor (e_x + 1 - \varrho)/2 \rfloor, \lfloor e_x/2 \rfloor) \\
&= \lfloor (e_x + 1 - \varrho)/2 \rfloor.
\end{aligned}
$$

As a result, $e_x - 2e$ is equal to either $\varrho - 1$ or ϱ. So the FP operator computes the square root of an integer with 2ϱ digits at most. The resulting significand m has ϱ digits and does not need any truncation. It might still have to be increased by one to account for the rounding direction, though.

This concludes the section on effectively performing basic FP computations in Coq and proving the correctness of the corresponding algorithms. These computations are not meant to be efficient, yet for multiplication, division, and square root, the algorithms have a reasonable complexity for standard formats, since they do not compute an intermediate result that is much more accurate than needed.

3.4. IEEE-754 binary formats and operators

Up to now, we have ignored exceptional values by assuming they could not occur (e.g. infinities) or that they were not meaningful (e.g. signed zeros). We now consider binary IEEE-754-compliant formats which are represented either as inductive types or as strings of bits (that is, as integers).

Section 3.4.1 uses an inductive type to account for exceptional values: signed infinities, signed zeros, NaNs and their payload. Finite nonzero numbers are represented by their sign, integer significand, and exponent. Finite numbers can no longer be arbitrarily large; only the values that can ultimately be stored are considered. So the real numbers that can be represented are now just a subset of those of an FLT format with $\beta = 2$. We do not consider decimal formats, which could be handled in sensibly the same way, except that significands would no longer be normalized.

The finite and exceptional cases are described by the inductive type `binary_float` that will be detailed below. Such a representation is close to the one given by level 3 of the IEEE-754 standard, but there is a difference. We do not make any distinction between quiet and signaling NaNs, as we do not formalize exception throwing and catching. As a result, the distinction between the two kinds of NaNs is simply part of their payload and we just ignore this distinction.

Section 3.4.2 then explains how to represent FP numbers as bounded nonnegative integers, which is the way they would be represented in memory for instance; this corresponds to level 4 of the IEEE-754 standard. Finally section 3.4.3 makes use of the results of section 3.3 to define IEEE-754-compliant arithmetic operators. In both sections 3.4.2 and 3.4.3, the functions are effective, that is, they can be used to perform computations within Coq.

3.4.1. *Formalizing IEEE-754 binary formats*

The three IEEE-754 binary formats, *binary32*, *binary64*, and *binary128*, and the generic interchange format are all built around the same principles. They can be characterized from just two constants: ϱ and $e_{\max}$. The first one describes the range of the integer significands, while the second one describes the range of the exponents.

Variable prec emax : Z.

In Flocq's formalism, $e_{\max}$ is chosen so that $2^{e_{\max}}$ is the smallest power of two that is too large to be representable. Note that this is one more than the definition of the maximal exponent in the IEEE-754 standard. From $e_{\max}$, we deduce the exponent $e_{\min}$ such that $2^{e_{\min}}$ is the smallest power of two that can be represented.

Let emin := (3 - emax - prec)%Z.

Note that the smallest positive normal number is thus $2^{e_{\min}+\varrho-1} = 2^{2-e_{\max}}$.

The choice of the parameters ϱ and $e_{\max}$ is mostly free. We only require the precision to be positive and $e_{\max}$ to be strictly larger than ϱ. The second requirement ensures that the square root of any number, even a small one, is a normal number, since it is equivalent to $e_{\min}/2 \geq 2 - e_{\max}$. The corresponding parameters for the IEEE-754 formats can be found in Table 3.1.

Hypothesis prec_gt_0_ : (0 < prec)%Z.
Hypothesis Hmax : (prec < emax)%Z.

Before we can define the set of FP numbers, we need a few auxiliary definitions. First of all, given a positive integer m and an integer e, the bounded function returns *true* when the following two conditions hold. One, $m \cdot 2^e$ (seen as a pair of integers) should be the canonical way of representing $m \cdot 2^e$ (seen as a real number). Two, $m \cdot 2^e$ should not exceed the overflow threshold. Note that, if the first condition holds, then we have $|m| < 2^\varrho$, so the second condition can be simplified to $e \leq e_{\max} - \varrho$ in that case.

Definition bounded (m : positive) (e : Z) :=
  andb (canonical_mantissa m e) (Zle_bool e (emax - prec)).

The call to canonical_mantissa checks whether m is in the integer range mandated by the value of e. (Note that, when e is smaller than e_{min}, this range is empty.) It first computes the magnitude of $m \cdot 2^e$ by calling digits2_pos to count the number of bits of m and by adding e to it. This magnitude is then passed to the FLT_exp function (section 3.1.3.4). This function returns the exponent that should have been used to represent $m \cdot 2^e$, so canonical_mantissa just has to check that it matches e.

Definition canonical_mantissa (m : positive) (e : Z) :=
Zeq_bool (FLT_exp emin prec (Zpos (digits2_pos m) + e)) e.

We also need to define the set of payloads that are allowed for NaNs. These payloads are positive integers that can fit into the significand. Note that the zero payload is reserved for representing infinities and thus is implicitly disallowed here by using the positive type.

Definition nan_pl (pl : positive) :=
Zlt_bool (Zpos (digits2_pos pl)) prec.

We can now define an IEEE-754 binary format as an inductive type binary_float. It is implicitly parameterized by ϱ and e_{max}, as is the function bounded it uses. It contains four branches, depending on whether a number is a signed zero, a signed infinity, a NaN (also signed, for uniformity), or a finite nonzero number. In all these branches, the Boolean value encodes the sign of the FP number: it is *true* whenever the number is negative.

Inductive binary_float :=
| B754_zero (s : bool)
| B754_infinity (s : bool)
| B754_nan (s : bool) (pl : positive) :
nan_pl pl = true -> binary_float
| B754_finite (s : bool) (m : positive) (e : Z),
bounded m e = true -> binary_float.

Notice that the branches for NaNs and for finite numbers store proofs that their arguments are not out of range. Thus one cannot create an object of type binary_float if these properties are not proved. As it might be useful to temporarily represent FP numbers that are not yet proved to satisfy these properties, a separate type full_float is provided. It contains the same four branches but without any kind of proof.

Inductive full_float :=
| F754_zero (s : bool
| F754_infinity (s : bool)
| F754_nan (s : bool) (pl : positive)

```
| F754_finite (s : bool) (m : positive) (e : Z).
```

There is also a valid_binary predicate that checks whether a value of type full_float can be converted to binary_float. The conversion itself is named FF2B. The definition of an FP number can thus be done in three steps: first, create a value x of type full_float, then check that it is well formed by proving valid_binary x = true and finally convert it to binary_float using FF2B.

```
Definition valid_binary (x : full_float) : bool :=
  match x with
  | F754_finite _ m e => bounded m e
  | F754_nan _ pl => nan_pl pl
  | _ => true
  end.
```

```
Definition FF2B : forall x : full_float,
  valid_binary x = true -> binary_float.
```

The converse operation B2FF removes the proofs from a binary_float value and returns a full_float value.

The behavior of the operators depends on the class of their inputs. To ease their specification, we define a few Boolean functions to classify FP numbers. For instance, is_finite_strict returns *true* if and only if the FP number is finite and nonzero, while is_finite also returns *true* when the number is zero. Another classifier that will be useful for formalizing operators is is_nan. All these functions are defined by pattern-matching over the input FP number. Here is the example of is_finite:

```
Definition is_finite (f : binary_float) : bool :=
  match f with
  | B754_finite _ _ _ _ => true
  | B754_zero _ => true
  | _ => false
  end.
```

The is_finite_FF function behaves similarly for full_float inputs. Two other useful functions are Bsign and sign_FF, which return the sign of their inputs of type binary_float and full_float, respectively.

An important function is B2R of type binary_float -> R. Given a finite FP number, it returns the corresponding real number using the F2R function (see section 3.1.1). Otherwise, it returns zero (even for infinite inputs). In other words, this function relates concrete values of type binary_float with more abstract ones of type R, with exceptional values such as infinities and NaNs being flushed to zero.

```
Definition B2R (f : binary_float) :=
  match f with
  | B754_finite s m e _ =>
    F2R (Float radix2 (cond_Zopp s (Zpos m)) e)
  | _ => 0
  end.
```

First, B2R extracts the sign s of the finite number, its positive integer significand m, and its exponent e. Note that the proof that (m, e) is the canonical representation of the number is ignored as it does not have any impact on the represented real number. Then, the function constructs a signed integer significand using the cond_Zopp function which negates its second argument when the first one evaluates to *true*. Then, it constructs an object of type float radix2, with radix2 representing $\beta = 2$. Finally, it converts that object to a real number using F2R. The FF2R function behaves similarly for full_float inputs.

Note that, except for $+0$ and -0, no two finite FP numbers can represent the same real number. As a result, the B2R function is injective for nonzero finite numbers. In other words, to prove that two such numbers are equal, we just have to prove that the real values they represent are equal and then apply the following theorem.

```
Theorem B2R_inj:
  forall x y : binary_float,
  is_finite_strict x = true ->
  is_finite_strict y = true ->
  B2R x = B2R y ->
  x = y.
```

3.4.2. *Bit-level representation*

The B2R function makes it possible to convert an FP number into a real number. There are some other conversion functions that make it possible to go to and from the bit-level representation:

```
Definition bits_of_binary_float : binary_float -> Z.
Definition binary_float_of_bits : Z -> binary_float.
```

These functions need to know how many bits are reserved for the exponent and how many bits are reserved for the significand.

```
Variable mw ew : Z.
Hypothesis Hmw : (0 < mw)%Z.
Hypothesis Hew : (0 < ew)%Z.
```

From these values *mw* and *ew*, we can infer the precision and the maximal exponent that implicitly parameterize the `binary_float` type. Note that the most-significant bit is implicit for normal numbers (see section 1.3.2).

Let emax := Zpower 2 (ew - 1).
Let prec := (mw + 1)%Z.

Hence, by giving some widths *mw* for the mantissa and *ew* for the exponent, one can define an IEEE-754 format represented on $mw + ew + 1$ bits. The sign bit is the most significant one, followed by the biased exponent (see section 1.3.2), and finally the significand.

Theorem bits_of_binary_float_range:
  forall x : binary_float,
  (0 <= bits_of_binary_float x < 2^(mw+ew+1))%Z.

An important property of these functions is that they are effective: Given a value of type `binary_float`, one can produce the integer representing its bit pattern, and conversely. This effectiveness will be needed for defining some parts of the CompCert compiler: representing FP numbers as integer words (section 7.2.3) and fiddling with their representation when performing conversions (section 7.3).

Due to their effectiveness, it is possible to test that these functions comply with the IEEE-754 standard by running them on various inputs. Another way to increase the confidence in their compliance is to prove a few properties that are supposed to hold according to the IEEE-754 standard. For instance, the two conversion functions have to be inverse of each other:

Theorem binary_float_of_bits_of_binary_float :
  forall x : binary_float,
  binary_float_of_bits (bits_of_binary_float x) = x.

Theorem bits_of_binary_float_of_bits :
  forall x : Z,
  (0 <= x < 2^(mw+ew+1))%Z ->
  bits_of_binary_float (binary_float_of_bits x) = x.

3.4.3. IEEE-754 arithmetic operators

Section 3.3 explained how to effectively compute the rounded result of a basic arithmetic operator ($+$, $\times$, $\div$, $\sqrt{\cdot}$) applied to FP numbers with unbounded exponents. These algorithms can be extended to IEEE-754 binary formats. Given some `binary_float` numbers and a rounding mode, such an FP operator should return a binary_float result that complies with the IEEE-754 specification. Moreover, this operator should be effective; for instance, it should not rely on operations on real numbers, as they are abstract and thus prevent computations.

Section 3.4.3.1 explains how these operators are structured so as to handle exceptional inputs such as zeros, infinities, and NaNs. Section 3.4.3.2 then shows how to specify the behavior of the operators, in particular when the output is either finite or overflowing. Finally, section 3.4.3.3 gives some details on the implementation of these operators in Flocq.

3.4.3.1. *Exceptional inputs*

Due to the inductive nature of the binary_float type, an FP operator can be implemented in Coq using pattern-matching on its inputs. For instance, the following excerpt shows how the FP multiplication operator handles some of the exceptional cases: product of infinities, product of zeros, and product of finite numbers by either zeros or infinities. For all these cases, the result is either zero or infinity, and one just has to properly compute its sign.

```
Definition Bmult ... (x y : binary_float) : binary_float :=
  match x, y with
  | B754_infinity sx,  B754_infinity sy
=> B754_infinity (xorb sx sy)
  | B754_infinity sx,
B754_finite sy _ _ _ => B754_infinity (xorb sx sy)
  | B754_finite sx _ _ _, B754_infinity sy
=> B754_infinity (xorb sx sy)
  | B754_finite sx _ _ _, B754_zero sy     => B754_zero (xorb sx sy)
  | B754_zero sx,
B754_finite sy _ _ _ => B754_zero (xorb sx sy)
  | B754_zero sx,        B754_zero sy       => B754_zero (xorb sx sy)
  | ...
  end.
```

When computing the product of a zero by an infinity, the result is a quiet NaN according to the default exception handling of IEEE-754. Since the binary_float format embeds the type, sign, and payload of NaNs, the Bmult function has to provide them. Unfortunately, the IEEE-754 standard is under-specified with respect to what their value is. The only guidance is that it should be for diagnostic purposes. If we were to choose arbitrary values, we would no longer accurately model existing architectures. So the Bmult function has to be parameterized by a user-defined function for generating NaNs; this function takes the two inputs and returns an FP number and a proof that it is a NaN.

```
Definition Bmult
```

```
(mult_nan : binary_float -> binary_float ->
            { nan | is_nan nan = true })
(x y : binary_float) :=
match x, y with
| B754_nan _ _, _ | _, B754_nan _ _
| B754_infinity _, B754_zero _
| B754_zero _, B754_infinity _ =>
    proj1_sig (mult_nan x y)
| ...
end.
```

In order to complete the Bmult function, we have to handle the remaining case, that is, the product of two finite numbers. Except in the case of an exact non-overflowing FP result, the result depends on the rounding mode, which will be passed as an argument to Bmult. This mode is defined as an element of the enumeration of the five standard IEEE-754 modes:

Inductive mode :=
  mode_NE | mode_ZR | mode_DN | mode_UP | mode_NA.

These modes can be translated into the ones defined in section 3.2 by using the round_mode function. Given an input of type mode, it returns a function of type R -> Z that tells us how to round a scaled mantissa to an integer significand (see section 3.2.2).

We now have all the components needed to state the type of Bmult:

Definition Bmult
  (mult_nan : binary_float -> binary_float -> { nan | ... })
  (m : mode) (x y : binary_float) : binary_float := ...

The signature of the other arithmetic operators is similar: they take a function for generating NaNs, a rounding mode, some FP operands, and they return the FP result. Their body is also similar: they decompose their inputs into the various classes of FP numbers so as to handle the exceptional cases early.

3.4.3.2. *Specifications*

Before looking at the way Flocq's operators are implemented for finite inputs (section 3.4.3.3), let us have a look at the way in which they are specified. Their specifications do not deal with exceptional inputs, since that would require an abstract arithmetic over

$$\mathbb{R}^* \cup \{\pm\infty\} \cup \{\pm 0\} \cup \{\text{NaN}\},$$

as implicitly done in the IEEE-754 standard. The lack of such a specification is not much of a hindrance though, since the operators are designed for effective computations. For instance, to prove that the result of Bmult is infinite if the inputs are infinite, one can evaluate the operator applied to such inputs.

While we can also execute the operators on finite inputs, such an execution does not tell us much about the general properties of an operator. In particular, it does not tell us that the returned value is the closest (according to a rounding mode) from the infinitely-precise result. This is both a defining feature of the arithmetic operators and a property useful in verifying programs that use them. As a first approximation, a specification for Bmult could thus be stated as follows:

```
forall mult_nan (m : mode) (x y : binary_float),
is_finite x = true -> is_finite y = true ->
let rounded := round (round_mode m) (B2R x * B2R y) in
let result := Bmult mult_nan m x y in
B2R result = rounded.
```

In the property above, round is the rounding operator over generic formats (see section 3.2) instantiated with FLT_exp using the minimal exponent and the precision of the target binary format. There is a major drawback with that property though: it does not hold when the inputs are finite but large enough to cause an overflow. So the specification has to account for that case as well. To do so, we modify the last line B2R result = rounded so that the formula first compares the rounded real to $2^{e_{\max}}$:

```
if Rlt_bool (Rabs rounded) (bpow radix2 emax) then
  B2R result = rounded
else
  B2FF result = binary_overflow m (xorb (Bsign x) (Bsign y)).
```

The binary_overflow function takes a rounding mode and a Boolean indicating on which side of the real line the overflow occurs. It returns a full_float number that is either an infinite number or the largest positive or negative FP number.

The specification is now valid but it suffers from a few issues related to its lack of completeness. First of all, when the rounded value is zero, we have no idea what the actual result is, since B2R maps all the exceptional values to the real zero. So the specification does not prevent a broken implementation from returning an infinity in case of underflow. By unfolding Bmult and its subfunctions, the user could still prove that it is not broken, but it is unwieldy. So it is better if the specification actually states that the returned value is finite when the rounded result is in the range of finite numbers.

There is a minor related issue. While we can deduce from the specification that zero is returned in case of underflow, we do not know its sign yet. So the specification could state that this zero has the same sign as the infinitely-precise product when nonzero. While sufficient, it is overly restrictive. We can do a bit better by stating instead that the result always satisfies the rule of signs, even when zero. Note that this statement also tells us which is the sign of the result when one of the inputs is zero (though this could have been obtained by a simple computation). To fix these two issues, we replace the B2R result = rounded branch of the conditional with the following conjunction in the specification of Bmult:

```
B2R result = rounded /\
is_finite result = true /\
Bsign result = xorb (Bsign x) (Bsign y)
```

There is still a last issue though, in the sense that most of this specification holds even for non-finite inputs. Indeed, for non-finite inputs, the result is an exceptional FP number and the product (B2R x * B2R y) equals zero. As a result, we are in the *then* branch of the conditional and the property (B2R result = rounded) holds. The property about the sign of the result also holds, except when the result is NaN (since the standard is underspecified in that case). The property about the finiteness of the result no longer holds for non-finite inputs, but it can easily be modified so that is does. This gives the following specification for Bmult:

Theorem Bmult_correct :
```
    forall mult_nan (m : mode) (x y : binary_float),
    let rounded := round (round_mode m) (B2R x * B2R y) in
    let result := Bmult mult_nan m x y in
    if Rlt_bool (Rabs rounded) (bpow radix2 emax) then
      B2R result = rounded /\
      is_finite result = andb (is_finite x) (is_finite y) /\
      (is_nan result = false ->
        Bsign result = xorb (Bsign x) (Bsign y))
    else
      B2FF result = binary_overflow m (xorb (Bsign x) (Bsign y)).
```

The specification for the FP division operator Bdiv is similar. There is a small difference due to the division over real numbers not being defined for a zero denominator in Coq. As a result, the case B2R y = 0 has to be explicitly excluded from the specification.

In the case of the FP addition Bplus and subtraction Bminus, the situation is a bit more complicated. First, the sum (B2R x + B2R y) can be nonzero for non-finite inputs. So we cannot get rid of the requirements for the finiteness of the inputs. Second, the rule for the sign of the result is slightly more complicated, since it depends on the sign of the inputs, on the sign of the infinitely-precise result, and on the rounding mode (see section 1.3.3). Compliance with the IEEE-754 is thus expressed as follows.

Theorem Bplus_correct :
```
    forall plus_nan (m : mode) (x y : binary_float),
```

```
is_finite x = true ->
is_finite y = true ->
let rounded := round (round_mode m) (B2R x + B2R y) in
let result := Bplus plus_nan m x y in
if Rlt_bool (Rabs rounded) (bpow radix2 emax) then
  B2R result = rounded /\
  is_finite result = true /\
  Bsign result =
    match Rcompare (B2R x + B2R y) 0 with
    | Eq =>
      match m with
      | mode_DN => orb (Bsign x) (Bsign y)
      | _ => andb (Bsign x) (Bsign y)
      end
    | Lt => true
    | Gt => false
    end
else
  B2FF result = binary_overflow m (Bsign x) /\
  Bsign x = Bsign y.
```

Notice that, in case of overflow, the result is stated as having the same sign as the first operand. To simplify the use of this property, the specification also states that both operands have the same sign. So the user is only one rewriting away from knowing that the result also has the sign of the second operand in case of overflow.

The remaining FP operation provided by Flocq is the square root Bsqrt. Its specification is much simpler than the previous ones, since no overflow can ever occur. So, there is no conditional anymore but we still have the three usual properties: the real value of the final result is the rounded value of the infinitely-precise result, the result is finite as soon as the input is a nonnegative finite number (including -0), and the sign of the result is the same as the sign of the input as long as the result is not NaN (remember that $\sqrt{-0} = -0$).

Theorem Bsqrt_correct :
```
  forall sqrt_nan (m : mode) (x : binary_float),
  let rounded := round (round_mode m) (sqrt (B2R x)) in
  let result := Bsqrt sqrt_nan m x in
  B2R result = rounded /\
  is_finite result =
    match x with
    | B754_zero _ => true
    | B754_finite false _ _ _ => true
    | _ => false
```

```
    end /\
(is_nan result = false ->
 Bsign result = Bsign x).
```

3.4.3.3. Implementation

Finally, let us look at an implementation of the arithmetic operators that satisfies the specifications for finite inputs. Section 3.3 explained how to perform such computations in the case of generic formats, of which FLT formats are a special case. For instance, when performing a division or a square root (section 3.3.2), one can first compute an approximation of the infinitely-precise result and then round it. While the same approach can be used here, there is a major difference. Indeed, the implementation now has to account for overflow at the time of rounding. So let us focus on the rounding operation.

The main function is `binary_round_aux`. Given a rounding mode and an approximation represented by its sign, its positive significand, its exponent, and the location of the infinitely-precise result with respect to it, the function computes a `full_float` number representing the final result. The function first truncates the input significand as mandated by the rounding mode and by the φ function describing the format. It then adjusts the input exponent accordingly. If the resulting exponent is too large for the number to be representable, it calls `binary_overflow`. Otherwise it returns the FP number formed by the input sign, truncated significand, and adjusted exponent.

The specification of the function relates its result to the infinitely-precise value x which we are trying to round. First, the resulting number is well formed (according to `valid_binary`) and thus can later be converted to a `binary_float` number using FF2B. Second, the resulting number satisfies all the properties needed for Bdiv and Bsqrt to satisfy their specification. In other words, if the rounded value $\square(x)$ is in the range of finite numbers, the result of `binary_round_aux` satisfies the following properties: its real value is equal to the rounded value of the infinitely-precise result, it is finite, and it has the same sign as x. Otherwise, if $\square(x)$ falls outside the range of finite numbers, the result of `binary_round_aux` is given by `binary_overflow`.

This specification has two preconditions. First, the sign, significand, exponent, and location passed to `binary_round_aux` should be compatible with x, obviously. Second, the significand should have sufficiently many bits so that it can be truncated to an FP number satisfying the bounded predicate. Both preconditions are satisfied by the functions described in section 3.3.2.

Theorem binary_round_aux_correct :
```
  forall (m : mode) (x : R) mx ex lx,
  inbetween_float radix2 (Zpos mx) ex (Rabs x) lx ->
  (ex <= fexp (Zdigits radix2 (Zpos mx) + ex))%Z ->
```

```
let result := binary_round_aux m (Rlt_bool x 0) mx ex lx in
valid_binary result = true /\
let rounded := round (round_mode m) x in
if Rlt_bool (Rabs rounded) (bpow radix2 emax) then
  FF2R radix2 result = rounded /\
  is_finite_FF result = true /\
  sign_FF result = Rlt_bool x 0
else
  result = binary_overflow mode (Rlt_bool x 0).
```

The implementation of Bmult also uses the binary_round_aux function. The main difference is that the location is now loc_Exact, since it computes the significand and exponent of the infinitely-precise product. This makes the first precondition of binary_round_aux_correct trivially satisfied. As for the second one, it is a consequence of the fact that the inputs of Bmult have canonical significands, by definition of binary_float. Indeed, if either input is a normal number, then the product of the input significands has more than ϱ bits. Otherwise, they are both subnormal numbers, so the sum of the input exponents is so small that the size of the significand no longer matters. So the definition of Bmult is now straightforward in the finite case:

Definition Bmult mult_nan (m : mode) x y : binary_float :=
```
  match x, y with
  | ...
  | B754_finite sx mx ex _, B754_finite sy my ey _ =>
    binary_round_aux m (xorb sx sy) (mx * my) (ex + ey) loc_Exact
  end.
```

The case of Bplus and Bminus is a bit more complicated, since the significand of the infinitely-precise result might be the result of a cancellation, so it might not be possible to truncate it to a canonical representation due to the lack of significant bits. The binary_round function is a wrapper around binary_round_aux that makes sure the input significand can be truncated. To that effect, it assumes that it is past the infinitely-precise result (as a sign, significand, and exponent) instead of an approximation of it. As a result, its specification does not have any precondition.

Theorem binary_round_correct :
```
  forall (m : mode) sx mx ex,
  let result := binary_round m sx mx ex in
  valid_binary result = true /\
  let x := F2R (Float radix2 (cond_Zopp sx (Zpos mx)) ex) in
  let rounded := round (round_mode m) x in
  if Rlt_bool (Rabs rounded) (bpow radix2 emax) then
    FF2R radix2 result = rounded /\
```

```
  is_finite_FF result = true /\
  sign_FF result = sx
else
  result = binary_overflow m sx.
```

Bplus and Bminus do not use this function directly but through yet another wrapper: binary_normalize. Instead of a positive significand, this function takes a signed significand and a sign that should be used for the result if the significand is zero. This gives the following definition for Bplus:

Definition Bplus plus_nan (m : mode) x y : binary_float :=
```
  match x, y with
  | ...
  | B754_finite sx mx ex _, B754_finite sy my ey _ =>
    let ez := Zmin ex ey in
    let mz :=
      (cond_Zopp sx (Zpos (fst (shl_align mx ex ez))) +
       cond_Zopp sy (Zpos (fst (shl_align my ey ez))))%Z in
    binary_normalize m mz ez
      (match m with mode_DN => true | _ => false end)
  end.
```

More generally, binary_normalize can be used to convert any dyadic number to an FP number.

This chapter has presented the formalization of FP arithmetic we are going to use throughout this book. Its main characteristic is its ability to have both high-level properties such as a mathematical definition of rounding and some computational contents. This last feature will be described in more detail in the next chapter, which shows how to use computations to reduce the proof burden.

4

Automated Methods

Since FP numbers are often meant to approximate real numbers, it comes without surprise that proving the correctness of an FP algorithm might involve verifying some facts about real-valued expressions. In fact, due to the discrete nature of FP numbers, some of these facts might even involve only integer expressions. Formally verifying such facts often imposes a large proof burden on the user. Fortunately, Coq's standard library comes with several tactics that might help in discharging this kind of goal. Section 4.1 details some of the most useful tactics for automating the proof of integer and real properties. For instance, field can prove the equality between two rational fractions, while psatz can prove that a system of polynomial inequalities has no real solution. Most of the theorems presented in this book were formally proved in part due to such tactics.

When verifying FP algorithms, these tactics might not bring enough automation though, since they do not know about rounding errors and handle only basic arithmetic operators. For instance, one often needs to compute the actual range of some FP expressions, e.g. to prove that they do not underflow or overflow, or to deduce a bound on the absolute error that was introduced due to rounding. One might also need to combine and bound round-off errors so as to analyze the accuracy of an algorithm. Therefore this chapter also presents the interval and gappa tactics for automatically computing numerical enclosures of expressions and using these enclosures inside formal proofs. The correctness proofs of the examples presented in Chapters 6, 8, and 9 make heavy use of these two tactics (or of the associated tool in the case of gappa).

Both tactics are based on the principles of interval arithmetic, but they tackle different kinds of expressions. Section 4.2 shows the inner working of the interval tactic, which bounds real-valued expressions by using properties such as their differentiability. In particular, these expressions should not contain any rounding operator. Handling these operators is the purpose of the gappa tactic, described in section 4.3. While this tactic does not support as many basic functions as `interval`, it can make use of reasoning methods based on forward error analysis to compute round-off errors.

4.1. Algebraic manipulations

When proving properties about integers or real numbers, one often needs to verify equalities or inequalities between expressions. For instance, the verification of an FP algorithm might require proving that the inequalities $|x - y| \leq \varepsilon_1$ and $|y - z| \leq \varepsilon_2$ imply $|x - z| \leq \varepsilon_1 + \varepsilon_2$. One step of reasoning is to show that $x - z = (x - y) + (y - z)$. Coq's standard library provides several tactics to automate such proofs.

Whether on integers or on real numbers, if the conclusion of the goal is an equality between two expressions involving additions and multiplications, the `ring` tactic might be of help [GRÉ 05]. In case the expressions also involve some divisions of real numbers, there is the `field` tactic. Both `ring` and `field` mostly reason using only the conclusion of the goal. If the conclusion of the goal does not hold in isolation but is a consequence of some polynomial equalities present in the hypotheses, one should turn to the `nsatz` tactic [GRÉ 11]. These tactics are detailed in section 4.1.1.

If the hypotheses or the conclusion of the goal is not just about equalities over the real numbers, but also about inequalities, the previous tactics are no longer powerful enough. If equalities and inequalities are linear expressions, then the old `fourier` tactic or the more efficient `lra` tactic can be used. If the expressions are polynomials, then one should turn to the `psatz` tactic [BES 07]. These tactics are detailed in section 4.1.2.

Finally, for goals about linear inequalities over the integers, the `omega` [CRÉ 04], `romega`, and `lia` tactics might be of use, as detailed in section 4.1.3.

4.1.1. *Polynomial equalities*

The `ring` tactic tries to prove an equality between integer or real expressions using only the axioms of a commutative ring, that is, commutativity and associativity of addition and multiplication, distributivity, identity and inverse for addition. For instance, it can successfully discharge the following goal conclusion:

$$3 \cdot z \cdot z \cdot e^y \cdot 3 \cdot e^y - 4 = -(2 - 3 \cdot z \cdot e^y) \cdot (e^y \cdot z \cdot 3 + 2).$$

The tactic starts by considering the conclusion as an equality between two polynomials being evaluated at a point that encompasses any subexpression that is neither a sum nor a product nor a constant. In the equality above, the two polynomials are $3 \cdot x_1 \cdot x_1 \cdot x_2 \cdot 3 \cdot x_2 - 4$ and $-(2 - 3 \cdot x_1 \cdot x_2) \cdot (x_2 \cdot x_1 \cdot 3 + 2)$ of $\mathbb{Z}[x_1, x_2]$. The goal states that their values at point $\vec{x} = (z, e^y)$ are equal.

Once the tactic has turned the goal into an equality between two polynomial evaluations, it normalizes each of the polynomials, that is, it computes its unique representation as a linear combination of monomials for some arbitrary ordering of the monomials. Since the normalization of a polynomial does not alter its evaluation, the goal is proved if both normalized polynomials are equal. For the above example, both polynomials get normalized to $(-4) + 9 \cdot (x_1^2 \cdot x_2^2)$, which concludes the proof.

The field tactic extends ring by also recognizing and handling the inverse for multiplication. This time, the tactic does not use polynomials but rational fractions, that is, quotients of polynomials. Note that the normalization of a rational fraction might produce some side conditions which the user will have to prove. For instance, x_k/x_k and its normal form 1 have the same evaluation only if the k-th coordinate of the evaluation point is nonzero. So applying field to the goal $\sqrt{y}/\sqrt{y} = 1$ replaces it with the goal $\sqrt{y} \neq 0$.

There are two variants of the tactics above: ring_simplify and field_simplify. Instead of proving equalities, these tactics can be used to simplify subexpressions using the same normalization procedure. Note that it might be hard to predict what the normal form might be in all generality, so these simplification tactics are mostly useful when the polynomial expression ends up being reduced to a single monomial.

While the ring and field tactics can make use of some simple hypotheses to normalize the polynomials further, they are not suited for proving most algebraic identities involving the context. For instance, they cannot prove the following implication

$$x_1 + x_2 = 0 \Rightarrow x_1^2 = x_2^2.$$

That is the purpose of the nsatz tactic. This tactic first preprocesses the goal so as to turn all the polynomial equalities, be they hypotheses or conclusion, into equalities to zero. So, for some evaluation point $\vec{x}$, nsatz has to prove a goal of the form

$$\left(\bigwedge_{1 \leq k \leq N} P_k(\vec{x}) = 0 \right) \Rightarrow P(\vec{x}) = 0$$

with $(P_k)_{1 \leq k \leq N}$ the hypothesis polynomials and P the conclusion polynomial. The ring tactic would just ignore all the hypothesis polynomials. Then nsatz guesses some integers $c \neq 0$ and $r > 0$ and some polynomials $(Q_k)_{1 \leq k \leq N}$ such that the following equality holds:

$$c \cdot P^r = \sum_{1 \leq k \leq N} Q_k \cdot P_k. \qquad [4.1]$$

For the above example ($P = x_1^2 - x_2^2$ and $P_1 = x_1 + x_2$), the tactic guesses $c = 1$, $r = 1$, and $Q_1 = x_1 - x_2$. Since Coq does not trust any tactic, it verifies that equation [4.1] indeed holds for these constants and polynomials using the same machinery as ring. Since the context implies $\forall k$, $P_k(\vec{x}) = 0$, Coq deduces that $(c \cdot P^r)(\vec{x}) = 0$ holds by evaluating the polynomials of equation [4.1]. Finally, the tactic uses the fact that P is part of an integral domain to conclude that $P(\vec{x})$ is zero.

4.1.2. Linear and polynomial inequalities over the real numbers

The tactics above only handle the case of equalities. Other tactics can be used when the hypotheses or the conclusion are inequalities. Let us first consider the case where inequalities involve only affine expressions over the real numbers. In other words, given some vector $\vec{x}$ that encompasses the nonlinear expressions of the goal, we are trying to prove

$$A \cdot \vec{x} + \vec{b} \succ 0 \Rightarrow \bot$$

with A a matrix with rational coefficients and $\vec{b}$ a rational vector. Each entry of the vector $A \cdot \vec{x} + \vec{b}$ is compared to zero using either $\geq$ or $>$ and one has to prove that no instance of $\vec{x}$ can satisfy all these relations, denoted $\succ$ as a whole.

Both tactics fourier and lra are meant to handle this kind of goal. They do so by looking for some vector $\vec{y} \neq 0$ with nonnegative integer coefficients such that both $\vec{y} \cdot A = 0$ and $\vec{y} \cdot \vec{b} \not\succ 0$ hold.[1] The positiveness of $\vec{y}$ thus contradicts the hypothesis $A \cdot \vec{x} + \vec{b} \succ 0$. For example, one can prove that the following inequalities

$$5x_1 + 2x_2 + 4 \quad \geq \quad 0$$

$$-11x_1 + 4x_2 - 5 \quad > \quad 0$$

$$x_1 - 8x_2 - 3 \quad \geq \quad 0$$

are contradictory. Indeed, combining these inequalities using $\vec{y} = (2, 1, 1)$ would otherwise imply that $2 \cdot 4 + 1 \cdot (-5) + 1 \cdot (-3) = 0$ is positive. More generally, if there are no instances of $\vec{x}$ that satisfy a system of inequality, both tactics will be able to prove it. The main difference between them lies in the way they search for $\vec{y}$. The method used by lra is usually much more efficient.

---

1 When $\succ$ compares each pair of entries using $\geq$, the relation $\not\succ$ is simply $<$.

When the inequalities involve polynomials rather than just affine expressions, the psatz tactic can be used instead to discharge the goal. The tactic first preprocesses the goal into

$$\left(\bigwedge_{1 \le k \le n} P_k(\vec{x}) \ge 0 \right) \Rightarrow \bot.$$

The tactic then looks for a sequence $(Q_J)_{J \in \{0,1\}^n}$ of *sum-of-square* polynomials such that

$$\sum_{J \in \{0,1\}^n} Q_J \cdot \prod_k P_k^{J_k} = -1. \tag{4.2}$$

As with nsatz, Coq has to verify that equation [4.2] actually holds for the found polynomials using the ring machinery. Then, since a sum-of-square polynomial is expressed as a sum of squared polynomials, its evaluation is trivially nonnegative at any point. So the tactic can deduce from $\forall k$, $P_k(\vec{x}) \ge 0$ that the evaluation of the left-hand side of equation [4.2] at point $\vec{x}$ is nonnegative, which contradicts the fact that it is equal to -1.

Note that it is much faster for lra to find a vector $\vec{y}$ than it is for psatz to find a sequence (Q_J) of sum-of-square polynomials. So it might be worth trying, even for a polynomial problem, to use a linear decision procedure. The nra tactic does so by preprocessing the goal to find additional facts before calling lra. For instance, if x^2 occurs in the hypotheses or in the conclusion, the tactic adds $x^2 \ge 0$ as a hypothesis.

4.1.3. Linear inequalities over the integers

Variants of the tactics of section 4.1.2 can also be used when variables are not of type $\mathbb{R}$ but of type $\mathbb{Z}$. Indeed, if a set of equalities and inequalities does not admit any solution over the real numbers, then it does not admit any over the integers. Since the converse property is not true, these tactic variants might not be powerful enough. For instance, $\forall x \in \mathbb{Z}$, $2x \ne 1$ holds, but its *relaxation* to real numbers $\forall x \in \mathbb{R}$, $2x \ne 1$ does not.

So there are a few tactics dedicated to proving that systems of linear inequalities have no solution over the integers: omega, romega, and lia. Of these three tactics, lia is the most recent and the most powerful. Similarly to the real case, the goal to discharge is

$$A \cdot \vec{x} + \vec{b} \ge 0 \Rightarrow \bot,$$

except that matrix A and vectors $\vec{b}$ and $\vec{x}$ now have integer coefficients. Note that, since we are over the integers now, $\vec{b}$ can be modified so that only $\geq$ is used as a relation.

When the tactic finds some noninteger solution $\vec{x}$ to the system, it strengthens the constraints so as to exclude this solution. There are two methods do so. First, if some integer expression e is larger than or equal to a noninteger bound b, the tactic can add a new constraint $e \geq \lceil b \rceil$ to the system. Second, if some integer expression e is larger than or equal to an integer bound b, that tactic can create two new systems from the original one. One system extends it with the constraint $e = b$, while the other one extends it with $e \geq b + 1$. If both new systems imply a contradiction, so does the original one.

This concludes the overview of the tactics shipped with Coq for proving properties involving polynomials over the real numbers or integers. Most of these tactics are meant to be complete, that is, given enough time and memory, they can discharge any goal that lies in their logic fragment. The tactics presented in the next sections follow a different philosophy; they discharge goals by performing approximate (but guaranteed) numerical computations. Due to the approximations, they might fail to prove properties that would look like trivialities to the previous tactics. But approximate computations make it possible to enrich the formulas the tactics tackle some elementary functions such as log or some rounding operator.

4.2. Interval arithmetic

When proving properties about FP algorithms, it is common to have to bound the value of some expressions knowing some bounds on the variables that occur in them. For instance, one might want to simplify an error bound $\frac{23}{4}\varepsilon + 38\varepsilon^2$ by removing the quadratic term knowing that ε is small enough (see section 6.1 for the example this error bound originates from). This means that one needs to prove the following kind of inequality:

$$\forall \varepsilon \in [0; 0.005], \quad \frac{23}{4}\varepsilon + 38\varepsilon^2 \leq 6\varepsilon.$$

This property falls in the scope of the psatz tactic, so one could expect it to be automatically discharged. Unfortunately, the tactic does not succeed in a timely fashion. Proving this property by hand would be tedious, so we turn to interval arithmetic in the hope of proving it automatically [MOO 63]. Indeed, interval arithmetic is a computational approach and thus can be implemented in a reflective setting in Coq (see section 2.1.6). Rather than manipulating just real numbers, this approach manipulates connected closed sets of real numbers, i.e. *intervals*. For instance, to prove an inequality $u \leq v$, one can compute an enclosure $\mathbf{u} = [\underline{u}; \overline{u}]$ of u and an enclosure $\mathbf{v} = [\underline{v}; \overline{v}]$ of v. If $\overline{u} \leq \underline{v}$, then $u \leq v$. If $\underline{u} > \overline{v}$, then the inequality

$u \leq v$ is provably false. In the other cases, nothing can be deduced. So this approach is useful in practice only if it is able to prove tight enough enclosures that do not overlap.

The above example can be stated as follows in Coq. The interval tactic is able to prove the goal by computing a numerical enclosure of $23/4 \cdot \varepsilon + 38\varepsilon^2 - 6\varepsilon$ and verifying that its upper bound is no larger than zero when $\varepsilon \in [0; 0.005]$.

```
Goal forall eps : R, 0 <= eps <= 1/200 ->
  23/4 * eps + 38 * eps^2 <= 6 * eps.
Proof.
  intros eps He.
  interval with (i_bisect_diff eps).
Qed.
```

Section 4.2.1 shows how such bounds can be obtained, so that the goal above is proved automatically and formally in Coq. Intervals are represented using pairs of multi-precision FP numbers, on which an effective arithmetic is defined, in the spirit of section 3.3. This effective arithmetic also encompasses some elementary functions, e.g. exp. Unfortunately, naive interval arithmetic can lead to overestimated enclosures due to the *dependency effect*, thus making it useless to prove some properties. Section 4.2.2 explains what this effect is and shows several approaches to alleviate this issue: interval bisection, automatic differentiation, and polynomial approximations. Finally, section 4.2.3 presents an application of these approaches to the computation of enclosures of integrals.

4.2.1. *Naive interval arithmetic*

Let $\mathbb{I}$ denote the set of intervals. An *interval extension* of a real-valued function $f : \mathbb{R}^n \to \mathbb{R}$ is a function $\mathbf{f} : \mathbb{I}^n \to \mathbb{I}$ that satisfies the *containment* property:

$$\forall \mathbf{x_1} \in \mathbb{I}, \ldots, \mathbf{x_n} \in \mathbb{I}, \ \forall x_1 \in \mathbb{R}, \ldots, x_n \in \mathbb{R},$$

$$x_1 \in \mathbf{x_1} \wedge \ldots \wedge x_n \in \mathbf{x_n} \Rightarrow f(x_1, \ldots, x_n) \in \mathbf{f}(\mathbf{x_1}, \ldots, \mathbf{x_n}).$$

Note that the interval extension $\mathbf{f}$ is not uniquely defined. Given some interval extension $\mathbf{f_1}$ of f, any function $\mathbf{f_2}$ that satisfies the following property will also be an interval extension:

$$\forall \mathbf{x_1} \in \mathbb{I}, \ldots, \mathbf{x_n} \in \mathbb{I}, \ \mathbf{f_1}(\mathbf{x_1}, \ldots, \mathbf{x_n}) \subseteq \mathbf{f_2}(\mathbf{x_1}, \ldots, \mathbf{x_n}).$$

Interval extension $\mathbf{f_1}$ is said to be *tighter* than $\mathbf{f_2}$.

The *set extension* of f is a function of $\mathbb{I}^n \to \mathcal{P}(\mathbb{R})$ defined as follows:

$$f(\mathbf{x_1},\ldots,\mathbf{x_n}) \stackrel{\text{def}}{=} \{f(x_1,\ldots,x_n) \mid x_1 \in \mathbf{x_1} \wedge \ldots \wedge x_n \in \mathbf{x_n}\}.$$

Since $f(\mathbf{x_1},\ldots,\mathbf{x_n})$ is generally not a closed connected subset of $\mathbb{R}$, the set extension of f is not a proper interval extension. It can, however, be turned into the tightest interval extension by computing the closed convex hull:

$$\mathbf{f}_{\text{tightest}}(\mathbf{x_1},\ldots,\mathbf{x_n}) = \text{hull}(f(\mathbf{x_1},\ldots,\mathbf{x_n})).$$

On the contrary, the least useful interval extension returns the widest possible interval: for any input, we have

$$\mathbf{f}_{\text{worst}}(\mathbf{x_1},\ldots,\mathbf{x_n}) = \mathbb{R}.$$

In practice, we can get arbitrarily close to the tightest extension, but usually at the expense of an increase of computation time.

An interval can be represented by its bounds: $[\underline{u}; \overline{u}]$. The bounds are meant to be real numbers, but in order to represent half-bounded intervals or $\mathbb{R}$, we also allow $\underline{u} = -\infty$ and/or $\overline{u} = +\infty$. Section 4.2.1.1 shows how to compute these bounds and how to make use of FP arithmetic to improve the efficiency. Section 4.2.1.2 then gives more details about an implementation of basic arithmetic operators inside the logic of Coq. Finally, section 4.2.1.3 covers the FP implementation of elementary functions such as exp.

4.2.1.1. *Computing with bounds and outward rounding*

Let us now define some interval extensions of the basic arithmetic operators. The enclosures below are easily obtained by remembering that addition, subtraction, and multiplication are continuous functions with some monotony properties.

Lemma 4.1. Given two enclosures $u \in [\underline{u}; \overline{u}]$ and $v \in [\underline{v}; \overline{v}]$ with finite bounds, we have

$$
\begin{aligned}
u + v &\in [\underline{u} + \underline{v}; \overline{u} + \overline{v}], \\
u - v &\in [\underline{u} - \overline{v}; \overline{u} - \underline{v}], \\
u \cdot v &\in [\min(\underline{u} \cdot \underline{v}, \underline{u} \cdot \overline{v}, \overline{u} \cdot \underline{v}, \overline{u} \cdot \overline{v}); \max(\underline{u} \cdot \underline{v}, \underline{u} \cdot \overline{v}, \overline{u} \cdot \underline{v}, \overline{u} \cdot \overline{v})].
\end{aligned}
$$

Note that the enclosures of lemma 4.1 are meaningful only for finite bounds. In the case of addition and subtraction, they are easily extended to infinite bounds, since the undefined operation $\infty - \infty$ cannot occur. For multiplication, any occurrence of the product $0 \cdot \infty$ in the formula above should be evaluated as zero.

Let us see how the enclosures of lemma 4.1 can be used to prove an inequality by performing operations only on bounds. Suppose we want to prove that $x \cdot (y + \sqrt{y}) \geq -11$ for $x \in [-3; 4]$ and $y \in [1; 2]$. We compute an enclosure of the expression as follows:

$$
\begin{aligned}
x \cdot (y + \sqrt{y}) &\in [-3; 4] \times ([1; 2] + \sqrt{[1; 2]}) \\
&\in [-3; 4] \times ([1; 2] + [1; \sqrt{2}]) \\
&\in [-3; 4] \times [2; 2 + \sqrt{2}] \\
&\in [-6 - 3\sqrt{2}; 8 + 4\sqrt{2}].
\end{aligned}
$$

Since $-6 - 3\sqrt{2} \geq -11$, we can conclude the proof.

For the sake of efficiency, we might not want to obtain enclosures by performing computations on bounds representing arbitrary real numbers. It might be better to restrict finite bounds to a subset of $\mathbb{R}$, e.g. to FP numbers. As mentioned before, enlarging the output intervals of an interval extension still gives an interval extension. This means that computing the lower bound using an FP operator with rounding toward $-\infty$ and the upper bound with rounding toward $+\infty$ gives a result suitable for an interval extension. The resulting interval extensions are not the tightest ones, but their implementations are effective and fast. For addition and subtraction, the definitions are as follows:

$$
[\underline{u}, \overline{u}] + [\underline{v}, \overline{v}] \overset{\text{def}}{=} [\triangledown(\underline{u} + \underline{v}), \triangle(\overline{u} + \overline{v})],
$$

$$
[\underline{u}, \overline{u}] - [\underline{v}, \overline{v}] \overset{\text{def}}{=} [\triangledown(\underline{u} - \overline{v}), \triangle(\overline{u} - \underline{v})].
$$

Let us revisit the example above. We still want to prove that $x \cdot (y + \sqrt{y}) \geq -11$ for $x \in [-3; 4]$ and $y \in [1; 2]$. But this time, we use only a decimal FP arithmetic with a precision of two digits:

$$
\begin{aligned}
x \cdot (y + \sqrt{y}) &\in [-3; 4] \times ([1; 2] + \sqrt{[1; 2]}) \\
&\in [-3; 4] \times ([1; 2] + [1; 1.5]) \\
&\in [-3; 4] \times [2; 3.5] \\
&\in [-11; 14].
\end{aligned}
$$

Since $-11 \geq -11$, we can conclude the proof. Note that if we had been using only one digit of precision, we would have obtained $[-20; 20]$ as the final interval, which would have prevented us from proving $x \cdot (y + \sqrt{y}) \geq -11$.

The interval tactic implements this approach based on interval arithmetic to automatically prove inequalities in Coq [MEL 08, MAR 16]. The above example is translated into a Coq script as follows. The tactic uses a radix-2 FP arithmetic to perform the computations on the bounds and, by default, the precision is 30 bits, so the proof below is successfully checked by Coq.

```
Goal forall x y : R,
  -3 <= x <= 4 -> 1 <= y <= 2 ->
  x * (y + sqrt y) >= -11.
Proof.
  intros x y Hx Hy.
  interval.
Qed.
```

The interval_intro variant of the tactic makes it possible to compute bounds of expressions rather than verifying *a priori* bounds. For instance, the proof line

```
interval_intro (x * (y + sqrt y)) as H.
```

would introduce a hypothesis named H in the context:

```
H : -687371982 / 67108864 <= x * (y + sqrt y)
                <= 916495975 / 67108864
```

Notice how the bounds are radix-2 FP numbers ($67108864 = 2^{26}$). They underestimate and overestimate the bounds $-6 - 3\sqrt{2}$ and $8 + 4\sqrt{2}$, respectively. The interval_intro tactic can be used to linearize some expressions before lra is called to discharge a goal.

Sections 4.2.1.2 and 4.2.1.3 will give more details about the actual FP and interval arithmetics used by both interval and interval_intro tactics.

4.2.1.2. *Implementation of interval arithmetic*

For the tactic to be effective, it needs a way to perform interval arithmetic, that is, a way to effectively and efficiently compute interval bounds. For this purpose, it relies on the FP operators described in section 3.3. The FP arithmetic used by the tactic is not IEEE-754 compliant though, since it supports neither signed zeros nor signed infinities. Finite numbers can have arbitrarily large significands and exponents. The tactic also supports an extra element $\bot_F$ that propagates as if it were a NaN. A finite number is represented by a pair of integers (m, e) that is interpreted as the real number $m \cdot 2^e$. The extra element $\bot_F$ will be used whenever no finite bound is known for an interval enclosure. For instance, if an expression x is enclosed in an interval represented by the pair $\langle (3, -1), \bot_F \rangle$ then $x \geq 1.5$.

Before taking a look at the implementation of interval arithmetic, let us look at FP arithmetic. Most FP operators accept arbitrary FP numbers as inputs and return FP results at a given precision but with unbounded exponents. As such, their rounding behavior exactly matches rounding operators for the FLX format from Flocq (see section 3.1.2.2).

The FP operators also ensure that $\perp_F$ is propagated from input to output. Moreover, dividing by zero returns $\perp_F$, and so does taking the square root of a negative number. To express their specification, the arithmetic operators on real numbers have been extended to a type ExtendedR that contains all the real numbers and an extra element $\perp_{\mathbb{R}}$ that propagates like a NaN. For instance, Xadd extends the addition Rplus of real numbers. Similarly, the Xround function extends the round function that returns the closest FP number at a given precision and in a given direction (see section 3.2.2).

This gives the following signature and correctness statement for the FP addition operator Fadd. (It has been simplified a bit for clarity.) The function FtoX converts an FP number into the corresponding real number or $\perp_{\mathbb{R}}$. The statement about Fadd says that it behaves exactly as if the values of the input numbers were summed exactly and then rounded to the given precision and according to the given rounding mode, in the spirit of the IEEE-754 standard.

Definition FtoX : float -> ExtendedR.
Definition Fadd :
  rounding_mode -> precision -> float -> float -> float.
Lemma Fadd_correct :
  forall mode prec x y,
  FtoX (Fadd mode prec x y) =
  Xround radix mode prec (Xadd (FtoX x) (FtoX y)).

Note that, for the purpose of implementing an interval arithmetic, the rounding behavior of IEEE-754 is needlessly strong. Indeed, the result does not have to be the closest FP number; any number that satisfies the rounding direction would suffice. As a matter of fact, elementary functions of section 4.2.1.3 will not be correctly rounded yet they will still be suitable for computing interval bounds.

Note also that, for the sake of efficiency, the tactic does not implement the FP operators over the traditional type Z of relative integers but on the type BigZ. This type encodes large integers as binary trees of 31-bit integers that can be natively handled by the processor when using the *virtual machine* reduction engine of Coq [GRÉ 06, ARM 10].

Now that we have an interface that provides FP operators, we can build an arithmetic on intervals. They are represented by pairs of FP numbers (and $\perp_F$), but the tactic also supports an extra kind of interval denoted $\perp_I$ to represent the result of a partial function when some values of the input intervals might be outside the function domain. For instance, the differentiability relation for a function f will relate an input x to the derivative $f'(x)$. When f is not differentiable at x, there is no corresponding derivative. The interval version of f' will thus return $\perp_I$ whenever the input interval contains points at which f is not differentiable. This knowledge will be needed when performing automatic differentiation (see section 4.2.2.2). Note that interval functions are extended so that they return $\perp_I$ whenever an input interval is $\perp_I$. That way, even for composed function, we can detect whether input domains were respected.

The signature and the correctness statement of an interval operator look as follows. The contains predicate characterizes that a given interval encloses the given real; only $\perp_I$ contains $\perp_\mathbb{R}$. The lemma states that the computed interval satisfies the containment property, with ExtendedR denoting the set of real numbers extended with $\perp_\mathbb{R}$.

```
Definition Iadd : precision -> interval -> interval -> interval.
```

```
Lemma Iadd_correct :
  forall prec (ix iy : interval) (x y : ExtendedR),
  contains ix x -> contains iy y ->
  contains (Iadd prec ix iy) (Xadd x y).
```

As illustrated by the signature of Iadd, all the interval operators take a precision as argument, which is then passed to the FP operators used to compute interval bounds. The interval tactic does not automatically choose the best precision. As the example below shows, it is up to the user to select a suitable precision by passing the i_prec option to the tactic. If the precision is too small, the final interval might not be tight enough to prove the goal. If the precision is too large, computing the final interval might be too expensive. By default, the tactic uses a 30-bit radix-2 FP arithmetic.

```
Goal forall x y : R,
  -3 <= x <= 4 -> 1 <= y <= 2 ->
  x * (y + sqrt y) >= -11.
Proof.
  intros x y Hx Hy.
  interval with (i_prec 4).
    (* no need for a 30-bit precision *)
Qed.
```

Note that the tactic is not even semi-complete, that is, even if some proposition holds and is in the scope of the tactic, it might happen that no choice of the parameters will make the tactic succeed. For instance, due to rounding errors, any radix-2 enclosure of $1/3$ would lead to a non-point interval, and thus this is also true for the composite expression $(1/3) \cdot 6$. So the tactic cannot successfully prove that $(1/3) \cdot 6 \in [2; 2]$ holds. It can, however, prove $(1/3) \cdot 6 \in [2 - \varepsilon; 2 + \varepsilon]$ for a fixed yet arbitrarily small ε. The user just has to select an FP precision large enough so that the round-off errors do not exceed ε, assuming memory and time are not constrained.

4.2.1.3. Elementary functions

As shown in section 4.2.1.1, most interval operators simply perform directed FP operations on bounds, depending on the monotony of the real function. For addition, subtraction, multiplication, division, and square root, section 3.3 described how to compute the corresponding FP operations. What about elementary functions like exponential, sine, cosine, and so on?

The first step is to build FP approximations of the real functions and to prove their correctness. A notable point is that we cannot expect to easily compute correctly-rounded approximations of the elementary functions. For instance, we can compute an FP number y smaller than $\exp(x)$ for x an FP number, but it is much harder to get $y = \triangledown(\exp(x))$. Indeed, while there are simple algorithms for computing $\triangledown(\exp(x))$, proving their termination is quite difficult in Coq,[2] which prevents us from implementing them within Coq's logic. Fortunately, our goal is to perform interval arithmetic, so we are fine even if the results are not always the closest possible FP numbers. So let us look for a good enough approximation rather than the best one.

Since the precision is arbitrary, methods based on the evaluation of fixed polynomial approximants cannot be used. So we evaluate truncated power series instead. There are two difficulties: knowing at which point to truncate the series and computing a bound for the remainder. There are techniques to solve these two issues, as can be seen in MPFR [FOU 07], but they are still too complicated for our purpose. If we wanted to approximate arbitrary special functions, e.g. Airy or Bessel, we would have to apply such methods, but the interval tactic only supports elementary functions.

The main difference between special functions and elementary functions is that the latter have *argument reduction* identities. For instance, the identity

$$\cos x = 2 \left(\cos \frac{x}{2} \right)^2 - 1 \tag{4.3}$$

---

2 The idea of the proof is as follows. If such an algorithm were to not terminate, it would mean that $\exp(x)$ is a rational number, which is impossible when x is a nonzero FP number. Ideally, one should also compute a bound on the distance between $\exp(x)$ and the closest FP number at the target precision.

makes it possible to transform the input x into an input arbitrarily close to 0. (Note that, for radix $\beta = 2$, the computation of $x/2$ is trivial and does not introduce any error as the exponent range is unbounded.) Once x is close enough to 0, the process of evaluating the power series becomes much simpler.

Consider an elementary function f that we want to approximate at point x. Let us assume that it has a convergent Taylor series $f(x) = \sum_{k \geq 0} (-1)^k a_k x^k$ and that the sequence $k \mapsto a_k x^k$ is nonnegative and decreasing. Then we have the following inequalities:

$$0 \leq (-1)^n \left(f(x) - \sum_{k=0}^{n-1} (-1)^k a_k x^k \right) \leq a_n x^n.$$

These two inequalities solve both issues at once: they tell us that we can stop summing terms when $a_n x^n$ becomes small enough and that the truncated power series is off from the value of $f(x)$ by at most $a_n x^n$.

As for the evaluation of the truncated power series itself, we perform it using interval arithmetic. Note that this interval computation only succeeds when evaluating FP functions, not interval ones. Indeed, due to the alternated signs, the dependency effect is at its worst here (see section 4.2.2). Fortunately, for FP functions, the input interval is a point interval $[x; x]$, so the overestimation remains under control as long as the precision of the interval computations is large enough. But for interval extensions, inputs are usually not point intervals, so we need an additional mechanism, which will be presented at the end of this section.

We have seen a way to compute an elementary function when x is close to 0, but we still have to invert the argument reduction process for other inputs. Unfortunately, for a function like cos, the reconstruction of the result using equation [4.3] takes a number of arithmetic operations proportional to $\ln |x|$, and each of these operations incurs an additional overestimation. To alleviate this issue, when approximating a function with a precision p, the intermediate computations are performed at a precision $p' = p + \alpha \ln |x|$ with α a heuristically defined constant.

The process is similar for functions other than cos. First, we look for an interval of $\mathbb{R}$ where the function is approximated by an alternating series. The series should converge fast enough, so that few terms are needed to reach the target accuracy. Then, we look for an argument reduction that brings any input into this interval. Finally, we look for a recurrence that efficiently and accurately computes the series. More details on how arguments are reduced and how precision is tweaked can be found in [MEL 12].

Once we have FP approximations for elementary functions, we can devise interval extensions. Since exp, ln, and arctan are monotonic, extending them to intervals is straightforward. Note that these extensions are not guaranteed to return the tightest results, since the corresponding FP functions do not even have this property.

As for cos, sin, and tan, the interval extensions used by the tactic are quite naive. Indeed, because they are not monotone, the implementation and the proof of these extensions have been simplified by making their results meaningful only on a small domain around zero. For instance, the interval extension of sin just returns $[-1; 1]$ if its input is neither a point interval nor a subset of $[-2\pi; 2\pi]$. Therefore, if the tactic is applied on an expression defined on a different domain, the user should rewrite the expression using the following equation: $\sin x = \sin(x + k \cdot 2\pi)$ for some integer k.

4.2.2. Fighting the dependency effect

Computing bounds on expressions using interval arithmetic will serve to automatically prove bounds on approximation errors. Such an error is the signed distance between an ideal value, e.g. $y = \exp(x)$, and an approximation of it, e.g. $\tilde{y} = 1 + x + x^2/2$. This error might be expressed in an absolute way, $\tilde{y} - y$, or in a relative one, $\tilde{y}/y - 1$. In both cases, the interval evaluation of the error may suffer from the *dependency effect*.

Let us see what happens when bounding $\tilde{y} - y$ for $x \in \mathbf{x} = [0; 1]$:

$$
\begin{aligned}
\tilde{y} - y \quad &\in \quad ([1; 1] + \mathbf{x} + \mathbf{x}^2/2) - \exp(\mathbf{x}) \\
&\in \quad ([1; 1] + [0; 1] + [0; 0.5]) - [\exp 0; \exp 1] \\
&\in \quad [1; 2.5] - [1; e] \\
&\in \quad [1 - e; 2.5 - 1] \subseteq [-1.72; 1.5].
\end{aligned}
$$

So the interval evaluation makes it possible to automatically conclude that $|\tilde{y} - y| \leq 1.72$ for $x \in [0; 1]$. While correct, this upper bound is a gross overestimation of the supremum of $|\tilde{y} - y|$. Indeed, $\tilde{y} - y$ is equal to $\sum_{k \geq 3} x^k/k!$, which is an increasing function of $x \in [0; 1]$. So the extrema of $\tilde{y} - y$ are reached for $x = 0$ and $x = 1$, which makes it possible to manually prove that $|\tilde{y} - y| \leq 0.22$. Since the computed upper bound 1.72 is so far from the actual supremum 0.22, it is presumably useless for proving anything interesting.

The reason for the overestimation comes from the multiple occurrences of the input variable x in the difference $\tilde{y} - y$. Interval arithmetic happens to give tight enclosures of $y \in [1; e]$ and $\tilde{y} \in [1; 2.5]$, but these two enclosures do not carry enough information to compute a tight enclosure of $\tilde{y} - y$. Indeed, they do not show that the variations of y and $\tilde{y}$ are correlated, and thus that they do not reach their extrema independently from each other.

Section 4.2.2.1 shows an approach based on bisection to avoid the dependency effect. It is effective but slow, and thus should be used only as a last resort. Before using it, one can try to study the variations of the subexpressions to tighten bounds; this is the approach described in section 4.2.2.2. Rather than studying variations at the first order, one can also approximate subexpressions using polynomials and truncated Taylor series, as shown in section 4.2.2.3.

4.2.2.1. *Bisection*

Usually, interval operators satisfy an *isotonicity* property, that is, tighter input intervals lead to tighter output intervals. Ultimately, if the input intervals are point intervals, the output intervals will be point intervals too (assuming there is no need for outward rounding of bounds). So, using tighter input intervals tends to reduce the dependency effect. The simplest way to reduce this effect is thus to subdivide the input intervals into sub-intervals. For instance, to bound $\tilde{y} - y$ for $x \in [0; 1]$, one can bound it separately for $x \in [0; 0.5]$ and $x \in [0.5; 1]$:

$$\tilde{y} - y \in [-0.65; 0.63] \quad \text{when} \quad x \in [0; 0.5],$$
$$\tilde{y} - y \in [-1.10; 0.86] \quad \text{when} \quad x \in [0.5; 1],$$

and then merge the results by taking the union of the two enclosures:

$$\tilde{y} - y \in [-1.10; 0.86] \quad \text{when} \quad x \in [0; 1].$$

As expected, this new enclosure is tighter than the one obtained by interval evaluation on $[0; 1]$. If one wants to prove that $|\tilde{y} - y| \leq 0.65$, then the enclosure obtained for $x \in [0; 0.5]$ is sufficient. The one for $x \in [0.5; 1]$ is not; it should be refined further by considering sub-intervals of $[0.5; 1]$.

The interval tactic performs this process automatically. Whenever the target property is not verified by interval evaluation on a sub-interval, the tactic splits this sub-interval into two smaller intervals. This process keeps going until the original interval has been split into a covering of sub-intervals on which the target property holds, or until a given recursion depth has been reached. In the latter case, the tactic fails to prove the property.

Since the expression might involve several input variables, the variable v on which to perform the bisection has to be selected using the i_bisect option of the tactic. The default maximum depth of the process is 15, that is, the range of the selected variable will recursively be split into smaller sub-intervals at most 14 times. This limit can be modified using the i_depth option. Specifying (i_depth 1) prevents the original interval from being split. The following script shows how to automatically prove $|\tilde{y} - y| \leq 0.22$ when $x \in [0; 1]$.

```
Goal
  forall x : R, 0 <= x <= 1 ->
  Rabs ((1 + x + x^2 / 2) - exp x) <= 22/100.
Proof.
  intros x Hx.
  interval with (i_bisect x).
  (* same as (i_bisect x, i_depth 15) *)
  (* (i_bisect x, i_depth 11) would fail *)
Qed.
```

We have seen already that the tactic is not semi-complete, due to rounding errors when computing interval bounds. Even if there were no rounding error, it could still fail to prove a valid proposition. For instance, computing an enclosure of $x - x$ knowing an enclosure of x does not involve any rounding error in radix 2. Yet it is impossible to prove $x - x \in [0; 0]$ by bisection on the interval of x. The tactic can, however, succeed in proving $x - x \in [-\varepsilon; \varepsilon]$ for an arbitrarily small ε. One just has to set a large enough recursion limit so that the interval enclosing x gets split into sufficiently small sub-intervals.

The next two approaches for reducing the dependency effect are not meant to solve a larger class of problems (though they do so in some degenerate cases, e.g. $x - x \in [0; 0]$). They are designed to improve the efficiency of the tactic by reducing the amount of splitting usually needed to prove a proposition.

4.2.2.2. Automatic differentiation

While the bisection approach might succeed in proving properties suffering from the dependency effect, it is expensive. In the example of $\tilde{y}-y$, this is caused by $\tilde{y}$ and $-y$ having opposite variations with respect to x: when one increases, the other decreases. Naive interval arithmetic does not track this correlation between subexpressions. So a first way to improve the tightness of enclosures is not only to compute enclosures of expressions, but also enclosures of their derivatives.

Let us suppose that a function f is differentiable on an interval $\mathbf{x}$ and that its derivative on that interval is enclosed in an interval $\mathbf{f}'(\mathbf{x})$, that is to say, $f'(\mathbf{x}) \subseteq \mathbf{f}'(\mathbf{x})$. The Taylor-Lagrange theorem at order 0 (also known as the mean-value theorem) can make use of that information. For any $x_0 \in \mathbf{x}$ (in practice, we choose the midpoint of $\mathbf{x}$), it states

$$\forall x \in \mathbf{x}, \ \exists \xi \in \mathbf{x}, \ f(x) = f(x_0) + (x - x_0) \times f'(\xi),$$

Weakening it using the containment property, the theorem becomes

$$\forall x \in \mathbf{x}, \ f(x) \in \mathbf{f}([x_0, x_0]) + (\mathbf{x} - [x_0, x_0]) \times \mathbf{f}'(\mathbf{x}). \qquad [4.4]$$

By definition, the right-hand side is an interval extension of f. Note that this approach is not limited to order 0; higher orders of the theorem can also be turned into interval extensions. The tactic, however, uses a slightly different mechanism for building higher-order extensions (see section 4.2.2.3).

Let us consider anew the problem of enclosing $\tilde{y} - y$ with $y = \exp(x)$ and $\tilde{y} = 1 + x + x^2/2$ for $x \in [0, 1]$. Setting $f(x) = (1 + x + x^2/2) - \exp(x)$, we have $f'(x) = 1 + x - \exp(x)$. A naive interval evaluation with bounds represented by arbitrary real numbers gives

$$\mathbf{f}'(\mathbf{x}) = [1; 2] - [1; e] = [1 - e; 1]$$

for $\mathbf{x} = [0; 1]$. We take $x_0 = 0.5$, so $f(x_0) = 1.625 - \exp(0.5)$. The Taylor-Lagrange formula thus gives the following enclosure of $f(x)$ for $x \in [0, 1]$:

$$\begin{aligned} f(x) &\in (1.625 - \exp(0.5)) + [-0.5; 0.5] \times [1 - e; 1] \\ &\in [-0.89; 0.84]. \end{aligned}$$

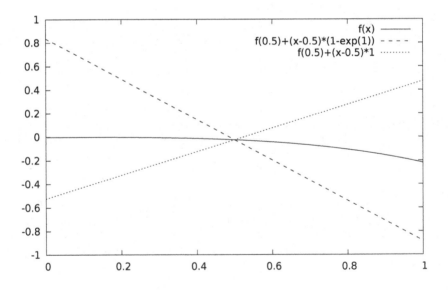

Figure 4.1. *Plot of f and of the two lines enclosing its values.*

Figure 4.1 shows a possible interpretation of the Taylor-Lagrange formula. The curved plot represents the graph of f while the two straight lines delimit the zone in which f can take its values, given the values of $f(0.5)$ and $\mathbf{f}'([0; 1])$. Note that dependencies were ignored when computing $\mathbf{f}'([0; 1])$, so the enclosure of $f'([0; 1])$ is grossly overestimated. Yet the resulting enclosure is tighter than the earlier enclosure $\tilde{y} - y \in [-1.72; 1.5]$ obtained by naive interval evaluation.

To automate this approach, we need some way to compute $\mathbf{f}'(\mathbf{x})$ given a function f and an interval $\mathbf{x}$. As manually done above, we could first symbolically compute an expression equal to the derivative of f and then evaluate it using intervals. Methods based on automatic differentiation do not use such a two-stage approach. They instead rely on arithmetic operators that propagate enclosures of expressions and their derivatives. Here are a few examples of such operators:

$$(\mathbf{u}, \mathbf{u}') + (\mathbf{v}, \mathbf{v}') \quad \overset{\text{def}}{=} \quad (\mathbf{u} + \mathbf{v}, \mathbf{u}' + \mathbf{v}'),$$

$$(\mathbf{u}, \mathbf{u}') \times (\mathbf{v}, \mathbf{v}') \quad \overset{\text{def}}{=} \quad (\mathbf{u} \times \mathbf{v}, \mathbf{u}' \times \mathbf{v} + \mathbf{u} \times \mathbf{v}'),$$

$$\exp(\mathbf{u}, \mathbf{u}') \quad \overset{\text{def}}{=} \quad (\exp(\mathbf{u}), \mathbf{u}' \times \exp(\mathbf{u})).$$

As with naive interval arithmetic, there is a containment property satisfied by these pairs of intervals. It is preserved by composition, so we get $(\mathbf{f}(\mathbf{x}), \mathbf{f}'(\mathbf{x}))$ at the end of the computation. Once we have $\mathbf{f}'(\mathbf{x})$, we compute $\mathbf{f}([x_0; x_0])$ and we apply the interval version of the Taylor-Lagrange formula given by equation [4.4]. The first member of the pair could be discarded at that point, but we use it to further refine the result:

$$f(x) \in \mathbf{f}(\mathbf{x}) \cap (\mathbf{f}([x_0; x_0]) + (\mathbf{x} - [x_0; x_0]) \times \mathbf{f}'(\mathbf{x})).$$

In some cases, we can do slightly better. If the interval $\mathbf{f}'(\mathbf{x})$ happens to be of constant sign, then we can prove that f is monotone on $\mathbf{x}$. In that case, we can just compute $\mathbf{f}([\underline{x}; \underline{x}])$ and $\mathbf{f}([\overline{x}; \overline{x}])$, take their convex hull, and thus obtain a superset of $f(\mathbf{x})$. Since the input intervals are point intervals, the dependency effect is minimal and the enclosure is tight. As a result, when combining automatic differentiation with bisection, no more splitting is needed once the interval evaluation $\mathbf{f}'$ has constant sign on each of the sub-intervals.

On the previous example, bisection with naive interval arithmetic required a bisection depth 12, that is, some part of the input interval had to be split down to a relative size of 2^{-11}. With automatic differentiation, the bisection depth does not exceed 3, that is, the relative size is no less than 2^{-2}. More precisely, the successful sub-intervals are $[0; 0.5]$, $[0.5; 0.75]$, and $[0.75; 1]$. The Coq script looks as follows.

Goal
```
forall x : R, 0 <= x <= 1 ->
Rabs ((1 + x + x^2 / 2) - exp x) <= 22/100.
```
Proof.

```
    intros x Hx.
    interval with (i_bisect_diff x, i_depth 3).
    (* (i_bisect_diff x, i_depth 2) would fail *)
Qed.
```

4.2.2.3. Polynomial approximations using Taylor models

The approach based on automatic differentiation can tackle more proofs than just naive interval arithmetic, but it might still need too many subdivisions for the most complicated approximation problems. The following example is inspired by the FP approximation of the exponential function proposed by Cody and Waite [COD 80, p. 65]. After an argument reduction that brings the input x into the interval $[-0.35; 0.35]$ (see section 6.2), the FP implementation approximates $\exp(x)$ using the rational function

$$r(x) = 2 \cdot \left(\frac{x \cdot p(x^2)}{q(x^2) - x \cdot p(x^2)} + 0.5 \right),$$

where p and q are two polynomials with FP coefficients. For an accurate *binary32* approximation, polynomials of degree 1 are sufficient:

$$p(x) = 0.25 + 1116769 \cdot 2^{-28} \cdot x,$$
$$q(x) = 0.5 + 13418331 \cdot 2^{-28} \cdot x.$$

Note that, unlike modern approaches to approximating the exponential [MUL 16], this one involves a division.

When evaluating the rational function r using FP arithmetic, some round-off errors will occur. We are not yet interested in bounding them; section 4.3 will later show how to do this. For now, we just want to prove that $r(x)$ is close to $\exp(x)$ when using real arithmetic. In other words, we want to bound the method error. Figure 4.2 plots the relative error.

The following Coq script proves that the relative error between $r(x)$ and $\exp(x)$ is less than $17 \cdot 2^{-34} \simeq 9.9 \cdot 10^{-10}$ for any $x \in [-0.35; 0.35]$. Constants of the form $m \cdot 2^e$ are represented using the notation m * pow2 e.

```
Goal forall x : R, Rabs x <= 35/100 ->
    let r := 2 *
      (x * p (x^2) / (q (x^2) - x * p (x^2)) + 1/2) in
    Rabs ((r - exp x) / exp x) <= 17 * pow2 (-34).
Proof.
    intros x Hx. simpl.
```

interval with (i_prec 40, i_bisect_taylor x 5).
Qed.

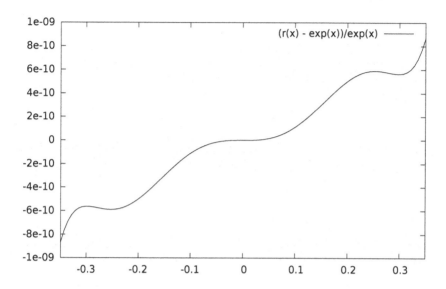

Figure 4.2. *Relative error between the rational function r and the exponential.*

As can be seen from the script, the proof amounts to calling the interval tactic. Since we are interested in proving a relative error of about 2^{-30}, the precision setting for the FP computations performed by the tactic cannot be left to its default value of 30, since the resulting round-off errors would incur too much of a perturbation. So we increase the precision to 40 bits using (i_prec 40).

If we were to use the method (i_bisect x) to reduce the dependency between $r(x)$ and $\exp(x)$, the proof could take hours to complete. Even using (i_bisect_diff x), the proof still takes about five minutes, which is much too long in an interactive setting. So we select a different method for completing this proof automatically. The option (i_bisect_taylor x 5) tells the tactic that it should approximate expressions using degree-5 polynomials in x. As with (i_bisect x) and (i_bisect_diff x), the option also tells the tactic that it should subdivide along x if it fails to prove the property for $x \in [-0.35; 0.35]$. This brings the proof time under one second. Degree 5 was chosen so as to minimize the proof time, but any degree between 3 and 13 would have been sufficient to make this example script complete in a few seconds.

With naive interval arithmetic, enclosures were represented by a single interval; with automatic differentiation, they were represented by a pair of intervals. Now they are represented by a pair $(\vec{\mathbf{p}}, \boldsymbol{\Delta})$ of a vector $\vec{\mathbf{p}}$ of intervals and an interval $\boldsymbol{\Delta}$. Given an interval $\mathbf{x}$ and a point x_0, $(\vec{\mathbf{p}}, \boldsymbol{\Delta})$ is a polynomial enclosure of a function f if there exists a polynomial

$$p(x) = \sum_{i < \text{size}(\vec{\mathbf{p}})} p_i \cdot (x - x_0)^i$$

with real coefficients $p_i \in \mathbf{p}_i$ for $i < \text{size}(\vec{\mathbf{p}})$, and such that

$$\forall x \in \mathbf{x}, \quad f(x) - p(x) \in \boldsymbol{\Delta}.$$

In practice, the interval coefficients $\mathbf{p}_i$ are tight, since they are obtained from computations on point intervals and thus do not suffer much from the dependency effect. So the tightness of a polynomial enclosure is mostly defined by the tightness of its error interval $\boldsymbol{\Delta}$.

As with naive interval arithmetic and automatic differentiation, we define an arithmetic over this new kind of enclosure. For instance, let us assume that $(\vec{\mathbf{p}}_f, \boldsymbol{\Delta}_f)$ and $(\vec{\mathbf{p}}_g, \boldsymbol{\Delta}_g)$ enclose the functions f and g on the interval $\mathbf{x}$ (using point x_0). Let us also assume that $\vec{\mathbf{p}}_f$ and $\vec{\mathbf{p}}_g$ have the same size s for simplicity. The enclosures of f and g mean that there are some coefficients p_{fi} and p_{gi} such that

$$\forall i < s, \quad p_{fi} \in \mathbf{p}_{fi} \quad \text{and} \quad p_{gi} \in \mathbf{p}_{gi},$$

$$\forall x \in \mathbf{x}, \quad \begin{cases} f(x) - \sum_{i<s} p_{fi} \cdot (x - x_0)^i & \in \quad \boldsymbol{\Delta}_f, \\ g(x) - \sum_{i<s} p_{gi} \cdot (x - x_0)^i & \in \quad \boldsymbol{\Delta}_g. \end{cases}$$

By summing these two enclosures, we deduce that the polynomial

$$p(x) = \sum_{i<s} (p_{fi} + p_{gi}) \cdot (x - x_0)^i$$

is such that

$$\forall x \in \mathbf{x}, \quad (f(x) + g(x)) - p(x) \in \boldsymbol{\Delta}_f + \boldsymbol{\Delta}_g.$$

So, if we define the vector $\vec{\mathbf{p}}$ by $\mathbf{p}_i = \mathbf{p}_{fi} + \mathbf{p}_{gi}$ for any $i < s$, the polynomial approximation $(\vec{\mathbf{p}}, \boldsymbol{\Delta}_f + \boldsymbol{\Delta}_g)$ is an enclosure of $f + g$ on the interval $\mathbf{x}$ (using point x_0).

The formula is similar for subtraction. For multiplication, the straightforward approach would produce a model of size $2s$. To avoid polynomials growing in size (and thus subsequent arithmetic operations getting costlier), the result of a multiplication is usually truncated to a polynomial approximation of size s by enlarging the error interval accordingly. For an elementary function, one would compute the Taylor series of the function, compose it with the input polynomial, and truncate it to get an enclosure. Note that, due to the use of Taylor series for handling elementary functions, these polynomial approximations are also called Taylor models [JOL 11, BRI 12].

Using this arithmetic on polynomial approximations, the tactic inductively computes a polynomial enclosure of the whole expression f on $\mathbf{x}$, taking for x_0 the midpoint of $\mathbf{x}$. Then it computes an interval enclosure of the polynomial part over $\mathbf{x}$ and adds the error interval $\boldsymbol{\Delta}$ to get an interval enclosure of the expression f. The degree of the polynomial approximations is selected using the i_bisect_taylor option of the tactic. (The size is equal to the degree plus one.) The larger the degree, the longer the computations on each sub-interval take, but the dependency effect gets hopefully reduced, so fewer subdivisions are needed. On the example of $f(x) = (1 + x + x^2/2) - \exp(x)$ for $x \in [0;1]$, verifying the bounds using polynomial approximations of degree 0 needs the same bisection depth as naive interval arithmetic. The use of polynomial approximations of degree 1 needs a similar depth as automatic differentiation. Polynomial approximations of degree 2 and more do not need any bisection to complete the proof.

To summarize, the downside of using polynomial approximations is an increase in the time needed for handling each sub-interval, with respect to naive interval arithmetic and automatic differentiation. We also lose the ability to detect monotone functions. The upside is that polynomial approximations usually reduce the impact of the dependency effect further than naive interval arithmetic (which can be seen as computing with polynomial approximations of degree 0) and automatic differentiation as presented in section 4.2.2.2 (which can be seen as computing with polynomial approximations of degree 1). As a result, using polynomial approximations of high enough degree should reduce the number of sub-intervals needed to complete the formal proof, to the point that the increased cost per sub-interval no longer matters.

4.2.3. Enclosing integrals

We have been using interval arithmetic to automatically compute bounds on expressions, but we can also use it to compute bounds on integrals [MAH 16]. Let us start with the following easy lemma.

Lemma 4.2. Let $f : \mathbb{R} \to \mathbb{R}$ be a function with an interval extension $\mathbf{f}$. Let u and v be two points such that f is integrable between them. If $u \in \mathbf{u}$ and $v \in \mathbf{v}$, then

$$\int_u^v f \in (\mathbf{v} - \mathbf{u}) \cdot \mathbf{f}(\text{hull}(\mathbf{u}, \mathbf{v})).$$

Proof. Without loss of generality, let us assume $u \leq v$. Let $[\underline{m}; \overline{m}]$ denote the interval $\mathbf{f}(\text{hull}(\mathbf{u}, \mathbf{v}))$. For any $x \in [u; v]$, we have $\underline{m} \leq f(x) \leq \overline{m}$. Since f is integrable between u and v, we have $(v - u) \cdot \underline{m} \leq \int_u^v f \leq (v - u) \cdot \overline{m}$. This property can be extended to $\mathbf{u}$ and $\mathbf{v}$ by using the properties of interval arithmetic, which concludes the proof. ∎

The previous lemma can only be applied if we verify beforehand that the function f is integrable. It would be better if the integrability property could also be derived from the computation $\mathbf{f}(\text{hull}(\mathbf{u}, \mathbf{v}))$. This is indeed the case due to two properties of the implementation of the `interval` tactic. First, whenever an expression is evaluated outside its definition domain, the resulting enclosure is $\perp_I$. Second, all the functions handled by the tactic happen to be continuous on their definition domain. As a result, once the tactic has computed $\mathbf{f}(\text{hull}(\mathbf{u}, \mathbf{v}))$, if the result is not $\perp_I$, it formally deduces that the function is continuous and thus integrable between u and v.

This gives the `interval` tactic a first procedure for computing bounds on integrals. Unfortunately, the bounds obtained that way are hardly tight, unless the function is close to constant. For instance, consider the identity function. Its enclosure on $[2; 4]$ is $[2; 4]$, so we get $\int_2^4 x \, dx \in (4 - 2) \cdot [2; 4] = [4; 8]$ which is a poor enclosure of $\int_2^4 x \, dx = 6$. Note that this overestimation is unrelated to the dependency effect, since the enclosure of f is optimal here. It is just that this approach does not take into account the variations of the integrand but only its extrema.

Using Chasles' relation to split the interval $[u; v]$ into sub-intervals, one can tighten the enclosure of the integral. For instance, instead of the enclosure $\int_2^4 x \, dx \in [4; 8]$, we get

$$\int_2^4 x \, dx = \int_2^3 x \, dx + \int_3^4 x \, dx \in (3 - 2) \cdot [2; 3] + (4 - 3) \cdot [3; 4] \subseteq [5; 7].$$

Using Chasles' relation to compute an integral is akin to performing a bisection to reduce the dependency effect. In other words, it tightens bounds but it is computationally intensive. So let us see how we can make use of polynomial approximations to compute tight bounds more efficiently. We rely on the following lemma.

Lemma 4.3. Let $f : \mathbb{R} \to \mathbb{R}$. Let p be a polynomial such that $\forall x \in [u; v]$, $f(x) - p(x) \in \Delta$ for some interval Δ. If f is integrable on $[u; v]$, then for any primitive P of p,

$$\int_u^v f \in P(v) - P(u) + (v - u) \cdot \Delta.$$

Proof. By hypothesis, the constant function Δ is an interval extension of $f - p$ on $[u; v]$, so $\int_u^v (f - p) \in (v - u) \cdot \Delta$. Since $\int_u^v f - (P(v) - P(u)) = \int_u^v (f - p)$, the lemma holds. ∎

Thus, if we can compute a polynomial approximation $(\vec{\mathbf{p}}, \boldsymbol{\Delta})$ of a function f on an interval, this lemma gives us a way to compute an enclosure of the integral of f. This method is implemented in the `interval` tactic to enclose by a constant interval each integral occurring in an expression. Let us consider the following integral taken from [AHM 15].

$$\int_0^1 \frac{\arctan \sqrt{x^2 + 2}}{\sqrt{x^2 + 2} \cdot (x^2 + 1)}\, dx = \frac{5\pi^2}{96}.$$

To numerically yet formally prove that this equality holds up to 10^{-15}, one can use the following Coq script. It makes use of Coquelicot's `RInt` operator described in section 9.2.1 to express the value of the integral.

Goal
```
  let f x := atan (sqrt (x*x + 2)) /
            (sqrt (x*x + 2) * (x*x + 1)) in
  Rabs (RInt f 0 1
         - 5/96*PI*PI) <= 1/10^15.
```
Proof.
```
  unfold f.
  interval with (i_integral_prec 49,
    i_integral_depth 4,
    i_integral_deg 10, i_prec 55).
```
Qed.

Coq checks the script above in a few seconds. Here is a description of the options passed to the tactic:

– Option `i_integral_prec` tells the tactic to compute enclosures with a relative accuracy of 49 bits. It means that the width of the resulting enclosure should be no larger than 2^{-49} times the magnitude of a straightforward evaluation of the integral. Default is 10 bits.

– Option `i_integral_depth` tells the tactic to recursively split an integration domain into two subdomains at most 4 times. This splitting happens when the target accuracy is not reached on an integration domain. Default is 3.

– Option `i_integral_deg` tells the tactic to use degree-10 polynomial approximations. This is the default.

– Option `i_prec` is not specific to enclosing integrals; it sets the precision of the underlying FP arithmetic to 55 bits (see section 4.2.1.2).

This application of interval arithmetic to computing and proving enclosures of univariate proper integrals completes this section dedicated to automating the proof of bounds of real-valued expressions by performing formally verified numerical computations. We will now see how automation can also be achieved if the expressions also contain FP operators.

4.3. Bounds on round-off error

In section 4.2.2.3, we have seen how to bound the method error between $\exp(x)$ and the following rational approximation:

$$r(x) = 2 \cdot \left(\frac{x \cdot p(x^2)}{q(x^2) - x \cdot p(x^2)} + 0.5 \right).$$

Let us now see how to bound the round-off error between $r(x)$ and the value computed by the following C code:

```
float X2 = x * x;
float PX2 = 0x1p-2f + X2 * 0x110A61p-28f;
float QX2 = 0x1p-1f + X2 * 0xCCBF5Bp-28f;
float R = 2.0f * ((x * PX2) / (QX2 - x * PX2) + 0.5f);
```

Note that all the FP operations in the code above are meant to be performed with the precision of the *binary32* format. Moreover, the FP coefficients that appear in the code are exactly representable in *binary32* and they are the same as the coefficients appearing in the polynomials p and q used for computing the method error.

The Gappa tool can be used to bound the difference between $r(x)$ and the computed value R. It is available both as a standalone tool [DAU 10] and as a Coq tactic [BOL 09c]. It takes as input a logical proposition about real-valued expressions. These expressions may contain basic arithmetic operators on real numbers $(+, \times, \div, \sqrt{\cdot})$ and rounding operators for binary floating- or fixed-point arithmetic. The tool tries to verify this logical proposition, generating a formal proof along the way, which can then be verified by Coq.

Section 4.3.1 presents the standalone tool and gives an overview of its input syntax. Internally, the tool performs naive interval arithmetic, which means that the dependency effect plagues its computations; section 4.3.2 shows how methods inspired from forward error analysis can alleviate this issue. Some proofs can be found faster or made shorter when some specialized variants of the enclosure property are introduced; they are presented in section 4.3.3. Still, all these properties only deal with enclosures and thus do not capture the discreteness of values computed by FP algorithms, so section 4.3.4 details how Gappa deals with this property. Section 4.3.5 explains how Gappa handles FP formats and rounding operators. Sometimes, enclosures are known not just on the inputs of an algorithm, but also on some larger expressions; section 4.3.6 shows how such additional constraints can be used to refine the knowledge on the inputs. Finally, since all the approaches mentioned so far might still not be sufficient for an automatic proof, some mechanisms described in section 4.3.7 make it possible for the user to pass extra knowledge to the tool.

4.3.1. The Gappa tool

Let us start with a very simple example that involves neither rounding operators nor round-off errors. The goal is to prove the following proposition:

$$\forall u, v \in \mathbb{R}, \quad u \in [2; 3] \wedge v \in [-1; 1] \Rightarrow u \cdot v \in [-3; 3].$$

A Coq script that makes use of the gappa tactic to automatically prove that proposition looks as follows.

```
Goal forall u v : R,
  2 <= u <= 3 /\ -1 <= v <= 1 ->
  -3 <= u * v <= 3.
Proof.
  gappa.
Qed.
```

The gappa tactic simply converts the current goal into a Gappa input file and passes it to the standalone tool. If the tool successfully verifies the proposition, it generates a Coq proof that is sent back to the tactic. Coq then checks that the proof indeed discharges the goal. For the above example, the corresponding input for the standalone Gappa tool is as follows. Note that any variable that appears in the logical proposition is universally quantified over $\mathbb{R}$.

```
{ u in [2,3] /\ v in [-1,1] -> u * v in [-3,3] }
```

Given this input, Gappa does not display anything. Only its return code shows that it found a proof. By passing the -Bcoq option to the tool, we get a 60-line Coq script that proves the property. This proof script relies on the Flocq library; it does not rely on any axiom other than the ones from Coq's standard library about real numbers. The approach the proof follows is quite close to the naive interval arithmetic described in section 4.2.1.

On such an example, the gappa tactic is hardly more powerful than the interval tactic. In fact, it is *a priori* less useful for expressions containing only arithmetic operators, since the tool does not know about automatic differentiation or polynomial approximations in order to reduce the dependency effect. So, let us go back to the example of the exponential of Cody and Waite. It involves FP operations and thus cannot be completely handled by the interval tactic.

As explained in section 1.1.4, by composing arithmetic operators with rounding operators, we get an exact description of the operations computed by an FP unit, as long as the computations produce no infinities nor NaN. This restriction also applies to Gappa, as the tool assumes that formats have no upper bound and that all the FP numbers are finite (possibly subnormal). As a result, it is of limited use for proving properties about programs that would experience exceptional behaviors such as overflows. (It can, however, be used to prove that a program cannot experience any such exceptional behavior.)

Let us see how Cody and Waite's code translates to the syntax of the standalone tool. Gappa's syntax makes it possible to give names to some expressions so as to simplify the writing of the logical proposition. So the input script for Gappa could start with the following four lines so as to define the infinitely-precise rational approximation $r(x)$ of $\exp(x)$:

```
x2 = x * x;
px2 = 0x1p-2 + x2 * 0x110A61p-28;
qx2 = 0x1p-1 + x2 * 0xCCBF5Bp-28;
r = 2 * ((x * px2) / (qx2 - x * px2) + 0.5);
```

Now, let us express the computed value. We need a unary operator for rounding to nearest (tie breaking to even) to the *binary32* format: float<24,-149,ne>. The float identifier indicates it is an FP rounding operator; its first argument is the precision; the second one is the exponent of the smallest positive FP number (see Table 3.1); the third one is the rounding direction (see section 4.3.5). In Flocq's formalism (see section 3.2), this rounding operator corresponds to

```
round radix2 (FLT_exp (-149) 24) ZnearestE.
```

Rather than providing the format constants, one can use a predefined name instead: float<ieee_32,ne>. The first operation of the C code can thus be expressed as

```
X2 = float<ieee_32,ne>(x * x);
```

Since applying this rounding operator after every operation might be tedious, there are two features to make a Gappa script shorter and more readable. First, one can give a shorter name to the operator:

```
@rnd = float<ieee_32,ne>;
```

Second, one can put a rounding operator before the = operator used for assigning names to expressions. This rounding operator will then be applied to each result of an arithmetic operation appearing on the right-hand side. This gives a compact way of expressing the value R actually computed by the C code using FP arithmetic:

```
X2 rnd= x * x;
PX2 rnd= 0x1p-2 + X2 * 0x110A61p-28;
QX2 rnd= 0x1p-1 + X2 * 0xCCBF5Bp-28;
R rnd= 2 * ((x * PX2) / (QX2 - x * PX2) + 0.5);
```

Note that, when using the rnd= assignment in the code above, the rounding operator rnd is only applied to addition, subtraction, multiplication, and division operators. It would also have been applied to square root, but neither to opposite and absolute values nor to constants. Note that this makes it different from the $\circ[\cdots]$ notation. For the example above, this restriction does not cause any issue since all the constants are representable exactly as *binary32* numbers. If that had not been the case, we should have added explicit rounding operators to the Gappa script. For example, instead of using a *binary32* hexadecimal literal in the definition of QX2, we could have used a non-representable decimal constant by explicitly rounding it to get a *binary32* number:

```
QX2 rnd= 0x1p-1 + X2 * rnd(4.99871783e-2);
```

Similarly, if mixed-precision opposite and absolute values had been in use, rounding operators should have been explicitly added, since these operations are generally not exact.

We are still missing a piece of information (though it will not make much of a difference here). We have not yet said that x itself is a *binary32* FP number. There are several ways to express this property. It might be done as part of the logical proposition, using the predicates from section 4.3.4. But there is a shorter way: one can just tell the tool that x is itself the rounded value of some real number. Indeed, x is its own rounded value, by idempotence of the rounding operator. But the actual identity of this real number does not matter, as long as it exists. Let us call it x_. We can thus add a new line to the script before the first usage of x.

```
@rnd = float<ieee_32,ne>;
x = rnd(x_);
X2 rnd= x * x;
```

Now that we have an expression named r that models the infinitely-precise computation and an expression named R that models the value computed by the C code, we can express the property that we want to prove about the bounds on the relative error between R and r. If some of these bounds are not known beforehand, we can use question marks in place of the intervals. In that case, Gappa tells us which property it is able to prove:

```
{ x in [-0.35,0.35] -> (R - r) / r in ? }
```

Gappa answers that, without any further guidance, it can provide a proof that the relative error between the computed value R and the exact value r is bounded by 1.22. This is a gross overestimation, as we would expect it to be no larger than a few ulps for the implementation to be somehow accurate. Note that this poor result is mostly due to the use of the question mark, since Gappa would otherwise be guided by the target enclosure. If we had told Gappa that we were interested in proving a much smaller bound, say 2^{-22}, the tool would have computed for a longer time and it would have answered that it can indeed find a proof for such a smaller bound. In particular, if Gappa had been called from Coq, the goal would have been fully specified, so it would have been automatically proved. But for now, let us assume we are still in the process of analyzing the code and we would like to understand where the overestimation comes from.

Since a division is involved, Gappa presumably tries to bound the relative round-off errors of its numerator and denominator. If an underflow occurs when computing either of these operands, then the corresponding round-off error will not be bounded, which will hamper the tool. The -Munconstrained option tells Gappa to assume that no division by zero nor underflow can occur. Under this assumption, the tool can now prove that the relative error between R and r is less than 2^{-22}. The option, however, prevents Gappa from generating a correct proof, as it contains holes in a dozen places where the tool used the assumption. Let us now massage the problem statement to see if we can convince Gappa that underflow is either absent or without any consequence. Indeed, if some parts of the proof only care about the absolute error, whether FP operations underflow or not makes no difference.

We could ask Gappa which of the numerator or denominator gets too close to zero by using the following proposition:

```
{ x in [-0.35,0.35] -> x * PX2 in ? /\ QX2 - x * PX2 in ? }
```

The answer is obvious though: since x crosses 0, so does the numerator $x \cdot PX2$. As a result, the relative round-off error of the FP product $\circ(x \cdot PX2)$ is not bounded, which prevents Gappa from proving that the relative error between R and r is small. Note that this does not mean that the relative error is large; it just means that Gappa cannot yet prove it is small. Since the issue comes from x crossing 0, let us weaken our logical proposition by assuming it does not. We do so by excluding some small interval enclosing zero from the range of x:

```
{ x in [-0.35,0.35] /\ not x in [-1b-30,1b-30] ->
  (R - r) / r in ? }
```

Gappa is now able to prove that the relative error is smaller than $2^{-22.4}$, which is satisfactory. The statement of the problem, however, is no longer satisfactory, since we have no idea what happens for x close to zero. Let us ask Gappa:

```
{ x in [-1b-30,1b-30] -> (R - r) / r in ? }
```

The tool answers that the relative error is bounded by 2^{-30}. This small bound is a consequence of the following fact: when x is close to zero, the summand 0.5 at the end of the computation tends to absorb any error.

We could tell Gappa to consider separately the cases $x \leq -2^{-30}$, $x \geq 2^{-30}$, and $x \in [-2^{-30}; 2^{30}]$. This user hint is written after the logical proposition, which is now back to its original form:

```
{ x in [-0.35,0.35] -> (R - r) / r in ? }
x in (-1b-30,1b-30);
```

As mentioned earlier, if the proposition does not contain any question mark, the user hint is not needed to solve this example. Indeed, Gappa is able to guess that x is an important variable. So it automatically splits the domain of x into smaller and smaller sub-intervals until it succeeds in proving the logical proposition on all the sub-intervals. Algorithm 4.1 presents the complete script for instructing Gappa to prove the round-off error bound. Lines starting with a # symbol are comments.

Let us assume for a moment that we are working in Coq and that Gappa has failed to prove the goal. For instance, it might have considered another variable was more important than x. We need some way to pass to the tool our knowledge that excluding $x \in [-2^{-30}; 2^{-30}]$ is helpful. The simplest way is to add the following tautology as a hypothesis to the proposition:

$$x \in [-2^{-30}; 2^{-30}] \vee x \notin [-2^{-30}; 2^{-30}].$$

There have been various uses of Gappa as a standalone tool to prove the correctness of algorithms. Here are a few examples of mathematical functions partly verified using Gappa: the logarithm [DIN 11], the square root [JEA 11], and some trigonometric functions [JEA 12]. More generally, it has also been used as a component of code generators for mathematical functions [MOU 11, KUP 14].

4.3.2. *Dependency effect and forward error analysis*

Similarly to the `interval` tactic, the Gappa tool internally performs naive interval arithmetic with multiprecision FP bounds. As such, it also suffers from the dependency effect. This is especially noticeable when bounding errors between approximated and exact terms. By using methods inspired from forward error analysis, Gappa can mitigate the impact of this kind of dependencies. Section 4.3.2.1 illustrates this approach on the example of bounding the relative error between two products. Section 4.3.2.2 generalizes the approach to other arithmetic operators. Since addition and relative error do not mix well, section 4.3.2.3 shows how Gappa handles this specific case.

Algorithm 4.1 Gappa script for bounding the round-off error of Cody and Waite's polynomial evaluation

---

```
# value computed using a binary32 arithmetic
@rnd = float<ieee_32,ne>;
x = rnd(x_);
X2 rnd= x * x;
PX2 rnd= 0x1p-2 + X2 * 0x110A61p-28;
QX2 rnd= 0x1p-1 + X2 * 0xCCBF5Bp-28;
R rnd= 2 * ((x * PX2) / (QX2 - x * PX2) + 0.5);

# infinitely-precise result
x2 = x * x;
px2 = 0x1p-2 + x2 * 0x110A61p-28;
qx2 = 0x1p-1 + x2 * 0xCCBF5Bp-28;
r = 2 * ((x * px2) / (qx2 - x * px2) + 0.5);

# logical proposition to be proved
{ x in [-0.35,0.35] -> |(R - r) / r| <= 3b-24 }
```

4.3.2.1. *Example: relative error of the product*

Let us consider a short example. We have four values u, v, $\tilde{u}$, and $\tilde{v}$, which satisfy the following properties:

$$u \in [1; 10] \quad \wedge \quad \tilde{u} \in [1; 10],$$

$$v \in [5; 20] \quad \wedge \quad \tilde{v} \in [5; 20],$$

$$\left| \frac{\tilde{u} - u}{u} \right| \leq 10^{-2} \quad \wedge \quad \left| \frac{\tilde{v} - v}{v} \right| \leq 10^{-3}.$$

We wish to bound the relative error between the two exact products $\tilde{u} \cdot \tilde{v}$ and $u \cdot v$. Naive interval arithmetic would lead to the following bounds:

$$\frac{\tilde{u} \cdot \tilde{v} - u \cdot v}{u \cdot v} \in \frac{[1; 10] \cdot [5; 20] - [1; 10] \cdot [5; 20]}{[1; 10] \cdot [5; 20]} = \frac{[-195; 195]}{[5; 200]} = [-39; 39].$$

Such large bounds on the relative error are pointless. The fact that only the bounds on the values have been taken into account, but not their closeness, hints to a solution. We have to make explicit the dependencies between $\tilde{u}$ and u and between $\tilde{v}$ and v. We can improve the tightness of the interval evaluation by first rewriting the relative error as follows:

$$\frac{\tilde{u} \cdot \tilde{v} - u \cdot v}{u \cdot v} = \frac{\tilde{u} - u}{u} + \frac{\tilde{v} - v}{v} + \frac{\tilde{u} - u}{u} \cdot \frac{\tilde{v} - v}{v}. \qquad [4.5]$$

There are still some dependencies: the same relative errors appear twice in the rewritten expression. These dependencies still cause the lower bound to be a bit underestimated, but at least the final bounds now have the correct order of magnitude:

$$\frac{\tilde{u} \cdot \tilde{v} - u \cdot v}{u \cdot v} \in [-10^{-2}; 10^{-2}] + [-10^{-3}; 10^{-3}]$$
$$+ [-10^{-2}; 10^{-2}] \cdot [-10^{-3}; 10^{-3}]$$
$$\in [-1.101 \cdot 10^{-2}; 1.101 \cdot 10^{-2}].$$

Notice that the bounds on u, v, $\tilde{u}$, and $\tilde{v}$ are not even used anymore during the interval computations above. The bounds on u and v are still useful though, since they serve to prove that $u \cdot v$ is nonzero and thus that the rewriting is legit. section 4.3.3.1 will later explain how even these bounds can be avoided, but for now, let us keep the bounds on u and v around. Once the bounds on $\tilde{u}$ and $\tilde{v}$ have been ignored, the Gappa script is as follows, with ut and vt denoting $\tilde{u}$ and $\tilde{v}$:

```
{ u in [1,10] /\ v in [5,20] ->
  |(ut - u) / u| <= 1e-2 ->
  |(vt - v) / v| <= 1e-3 ->
  (ut * vt - u * v) / (u * v) in ? }
```

Gappa automatically tries the above rewriting step to reduce the dependency effect and answers that the relative error between $\tilde{u} \cdot \tilde{v}$ and $u \cdot v$ is enclosed in $[-1.099 \cdot 10^{-2}; 1.101 \cdot 10^{-2}]$, which is a bit tighter than what the manual computation above gave. In fact, both computed bounds are now optimal. The tightened lower bound comes from the following observation. Given $\varepsilon_1 \geq -1$ and $\varepsilon_2 \geq -1$, the expression $\varepsilon_1 + \varepsilon_2 + \varepsilon_1 \cdot \varepsilon_2$ is increasing with respect to both ε_1 and ε_2, as can be seen by computing its partial derivatives. As a result, if we have the enclosures $\varepsilon_1 \in [\underline{\varepsilon}_1; \overline{\varepsilon}_1]$ and $\varepsilon_2 \in [\underline{\varepsilon}_2; \overline{\varepsilon}_2]$ with $\underline{\varepsilon}_1 \geq -1$ and $\underline{\varepsilon}_2 \geq -1$, we get

$$\varepsilon_1 + \varepsilon_2 + \varepsilon_1 \cdot \varepsilon_2 \in [\underline{\varepsilon}_1 + \underline{\varepsilon}_2 + \underline{\varepsilon}_1 \cdot \underline{\varepsilon}_2; \overline{\varepsilon}_1 + \overline{\varepsilon}_2 + \overline{\varepsilon}_1 \cdot \overline{\varepsilon}_2]. \qquad [4.6]$$

Since both lower bounds $\underline{\varepsilon}_1 = -10^{-2}$ and $\underline{\varepsilon}_2 = -10^{-3}$ are negative, their product is positive, which mechanically tightens the lower bound on the relative error of the product, compared to the bounds obtained by interval arithmetic.

Another way to obtain the tightened bounds would have been to rewrite $\varepsilon_1 + \varepsilon_2 + \varepsilon_1 \cdot \varepsilon_2$ into $(1 + \varepsilon_1) \cdot (1 + \varepsilon_2) - 1$ before doing the interval evaluation. The main issue with such a rewriting is that it induces a large cancellation when computing the bounds. For example, if ε_1 and ε_2 are bounded by 2^{-53}, getting a final bound of about 2^{-52} would require an internal precision of about 110 bits (instead of just a few bits), thus making the generated Coq proof much longer to verify.

This example has shown how rewriting a relative error using equation [4.5] and how some care in the interval computations, as in equation [4.6], make it possible for Gappa to compute tight error bounds when a product is involved. Let us now see how this approach can also be applied to other operators and to the absolute error.

4.3.2.2. Forward error analysis

Whenever Gappa has to bound an error between two expressions $\tilde{u} \diamond \tilde{v}$ and $u \diamond v$, it tries to express it from the errors between $\tilde{u}$ and u and between $\tilde{v}$ and v. In the equations below, evaluating the right-hand sides using interval arithmetic often gives much tighter bounds than evaluating the left-hand sides. These reformulations are inspired from the rules used when performing forward error analysis:

1) "The absolute error of the sum is the sum of the absolute errors."

$$(\tilde{u} + \tilde{v}) - (u + v) = (\tilde{u} - u) + (\tilde{v} - v).$$

$$(\tilde{u} - \tilde{v}) - (u - v) = (\tilde{u} - u) - (\tilde{v} - v).$$

2) "The relative error of the product is the sum of the relative errors, at first order." In the formula below, the rightmost term is the second-order term. While often negligible, it cannot be ignored when doing a formal proof.

$$\frac{\tilde{u} \cdot \tilde{v} - u \cdot v}{u \cdot v} = \frac{\tilde{u} - u}{u} + \frac{\tilde{v} - v}{v} + \frac{\tilde{u} - u}{u} \cdot \frac{\tilde{v} - v}{v}. \qquad \text{[4.5 again]}$$

As explained at the end of section 4.3.2.1, Gappa performs interval computations in a way such that the multiple occurrences of the relative errors between $\tilde{u}$ and u and between $\tilde{v}$ and v do not cause any overestimation.

3) "The relative error of the quotient is the sum of the relative errors, at first order." Note that, since Gappa manipulates signed errors, the formula actually computes a difference rather than a sum:

$$\frac{\tilde{u}/\tilde{v} - u/v}{u/v} = \left(\frac{\tilde{u} - u}{u} - \frac{\tilde{v} - v}{v} \right) \cdot \left(1 + \frac{\tilde{v} - v}{v} \right)^{-1}.$$

As with the product, the right-hand side is chosen so as to avoid cancellations. Moreover, the interval computations can again be arranged so that the two occurrences of $(\tilde{v} - v)/v$ do not cause any overestimation. Indeed, the right-hand side is a homographic function with respect to $(\tilde{v} - v)/v$ and thus monotone on both parts of its definition domain.

4) "The square root halves the relative error, at first order."

$$\frac{\sqrt{\tilde{u}} - \sqrt{u}}{\sqrt{u}} = \sqrt{1 + \frac{\tilde{u} - u}{u}} - 1.$$

The halving of the relative error is due to square root being of derivative $1/2$ at point 1. Note that this formula is not as efficient as the previous ones, since it requires a large precision due to the cancellation that occurs when computing the right-hand side. So it is better rewritten as follows, even if it introduces a dependency:

$$\frac{\sqrt{\tilde{u}} - \sqrt{u}}{\sqrt{u}} = \frac{\tilde{u} - u}{u} \cdot \left(\sqrt{1 + \frac{\tilde{u} - u}{u}} + 1 \right)^{-1}.$$

These four cases do not cover some combinations of arithmetic operators and absolute/relative errors. Gappa emulates the missing cases by transforming relative errors into absolute errors, and *vice versa*. But due to the prevalence of additions and multiplications in FP programs, Gappa also knows some rules for directly handling the absolute error of the product and the relative error of the sum. For the absolute error of the product, the rule is a simple variant of the one for the relative error:

$$\tilde{u} \cdot \tilde{v} - u \cdot v = (\tilde{u} - u) \cdot v + u \cdot (\tilde{v} - v) + (\tilde{u} - u) \cdot (\tilde{v} - v).$$

For the relative error of the sum, the situation is slightly more complicated because cancellations cause the relative error to grow arbitrarily large, so the following section is dedicated to this issue.

4.3.2.3. *Relative error of the sum*

A formula suitable for computing the relative error of the sum should not introduce any artificial cancellation when evaluated with intervals, even in presence of some dependency effect. Gappa deals with this issue in two steps. First, the following equality is used:

$$\frac{(\tilde{u} + \tilde{v}) - (u + v)}{u + v} = \frac{u}{u + v} \cdot \frac{\tilde{u} - u}{u} + \left(1 - \frac{u}{u + v} \right) \cdot \frac{\tilde{v} - v}{v}. \qquad [4.7]$$

As with some of the previous formulas, Gappa is able to evaluate the right-hand-side expression with intervals in a way such that the two occurrences of $u/(u + v)$ do not cause any overestimation. As a result, if the relative errors between $\tilde{u}$ and u, and between $\tilde{v}$ and v, are independent from u and v, the above formula gives a tight enclosure on the relative error of the sum, as long as Gappa is able to compute a tight enclosure of $u/(u + v)$.

Unfortunately, since u and v might be correlated, there is no guaranteed way to compute a tight enclosure of $u/(u + v)$. Nonetheless, Gappa tries to compute some bounds for the following expressions and keeps the tightest ones. Trying several expressions give more opportunities for a hypothesis of the problem to apply.

$$\frac{u}{u + v} = \left(1 + \frac{v}{u}\right)^{-1} = 1 - \frac{v}{u + v} = 1 - \left(1 + \frac{u}{v}\right)^{-1}. \qquad [4.8]$$

Let us consider a small but representative example for the relative error of the sum. The expression $c + x \cdot q$ represents the last step of a polynomial evaluation, with c a nonzero constant (see section 6.3 for a different approach). Moreover, let us assume that the only error comes from the evaluation of q, so we are interested in the error between $c + x \cdot \tilde{q}$ and $c + x \cdot q$. According to equation [4.7], this relative error ε can be expressed as

$$\varepsilon = \frac{c}{c + x \cdot q} \cdot 0 + \left(1 - \frac{c}{c + x \cdot q}\right) \cdot \frac{x \cdot \tilde{q} - x \cdot q}{x \cdot q}.$$

If equation [4.5] is used to bound the relative error between $x \cdot \tilde{q}$ and $x \cdot q$ and if the last variant of equation [4.8] is used to bound $c/(c + x \cdot q)$, computing an enclosure of ε using the expression above gives the same interval as if we were bounding the following expression of ε.

$$\varepsilon = \left(1 - \left(1 + \frac{x \cdot q}{c}\right)^{-1}\right) \cdot \frac{\tilde{q} - q}{q}.$$

Let $\mathbf{e_q}$ denote an interval enclosing the relative error ε_q between $\tilde{q}$ and q. Let $\mathbf{t} = [\underline{t}; \overline{t}]$ denote an interval enclosing the ratio $(x \cdot q)/c$. Assuming that the magnitude of $x \cdot q$ is small with respect to c, we have $\underline{t} > -1$. As a result, ignoring the rounding errors that occur when computing the bounds, Gappa obtains the following enclosure for the relative error:

$$\varepsilon \in \left[\frac{\underline{t}}{1 + \underline{t}}; \frac{\overline{t}}{1 + \overline{t}}\right] \cdot \mathbf{e_q}.$$

This enclosure shows that, when the addends c and $x \cdot q$ cancel, the $\underline{t}$ bound becomes close to -1, so the relative error ε_q is amplified by the final addition. If there is no cancellation however, then $\underline{t}$ is larger than $-\frac{1}{2}$ and the enclosure properly expresses that the magnitude of the relative error decreases. For instance, if $x \cdot q$ is two orders of magnitude smaller than the constant c, then the relative error ε is two orders of magnitude smaller than ε_q, which usually makes it negligible. This example shows how Gappa can recover some traditional properties obtained by error analysis using a combination of naive interval arithmetic and dedicated identities.

4.3.3. Variations around enclosures

Gappa does not handle just simple enclosures of expressions internally. It also supports a few other families of properties that mitigate some of the shortcomings of these enclosures. Section 4.3.3.1 presents the REL predicate which makes it possible to avoid quotients when dealing with relative errors. Section 4.3.3.2 presents the ABS and NZR predicates which make it simpler to handle excluded values such as zeros and subnormal numbers.

4.3.3.1. Enclosures of relative errors

When it comes to relative errors, expressions are possibly ill-formed when Gappa cannot prove that the denominator is nonzero. Yet the notion of a bounded relative error between two expressions $\tilde{u}$ and u can be given a meaning even if u is equal to zero. Indeed, $\tilde{u}$ just has to be zero too, in that case. (More generally, a signed relative error at least -1 guarantees that $\tilde{u}$ and u have the same sign.)

This boundedness property is also preserved for some arithmetic operations. For instance, let us suppose that the relative errors between $\tilde{u}$ and u and between $\tilde{v}$ and v are bounded. equation [4.5] tells us that the relative error between $\tilde{u} \cdot \tilde{v}$ and $u \cdot v$ is also bounded (by interval arithmetic), as long as $u \cdot v$ is nonzero. But if $u \cdot v$ is zero, then either u or v is zero, which means that either $\tilde{u}$ or $\tilde{v}$ is zero. So $\tilde{u} \cdot \tilde{v}$ is zero when $u \cdot v$ is zero. As a result, the relative error between $\tilde{u} \cdot \tilde{v}$ and $u \cdot v$ is bounded, whether $u \cdot v$ is zero or not.

This example tends to show that representing a bounded relative error by the enclosure $(\tilde{u} - u)/u \in \mathbf{e_u}$, while intuitive, would prevent some computations from succeeding when some information on u is missing. So Gappa relies on a slightly different representation. It internally uses the following ternary relation:

$$\mathrm{REL}(\tilde{u}, u, \mathbf{e_u}) \overset{\mathrm{def}}{=} \exists \varepsilon_u \in \mathbf{e_u},\ \tilde{u} = u \cdot (1 + \varepsilon_u).$$

The approach presented in section 4.3.2.2 does not change much with this new representation for enclosures of relative errors, since the same formulas can be obtained. As an example, let us rephrase equation [4.5] using REL. Given $\varepsilon_u \in \mathbf{e_u}$ and $\varepsilon_v \in \mathbf{e_v}$ such that

$$\tilde{u} = u \cdot (1 + \varepsilon_u) \quad \text{and} \quad \tilde{v} = v \cdot (1 + \varepsilon_v),$$

we deduce that

$$\tilde{u} \cdot \tilde{v} = (u \cdot v) \cdot (1 + (\varepsilon_u + \varepsilon_v + \varepsilon_u \cdot \varepsilon_v)).$$

So, given two properties $\mathrm{REL}(\tilde{u}, u, \mathbf{e_u})$ and $\mathrm{REL}(\tilde{v}, v, \mathbf{e_v})$, we deduce some property $\mathrm{REL}(\tilde{u} \cdot \tilde{v}, u \cdot v, \mathbf{e})$ with $\mathbf{e}$ an interval computed from $\mathbf{e_u}$ and $\mathbf{e_v}$ in a way similar to equation [4.5].

The -/ symbol can be used to express the REL ternary relation in Gappa scripts. The two syntaxes for expressing a property $\mathrm{REL}(approx, exact, \ldots)$ are as follows:

```
approx -/ exact in [bnd1, bnd2]
|approx -/ exact| <= bnd
    # equivalent to approx -/ exact in [-bnd, bnd]
```

Note that the -/ symbol is part of the ternary relation. It is not a binary operator on its own right, so its use inside a subexpression is meaningless and thus forbidden. Let us rephrase the example from section 4.3.2.1 using this syntax. It becomes as follows:

```
{ u in [1,10] /\ v in [5,20] ->
  |ut -/ u| <= 1e-2 ->
  |vt -/ v| <= 1e-3 ->
  ut * vt -/ u * v in ? # REL(ut*vt, u*v, ?)
}
```

Since the ternary relation REL makes it possible to relax the constraint that the denominator of a relative error has to be nonzero, the first hypothesis can be removed. So the following script works just as well and returns the same enclosure:

```
{ |ut -/ u| <= 1e-2 ->
  |vt -/ v| <= 1e-3 ->
  ut * vt -/ u * v in ? }
# result: ut * vt -/ u * v in [-0.01099, 0.01101]
```

The REL relation is not directly available when calling Gappa from Coq scripts. The gappa tactic instead recognizes the proposition $|\tilde{u} - u| \le b \cdot |u|$ and replaces it with $\mathrm{REL}(\tilde{u}, u, [-b; b])$ as long as b is a positive FP constant. Indeed the following equivalence holds:

$$\forall \tilde{u}, u \in \mathbb{R}, \quad (b \ge 0 \wedge |\tilde{u} - u| \le b \cdot |u|) \Leftrightarrow \mathrm{REL}(\tilde{u}, u, [-b; b]).$$

The following script shows a variant with different error bounds to the example about the relative error of the multiplication, using dyadic constants so that they are recognized by the tactic.

Goal forall ut u vt v : R,
  Rabs (ut - u) <= (1/1024) * Rabs u ->
  Rabs (vt - v) <= (2/1024) * Rabs v ->
  Rabs (ut * vt - u * v) <= (4/1024) * Rabs (u * v).
Proof.
  gappa.
Qed.

4.3.3.2. Zero and subnormal numbers

While Gappa internally uses the REL relation and offers some syntax for the user to express such a relation between expressions, it also recognizes enclosures of relative errors written as $(\tilde{x} - x)/x \in \mathbf{e_x}$, be they used as hypotheses or as conclusions. But Gappa is able to convert them into REL-based properties only if it is able to prove that x is nonzero from the hypotheses.

The main way for Gappa to prove that an expression x is nonzero is to compute an enclosure $x \in \mathbf{x}$ and to check that the interval $\mathbf{x}$ cannot contain zero. This is done by verifying that either the lower bound of the interval is positive or its upper bound is negative. Yet x might be nonzero while $\mathbf{x}$ contains zero. In that case, two other approaches can be tried by Gappa.

First of all, Gappa internally manipulates a predicate that precisely states that an expression is nonzero, whatever bounds can be computed on it:

$$\mathrm{NZR}(x) \overset{\text{def}}{=} x \neq 0.$$

The main occurrences of this predicate are as follows. First of all, if a hypothesis is stated as x <> 0, then Gappa directly handles it as $\mathrm{NZR}(x)$ rather than $\neg(x \in [0; 0])$. Second, the tool knows a few theorems that propagate the NZR predicate. For instance, if two expressions satisfy it, then their product satisfies it too. Third, if an expression is nonzero, any expression equal to it is nonzero. Finally, as already mentioned, if the enclosure of an expression does not contain zero, then Gappa deduces that it cannot be zero. For instance, Gappa can prove the following proposition even though it does not know any enclosure of v:

{ u in [1,2] /\ v <> 0 -> u * v <> 0 }

Gappa is able to deduce that an expression x is nonzero if its enclosure $\mathbf{x}$ does not contain zero, but this is a bit too coarse. Indeed, knowing that an enclosure of $|x|$ does not contain zero would be sufficient. For instance, an FP number in the normal range satisfies such a property. Separately solving the problem for positive and negative values makes it possible to notice that the expression is indeed nonzero at the expense of performing two proofs. Gappa can avoid this cost in some cases. Indeed, rather than just computing bounds of expressions, the tool also tries to compute bounds on absolute values of expressions. More precisely, Gappa internally supports the following kind of enclosures:

$$\text{ABS}(x, [\underline{x}; \overline{x}]) \stackrel{\text{def}}{=} \underline{x} \geq 0 \wedge |x| \in [\underline{x}; \overline{x}].$$

So Gappa can now deduce that x is nonzero when $\text{ABS}(x, [\underline{x}; \overline{x}])$ holds and $\underline{x}$ is positive. As with standard enclosures and the NZR predicate, the tool knows how to propagate the ABS relation along some arithmetic operations. For instance, since $|u \cdot v| = |u| \cdot |v|$, the product between expressions is easily handled by computing an interval multiplication:

$$\text{ABS}(u, \mathbf{u}) \wedge \text{ABS}(v, \mathbf{v}) \Rightarrow \text{ABS}(u \cdot v, \mathbf{u} \cdot \mathbf{v}).$$

For addition and subtraction, Gappa computes an interval enclosing both $|u + v|$ and $|u - v|$ using an algorithm inspired from the triangle inequalities

$$||u| - |v|| \leq |u \pm v| \leq |u| + |v|.$$

The following example exercises these inequalities.

```
{ |u| in [3,5] -> |v| <= 1 -> u + v <> 0 }
```

Gappa successfully produces a proof, without having to consider the cases $u \in [-5; -3]$ and $u \in [3; 5]$ separately. Indeed, from the hypotheses $\text{ABS}(u, [3; 5])$ and $\text{ABS}(v, [0; 1])$, Gappa deduces $\text{ABS}(u + v, [2; 6])$, and thus that $u + v$ is nonzero.

The ABS relation is useful for stating that an expression is nonzero, but it can also be used to state that an expression is not too close to zero. For instance, when rounding in an FP format, the relative error can be bounded according to the precision if the input is in the range of normal numbers. In other words, whenever Gappa has to prove a property like $\text{REL}(\Box(x), x, \mathbf{e_x})$, it looks for a property $\text{ABS}(x, [\underline{x}; \overline{x}])$ and checks whether $\underline{x}$ is large enough to ensure that x is in the range of normal numbers. The handling of rounding operators will be detailed in section 4.3.5.

4.3.4. Discreteness of FP numbers

In addition to the basic enclosure of an expression, we now have the following predicates and relations: NZR which is used to denote a nonzero expression, ABS which bounds the absolute value of an expression, and REL which bounds the relative error between two expressions. All of them can be seen as improvements of the enclosure relation that are primarily meant to simplify the search of a proof and reduce its final size. But there are also some relations that enlarge the expressiveness of Gappa: the FIX and FLT relations (not to be confused with the formats from Chapter 3).

These relations are meant to account for the discreteness property of FP formats. They relate an expression x and an integer k as follows:

$$\text{FIX}(x, k) \quad \overset{\text{def}}{=} \quad \exists m, e \in \mathbb{Z}, \; x = m \cdot 2^e \wedge k \leq e,$$

$$\text{FLT}(x, k) \quad \overset{\text{def}}{=} \quad \exists m, e \in \mathbb{Z}, \; x = m \cdot 2^e \wedge |m| < 2^k.$$

In other words, $\text{FIX}(x, k)$ means that the real x can be represented by a binary fixed-point number with a least significant bit of weight 2^k. This is thus equivalent to (FIX_format radix2 k x) using Flocq's definition (see section 3.1.2.3). As for $\text{FLT}(x, k)$, it means that x can be represented by a binary FP number with an integer significand of at most k bits. This is equivalent to (FLX_format radix2 k x) (see section 3.1.2.2). Note that, as with enclosures, only the expression x is meant to be symbolic; for Gappa to successfully use the FIX and FLT relations, the value of the integer k has to be known. Finally, to express (FLT_format radix2 emin prec x) in Gappa, one can use $\text{FIX}(x, e_{\min}) \wedge \text{FLT}(x, \varrho)$.

If a property $\text{FIX}(x, k)$ holds, any property $\text{FIX}(x, \ell)$ with ℓ smaller than k also holds. For FLT, it is the opposite: if a property $\text{FLT}(x, k)$ holds, any property $\text{FLT}(x, \ell)$ with ℓ larger than k also holds.

The following lemmas are used by Gappa to deduce such properties on expressions involving arithmetic operators:

$$\begin{aligned}
\text{FIX}(u, k) \wedge \text{FIX}(v, \ell) &\;\Rightarrow\; \text{FIX}(u + v, \min(k, \ell)), \\
\text{FIX}(u, k) \wedge \text{FIX}(v, \ell) &\;\Rightarrow\; \text{FIX}(u \cdot v, k + \ell), \\
\text{FLT}(u, k) \wedge \text{FLT}(v, \ell) &\;\Rightarrow\; \text{FLT}(u \cdot v, k + \ell).
\end{aligned}$$

The last lemma is not suitable when u (or v) is a power of two. Indeed, if v can be represented as an FP number with an integer significand of at most ℓ bits, this is also true for the real $2^e \cdot v$ for any integer e. Yet the above lemma only implies the weaker property $\text{FLT}(2^e \cdot v, 1 + \ell)$. So Gappa uses a variant of it when k or ℓ is equal to 1, so as to avoid this pessimization.

There are also some trivial lemmas to handle the unary opposite and the absolute value. More importantly, Gappa knows how to deduce some FLT property from a FIX property when it knows an enclosure, and *vice versa*. For instance, if x is a multiple of 2^{-3} yet $|x|$ is smaller than 13, then Gappa deduces that x can be represented by an FP number with an integer significand of at most 7 bits:

```
{ @FIX(x,-3) /\ |x| <= 13 -> @FLT(x,7) }
```

Symmetrically, if x can be represented by an FP number with an integer significand of at most 7 bits yet $|x|$ is larger than 13, then x is a multiple of 2^{-3}:

```
{ @FLT(x,7) /\ |x| in [13,10000] -> @FIX(x,-3) }
```

Finally, the last ingredient is the ability for Gappa to use a FIX or FLT property to sharpen an enclosure. For instance, Gappa is able to prove the following proposition:

```
{ @FLT(x,2) /\ x in [1.1,1.9] -> x in [1.5,1.5] }
```

This sharpening of bounds is implemented by rounding inward the bounds of the interval toward the corresponding FIX_format (for FIX) or FLX_format (for FLT). For instance, in the example above, the property FLT(x, 2) corresponds to the Coq property FLX_format radix2 2 x. Let us denote $\triangledown$ and $\triangle$ the directed rounding operators for this format. Since x is representable in this format, $\triangledown(x) = \triangle(x) = x$. So $x \geq 1.1$ implies $x = \triangle(x) \geq \triangle(1.1) = 1.5$ by monotony of $\triangle$. Similarly, $x = \triangledown(x) \leq \triangledown(1.9) = 1.5$. Thus $x = 1.5$.

4.3.5. Rounding operators

All the methods described so far are meant to handle expressions formed from the basic arithmetic operators. Let us see how Gappa handles rounding operators. They are defined in a way similar to section 3.2. Two families of rounding operators are supported: fixed and float.

A rounding operator in the fixed family describes how to round a real number to some FIX_format format (see section 3.1.2.3). It is parameterized by the exponent of the least significant bit and the rounding direction. Thus the Gappa expression fixed<e,dir>(x) is translated into the following Coq expression:

```
round radix2 (FIX_exp e) dir x.
```

A rounding operator in the float family describes how to round a real number to some FLT_format format (see section 3.1.2.1). It is parameterized by the precision, the exponent of the smallest positive power that can be represented, and the rounding direction. Thus the Gappa expression float<p,e,dir>(x) is translated into the following Coq expression:

```
round radix2 (FLT_exp e p) dir x.
```

The IEEE-754 rounding directions are written as follows in Gappa's syntax:

– ne rounds to nearest, with tie breaking to even;

– na rounds to nearest, with tie breaking away from zero;

– up rounds toward $+\infty$;

– dn rounds toward $-\infty$;

– zr rounds toward 0.

Let us now see what kind of properties Gappa is able to deduce for expressions involving rounding operators. First of all, it can obtain instances of the FIX and FLT properties introduced in section 4.3.4. For example, if $\square$ designates some float<p,e,dir> rounding operator, then both $\mathrm{FIX}(\square(x),e)$ and $\mathrm{FLT}(\square(x),p)$ hold. Moreover, if Gappa already knows an enclosure of $\square(x)$, it can sharpen it the same way it was done in section 4.3.4, that is, by rounding the bounds inward. The situation is similar for a fixed format.

From $\mathrm{FIX}(x,k)$, Gappa can deduce $\mathrm{FIX}(\square(x),k)$, whatever the rounding operator $\square$. Similarly, from $\mathrm{FLT}(x,\ell)$, it can deduce $\mathrm{FLT}(\square(x),\ell)$. Both deductions are instances of lemma 3.45.

For any rounding operator $\square$, given an enclosure $x \in [\underline{x};\overline{x}]$, Gappa can also compute an enclosure $\square(x) \in [\square(\underline{x});\square(\overline{x})]$, due to the monotony of $\square$.

As for rounding errors, from an enclosure of x, $\square(x)$, $|x|$, or $|\square(x)|$ (which ones are used depends on the actual rounding operator), Gappa can compute an enclosure of the absolute error $\square(x) - x$, or of the relative error as a $\mathrm{REL}(\square(x),x,\ldots)$ property. Here are a few examples:

$$\square(x) - x \in [-2^{e-1};2^{e-1}] \quad \text{when } \square \text{ is fixed<}e\text{,ne>,}$$
$$x \leq 0 \;\Rightarrow\; \square(x) - x \in [0;2^e] \quad \text{when } \square \text{ is fixed<}e\text{,zr>,}$$
$$|\square(x)| > 2^{e+p-1} \;\Rightarrow\; \mathrm{REL}(\square(x),x,[-2^{-p};2^{-p}]) \quad \text{when } \square \text{ is float<}p,e\text{,ne>.}$$

The last implication covers the situation where x is in the range of normal numbers. As such, it is not strong enough to prove that the relative rounding error of the sum of two representable numbers is always bounded. Indeed, the sum might well be in the subnormal range. Yet, in that case, the result is known to be computed exactly because the inputs are representable (see section 5.1.2.2). The following theorem handles that case by generalizing the property to any real number that is a multiple of 2^e:

$$\mathrm{FIX}(x,e) \;\Rightarrow\; \mathrm{REL}(\square(x),x,[-2^{-p};2^{-p}]) \quad \text{when } \square \text{ is float<}p,e\text{,ne>.}$$

Finally, on a related note, given suitable FIX and FLT properties on x, Gappa can also deduce that x is representable:

$$\mathrm{FIX}(x,e) \;\Rightarrow\; \square(x) = x \quad \text{when } \square \text{ is fixed<}e,\ldots\text{>,}$$
$$\mathrm{FLT}(x,p) \wedge \mathrm{FIX}(x,e) \;\Rightarrow\; \square(x) = x \quad \text{when } \square \text{ is float<}p,e,\ldots\text{>.}$$

4.3.6. *Backward propagation*

So far, all the theorems we have considered act as forward propagators: known facts on the leaves of an expression are inductively combined to obtain some new facts on the whole expression. This process is satisfactory when verifying that a round-off error is bounded, but not when some extra knowledge is known on some expressions. In particular, if we are trying to bound an expression that is not a round-off error, we are not making any use of the target bounds.

Let us illustrate the issue on an example. We have two real numbers x and y that are constrained in such a way that the 2D point of coordinates (x, y) is included in the triangle of vertices $(0, 1)$, $(1, 0)$, and $(0, -1)$. We would like to prove that this point is also included in the disk of center $(0, 0)$ and of radius $\sqrt{1.1}$. This can be expressed by the following Gappa script:

```
{ x in [0,1] /\ y in [-1,1] /\
  x+y in [-1,1] /\ x-y in [-1,1] ->
  x*x+y*y in [0,1.1] }
```

Plain interval arithmetic is just able to deduce $x^2 \in [0; 1]$ and $y^2 \in [0; 1]$, and then $x^2 + y^2 \in [0; 2]$, while the optimal enclosure is $x^2 + y^2 \in [0; 1]$. Note that this enclosure could be obtained by telling Gappa about the following equality, but that would defeat the point of this example.

$$x^2 + y^2 = \tfrac{1}{2}((x + y)^2 + (x - y)^2).$$

As already mentioned, Gappa computes the enclosure $x^2 + y^2 \in [0; 2]$, but since the interval is not included in $[0; 1.1]$, it cannot prove the logical proposition. It then tries to do some case splitting: it separately studies the cases $x^2 + y^2 \leq 1.1$ and $x^2 + y^2 \geq 1.1$. (Let us ignore the fact that 1.1 cannot be represented as a dyadic number; it does not matter for this example.) The hypothesis $x^2 + y^2 \leq 1.1$ makes the goal trivial to prove, given that $x^2 + y^2 \geq 0$ is already known. So we are left with the case $x^2 + y^2 \geq 1.1$, from which we deduce $x^2 + y^2 \in [1.1; 2]$.

In a forward setting, there is not much we can do with this enclosure. That is where backward propagation comes in, as its purpose is to deduce bounds on x and y given bounds on more complicated expressions. From $x^2 \in [0; 1]$ and $x^2 + y^2 \in [1.1; 2]$, we can deduce $y^2 \in [1.1; 2] - [0; 1] = [0.1; 2]$. Indeed, it is just a matter of expressing y^2 as $(x^2 + y^2) - x^2$ and then doing plain interval arithmetic. Since we had previously computed $y^2 \in [0; 1]$, we now have a tighter enclosure: $y^2 \in [0.1; 1]$. We can perform a similar propagation for x^2 and we get $x^2 \in [0.1; 1]$. Practically, whenever some subexpression $u + v$ appears in a logical proposition, Gappa automatically instantiates the following two equalities, which mean that, whenever Gappa gets some bounds on the right-hand side, it can use them for the left-hand side.

$$u = (u+v) - v,$$
$$v = (u+v) - u.$$

A similar mechanism is then used to refine the enclosure of x given the enclosure of x^2. First, new bounds on $|x|$ are computed using the equality $|x| = \sqrt{x^2}$. This approximately gives $|x| \in [0.3; 1]$. Second, backward propagation disposes of the absolute value, which gives $x \in [0.3; 1]$.

We can now make use of the hypotheses on $x + y$ and $x - y$, again by backward propagation:

$$y = x - (x - y) \quad \in \quad [0.3; 1] - [-1; 1] = [-0.7; 2],$$
$$y = (x + y) - x \quad \in \quad [-1; 1] - [0.3; 1] = [-2; 0.7].$$

We have thus proved $y \in [-0.7; 0.7]$. With this new enclosure on the input variable, we can now do a forward propagation to deduce $y^2 \in [0; 0.5]$. Then we can perform again all the steps of the previous backward propagation. This will give an even tighter enclosure of y. We iterate this process until the computed bounds of $x^2 + y^2$ contradict the original enclosure, which completes the proof.

If we had wanted to prove a tighter upper bound on $x^2 + y^2$, e.g. $x^2 + y^2 \leq 1.01$, a higher number of iterations would have been needed but it would still have succeeded. It does not work for the optimal bound $x^2 + y^2 \leq 1$, since Gappa would then run forever, computing tighter and tighter enclosures of y, but never reaching $y \in [0; 0]$, from which it could conclude the proof.[3]

This kind of infinite loop might prevent Gappa from exploring other proof paths. To detect them, a heuristic is used: whenever a computation improves an enclosure, Gappa quantifies this improvement; if it does not exceed some threshold, the tool discards the improved enclosure. Note that this might cause Gappa to fail to prove a proposition, if the discarded enclosure was actually mandatory for completing the proof. This hardly happens in practice, though.

---

3 Since Gappa does not perform infinitely-precise computations, it would eventually reach a fixed point and fail, but way too slowly to make a difference from the user point of view.

4.3.7. User hints

Some limitations of the Gappa tool we have mentioned can often be circumvented by providing it with some user knowledge by the way of hints. To illustrate this usage, let us assume that Gappa has to bound an error between $\tilde{u} \diamond \tilde{v}$ and $u \diamond v$ for some binary operator $\diamond$. As explained in section 4.3.2, it first tries to bound the errors between $\tilde{u}$ and u, and between $\tilde{v}$ and v. The process is recursively repeated on these errors between smaller expressions, until Gappa reaches elementary errors that it can bound. For instance, it might be an error between an expression and itself (that is, no error), or it might be an error given as a hypothesis by the user, or it might be an error that can be computed by naive interval arithmetic without much overestimation.

This process only succeeds if the way Gappa decomposes errors makes sense. For instance, if $\tilde{u}$ (respectively $\tilde{v}$) is close to u (respectively v), then Gappa can bound the error between $\tilde{u} \diamond \tilde{v}$ and $u \diamond v$, but it will not succeed in bounding the error between $\tilde{u} \diamond \tilde{v}$ and $v \diamond u$, even if $\diamond$ is a commutative operator. Indeed, the tool will only consider the errors between $\tilde{u}$ and v, and between $\tilde{v}$ and u; none of them are meaningful.

More generally, if Gappa has to bound an error between two expressions that do not have a similar structure, it will fail. In that case, the user has to help Gappa. Let us consider an example. We have six reals u, u_h, u_ℓ, v, v_h, and v_ℓ, such that $u_h + u_\ell$ (respectively $v_h + v_\ell$) approximates u (respectively v). We assume that u_ℓ (respectively v_ℓ) is negligible with respect to u_h (respectively v_h), that is, $u_h + u_\ell$ (respectively $v_h + v_\ell$) is a 2-expansion [DEK 71, SHE 97]. For optimization purpose, when doing the product of two 2-expansions, we choose to ignore the least significant partial product $u_\ell \cdot v_\ell$. We are interested in bounding the error between the partial expansion product $p = u_h \cdot v_h + u_\ell \cdot v_h + u_h \cdot v_\ell$ and the exact product $u \cdot v$.

Since we are interested in bounding an error between a sum and a product, none of the rules from forward error analysis can be used. So we have to give some indications to Gappa. First, we would like Gappa to decompose the relative error between p and $u \cdot v$ into two relative errors, the first one between p and $q = (u_h + u_\ell) \cdot (v_h + v_\ell)$, and the second one between q and $u \cdot v$. Indeed, p approximates q because a presumably negligible term has been removed, and q approximates $u \cdot v$ because it is the product of two expressions that approximate the factors of $u \cdot v$. Moreover, the second relative error is between two expressions with a similar structure, so Gappa will handle it properly.

The expressions p and q still do not have the same structure, so we have to help the tool further. Since we are interested in the relative error, let us rewrite it in a way that makes it simpler for the tool to bound it:

$$\frac{p - q}{q} = -\frac{u_\ell \cdot v_\ell}{q} = -\frac{u_\ell}{u_h + u_\ell} \cdot \frac{v_\ell}{v_h + v_\ell}. \qquad [4.9]$$

As explained in section 4.3.2.3, the tool happens to know how to bound an expression of the form $b/(a + b)$ given an enclosure of b/a and the fact that a and $a + b$ are nonzero. In other words, the tool just needs some enclosures of u_ℓ/u_h and v_ℓ/v_h, which come from the fact that u_ℓ (respectively v_ℓ) is negligible with respect to u_h (respectively v_h).

Let us consider a concrete instance of this example. We suppose that the relative error between $u_h + u_\ell$ and u is bounded by 10^{-13} and the relative error between $v_h + v_\ell$ and v is bounded by 10^{-14}. Let us also suppose that u_ℓ is smaller than u_h by a factor of 10^9, and similarly for v_ℓ and v_h. Finally we assume some artificial enclosures on u_h and v_h just so that Gappa can successfully prove that the various divisors are nonzero. Note that the only bounds that actually matter are the bounds 10^{-13} and 10^{-14} on the relative error, since the other ones have only a negligible effect on the final error bound.

At this point, Gappa would fail to bound the relative error between p and $u \cdot v$. So we also add equation [4.9] as a hypothesis. From this equality, Gappa infers not only a way to bound the relative error between p and q, but also that q is a potential intermediate expression when bounding the error between p and $u \cdot v$. This gives the following Gappa script.

```
p = uh * vh + uh * vl + ul * vh;
q = (uh + ul) * (vh + vl);

{ |uh + ul -/ u| <= 1e-13 /\ |vh + vl -/ v| <= 1e-14 ->
  |ul / uh| <= 1e-9     /\ |vl / vh| <= 1e-9     ->
  uh in [1,1000]        /\ vh in [1,1000]        ->
  (p - q) / q = - (ul / (uh + ul)) * (vl / (vh + vl)) ->
  p -/ u * v in ? }
```

Gappa answers that the relative error between p and $u \cdot v$ is bounded by $11 \cdot 10^{-14}$ (plus some second-order terms) which is the sum of the relative errors, as expected from a product. The added equality has properly captured the idea that $u_\ell \cdot v_\ell$ is negligible with respect to q and thus can be ignored. This equality appears as a hypothesis, so it has to be proved separately, e.g. using the field tactic described in section 4.1.1.

More generally, the user should add new hypotheses, either as equalities or as enclosures, whenever the tool is missing some information to complete a proof. Chapter 6 will give several examples of such user hints. Thanks to such hints, the tool can automatically produce a formal proof of the input proposition (with the hints as hypotheses).

5

Error-Free Computations and Applications

Even if FP computations are inexact most of the time, we can prove the exactness of some computations or retain some exactness using remainders as done in this chapter. We are interested in subcases where an FP computation is error-free, and will give several sufficient assumptions. When an FP computation is not exact, we moreover aim at computing the error it made, using FP operators only [DEK 71]. This kind of algorithm is called an error-free transformation (EFT). Given an operation $\diamond$ on two operands x and y, the corresponding EFT produces two FP numbers r_h and r_ℓ such that $r_h = \Box(x \diamond y)$ and $x \diamond y = r_h + r_\ell$. In other words, the EFT produces the rounding of the operation and the exact error of this operation. These operators will probably be recommended by the next version of the IEEE-754 standard (that will be published in 2018) for the addition (denoted as augmentedAddition) and the multiplication (denoted by augmentedMultiplication). When the error cannot be represented by an FP number (as in the case of division and square root), we will study the conditions needed to ensure that the remainders $x - \Box(x/y) \cdot y$ and $x - \Box(\sqrt{x})^2$ are representable by an FP number. These remainders can then be used in subsequent computations. The algorithms for computing these EFTs are well-known [DEK 71, SHE 97, BOL 03] and their formalization adds an extra level of guarantee. In particular, the requirements on the radix and the handling of underflow have to be specified.

This chapter is organized by operation. Section 5.1 deals with addition. This includes theorems ensuring exact addition/subtraction and how to compute the error of an addition. Section 5.2 deals with multiplication and gives two algorithms for its EFT, depending on whether an FMA is available. Section 5.3 is about the remainder for the integer division. Section 5.4 deals with the remainders of both the FP division and square root. Section 5.5 describes another exact FP computation: taking the square root of the square. Section 5.6 provides exact and approximated algorithms for computing the error of the FMA.

5.1. Exact addition and EFT for addition

The first operation we consider is FP addition. The EFTs for addition, known as 2Sum and Fast2Sum, are given in section 5.1.3. Before that, we focus on simpler but useful results that guarantee that a given addition or subtraction is exact. Section 5.1.1 describes the exact subtraction when the two inputs have the same magnitude, also known as Sterbenz' theorem. Section 5.1.2 ensures exact additions when the result is close to zero.

5.1.1. *Exact subtraction*

If two FP numbers are close to one another, their subtraction is exact. Its is usually attributed to Sterbenz [STE 74] even though it seems to be older. More precisely, it states that, if x and y are FP numbers such that $\frac{y}{2} \leq x \leq 2y$, then $x - y$ is computed without error, because it fits into an FP number. Note that the integer 2 occurring in the inequalities is not the radix; this lemma holds whatever the radix $\beta > 1$.

This exact computation is a cancellation: the magnitude of the result is much smaller than the magnitude of the inputs. In particular, it means that the previous errors are magnified by this cancellation, possibly making it catastrophic. Cancellations are often blamed for FP inaccuracy, although a cancellation is an exact operation.

Theorem 5.1 (Sterbenz). Assume a generic FP format described by a monotone function φ. Given two FP numbers x and y in the format such that

$$\frac{y}{2} \leq x \leq 2y,$$

$x - y$ is representable in the format.

The most interesting part is what is required on the FP format. As explained in section 3.1.3, the format is defined by a generic function φ. This function must be valid but theorem 5.1 requires φ to be monotone as well. Consider for example the FTZ (flush-to-zero) format, which is not monotone. The smallest positive number η and its successor are near enough one to another, but their difference is ulp(η), which is not an FP number in the FTZ format. Thus, Sterbenz' theorem does not hold in the FTZ format.

As for the proof of theorem 5.1, it relies on an interesting intermediate lemma similar to some existing result [FER 95]. We recall that mag(x) is the integer such that $|x| \in [\beta^{\mathrm{mag}(x)-1}; \beta^{\mathrm{mag}(x)})$.

Lemma 5.2 (generic_format_plus). Assume a generic FP format described by a monotone function φ. Given two FP numbers x and y in the format such that

$$|x + y| \leq \beta^{\min(\mathrm{mag}(x),\mathrm{mag}(y))},$$

then $x + y$ is representable in the format.

Proof. Under the previous assumptions, that is to say the validity and monotony of φ, and given two FP numbers x and y such that $|x + y| \leq \beta^{\min(\mathrm{mag}(x),\mathrm{mag}(y))}$, we consider the real value $x + y$. Either $x + y$ is exactly $\beta^{\min(\mathrm{mag}(x),\mathrm{mag}(y))}$, which is in the format, or it is smaller and then $\mathrm{mag}(x + y)$ is smaller than $\min(\mathrm{mag}(x), \mathrm{mag}(y))$. As φ is monotone, then $\mathrm{cexp}(x + y) = \varphi(\mathrm{mag}(x + y)) \leq \varphi(\min(\mathrm{mag}(x), \mathrm{mag}(y))) = \min(\varphi(\mathrm{mag}(x)), \varphi(\mathrm{mag}(y))) = \min(\mathrm{cexp}(x), \mathrm{cexp}(y))$. As x and y are FP numbers, they are representable with an exponent $\mathrm{cexp}(x)$ for x and $\mathrm{cexp}(y)$ for y. Therefore, $x + y$ can be represented with the exponent $\min(\mathrm{cexp}(x), \mathrm{cexp}(y))$, which was proved to be greater than or equal to $\mathrm{cexp}(x + y)$ above. Therefore, $x + y$ is in the format. ∎

5.1.2. *Exact addition when the result is close to zero*

There are other cases where we know there is no rounding error, in particular when the result is close to zero. The first result is when the result is zero and the second one is only valid in the FLT formats. See Table 3.2 for a comparison of some Flocq FP formats, including FLT.

5.1.2.1. *Exact addition when the result is zero*

We first need an auxiliary lemma corresponding to the existence of some kind of subnormal exponent, meaning an integer e such that $e \leq \varphi(e)$. Then all FP numbers in the format can be expressed as multiples of β^e.

Lemma 5.3 (subnormal_exponent). Assume a generic FP format described by a function φ such that $\forall z, \varphi(\varphi(z) + 1) \leq \varphi(z)$ (meaning it is a non-FTZ format, see also lemma 3.12). If there is an exponent e such that $e \leq \varphi(e)$, then for any FP number x in the format, $x = \lfloor x\beta^{-e} \rfloor \cdot \beta^e$.

Proof. This proof reduces to ensuring that $\lfloor x\beta^{-e} \rfloor = x\beta^{-e}$. It is enough to prove that $e \leq \mathrm{cexp}(x)$. This is automatically proved using omega (see section 4.1.3) from the properties of the function φ, namely, its validity and its not being flush-to-zero. ∎

We now focus on the addition with a zero result. Except in FTZ formats, it implies that the value to be rounded is zero too. We assume the format is described by a φ such that $\forall z, \varphi(\varphi(z) + 1) \leq \varphi(z)$ (see also lemma 3.12)

Lemma 5.4 (round_plus_eq_zero). Assume a generic FP format described by a function φ such that $\forall z, \varphi(\varphi(z) + 1) \leq \varphi(z)$ (meaning it is a non-FTZ format, see also lemma 3.12). Let x and y be FP numbers in this format. If $\square(x + y) = 0$, then $x + y = 0$.

Proof. We consider $e = \text{cexp}(x + y)$ and distinguish between two subcases. If $e \leq \varphi(e)$, then we apply lemma 5.3 both to x and y. Therefore, $x + y$ can be represented with exponent $e = \text{cexp}(x+y)$. Therefore it rounds to itself. So $0 = \square(x+y) = x+y$.

If $\varphi(e) < e$, we suppose $x + y \neq 0$ by contradiction. Then $\beta^{e-1} \leq |x + y| < \beta^e$. We have by monotony of rounding that $\square\left(\beta^{e-1}\right) \leq \square(|x + y|)$. But the contrary also holds: $\square(|x + y|) = 0 < \beta^{e-1} = \square\left(\beta^{e-1}\right)$, as β^{e-1} is in the format by lemma 3.3. ∎

5.1.2.2. *Exact addition when the result is subnormal*

Now, let us assume an FLT format with precision ϱ and minimal exponent $e_{\min}$. The smallest positive FP number is then $\beta^{e_{\min}}$ and the smallest positive normal FP number is $\beta^{\varrho-1+e_{\min}}$. In this format, all FP additions such that the result is subnormal (and slightly above) are exact [GOL 91].

Lemma 5.5 (FLT_format_plus_small). Given an FLT format and two FP numbers x and y such that

$$|x + y| \leq \beta^{\varrho+e_{\min}},$$

$x + y$ is an FP number in the same format.

Proof. The proof is very similar to the previous one, only simpler. As x and y are in the format, we apply lemma 5.3 twice with $e = e_{\min}$. Therefore, $x + y$ can be expressed as $m\beta^{e_{\min}}$. As $|x + y| \leq \beta^{\varrho+e_{\min}}$, then $|m| \leq \beta^\varrho$. If $|m| < \beta^\varrho$, then the FP number with integer significand m and exponent $e_{\min}$ is in the format by definition of the FLT format (see section 3.1.2.1). This number is also equal to $x + y$. Else $|m| = \beta^\varrho$, so $|x + y| = \beta^{\varrho+e_{\min}}$ is representable, and then $x + y$ is also representable. ∎

5.1.3. *Fast2Sum and 2Sum algorithms*

In the general case, an FP addition is not exact. Its error is a real number, that we may want to approximate. The most interesting case is when the rounding is to nearest. In this case, the error of an FP addition is an FP number. Moreover, there are algorithms to compute this error. These are called the 2Sum [KNU 98] and Fast2Sum [DEK 71] algorithms. We assume that the radix is 2 and that rounding is to nearest, with any symmetric tie-breaking rule τ. This means that $\circ^\tau(-x) = -\circ^\tau(x)$ and holds both for tie breaking to even and tie breaking away from zero.

5.1.3.1. *Fast2Sum algorithm*

Algorithm 5.1 is very simple as we compute r_h, the rounding of $x + y$, and another value r_ℓ, which should be 0 if there were no rounding errors. Unfortunately, it only works when $|y| \leq |x|$. It means we have to know beforehand which input has the largest magnitude before calling this algorithm.

Algorithm 5.1 Fast2Sum

Input: FP numbers x and y such that $|y| \leq |x|$
Output: FP numbers r_h and r_ℓ such that $r_h + r_\ell = x + y$

$$r_h = \circ^\tau(x + y)$$
$$r_\ell = \circ^\tau[y + (x - r_h)]$$

Theorem 5.6 (Fast2Sum_correct). Given a symmetric tie-breaking rule τ and two FP numbers x and y such that $|y| \leq |x|$, the results of algorithm 5.1 (Fast2Sum) are such that

$$r_h + r_\ell = x + y.$$

This means that r_ℓ is exactly the remainder of the FP addition. This proof is much more complex than the previous lemmas and not described here [DAU 01].

5.1.3.2. *2Sum algorithm*

For unknown inputs, we can test which one is the largest but this test may be hard to predict, thus leading to poor performance on modern architectures. It may be faster to use algorithm 5.2 which works whatever the magnitude of the inputs.

Algorithm 5.2 2Sum

Input: FP numbers x and y
Output: FP numbers r_h and r_ℓ such that $r_h + r_\ell = x + y$

$$r_h = \circ^\tau(x + y)$$
$$x' = \circ^\tau(r_h - x)$$
$$dx = \circ^\tau[x - (r_h - x')]$$
$$dy = \circ^\tau(y - x')$$
$$r_\ell = \circ^\tau(dx + dy)$$

Theorem 5.7 (TwoSum_correct). Given a symmetric tie-breaking rule τ and two FP numbers x and y, the results of algorithm 5.2 (2Sum) are such that

$$r_h + r_\ell = x + y.$$

This proof is also quite complex and it relies on theorem 5.6 [DAU 01].

In both algorithms, there is no problem with underflow, as underflowing additions are correct (see section 5.1.2.2). For more details about the (formal) proof, we refer the reader to [DAU 01].

These algorithms are quite robust. They behave well when double rounding occur [MAR 13] or when the rounding is not to nearest (hence the error may not be representable) [BOL 17]: this means they provide a good approximation of the FP remainder in all these cases. Moreover, these algorithms do not overflow (assuming $x + y$ does not) except in a single case [BOL 17].

5.2. EFT for multiplication

The addition case leads to the simplest theorem statements, since the rounding error is always an FP number. This is also the case when considering multiplication with an infinite exponent range. See Table 3.2 for a comparison of some Flocq FP formats, including FLX.

Lemma 5.8 (mult_error_FLX). Consider an FLX format and two FP numbers x and y. Then $\circ^\tau(x \cdot y) - x \cdot y$ is an FP number in the same format.

This is not true anymore when we consider an FLT format: the error term may be too small to be an FP number due to the limited exponent range. For instance, consider the product of β^{-2} and $\beta^{e_{\min}}$ (the smallest FP number). It is rounded to 0 in rounding to nearest. The error term is then $\beta^{e_{\min}-2}$, which has a too small exponent to be an FP number. This explains why our theorems have to handle underflow, either by preconditions or conditions on the outputs. A sufficient condition for the error term to be an FP number is given by the following lemma.

Lemma 5.9 (mult_error_FLT). Consider an FLT format and two FP numbers x and y such that

$$x \cdot y = 0 \quad \vee \quad |x \cdot y| \geq \beta^{e_{\min}+2\varrho-1},$$

then $\circ^\tau(x \cdot y) - x \cdot y$ is an FP number in the same format.

If we assume an architecture providing the FMA instruction, the EFT for multiplication is the rather trivial algorithm 5.3.

The computation of r_ℓ requires an FMA. This FMA computes exactly the rounding of the multiplication error. Therefore, if this error is an FP number, it is correctly computed. Provided that no underflow happens, this algorithms returns the expected result.

Algorithm 5.3 Computation of the error of a multiplication when an FMA is available

$$r_h = \circ^\tau (x \cdot y)$$
$$r_\ell = \circ^\tau (x \cdot y - r_h)$$

Now for underflow, we require preconditions on the inputs: if x or y is zero, or if $|x \cdot y| \geq \beta^{e_{\min}+2\varrho-1}$, the computed value r_ℓ of algorithm 5.3 is correct, meaning $r_h + r_\ell = x \cdot y$.

Even when underflow happens, algorithm 5.3 computes the rounding of the multiplication error (possibly not the multiplication error). In our initial example $\beta^{e_{\min}} \cdot \beta^{-2}$, it would return zero, which is the correct rounding of the FP error of the multiplication.

When no FMA is available, there is a more complicated algorithm to compute this error described in section 5.2.2. It relies on an auxiliary function that splits an FP number into its upper and lower parts, as described in section 5.2.1.

5.2.1. Splitting of an FP number

When no FMA is available, one may still compute the multiplication error. The first (and most important) step is the splitting of an FP number, meaning getting its upper and lower halves.

5.2.1.1. Algorithm and idea of the proof of Veltkamp's splitting

Veltkamp's algorithm [VEL 68, VEL 69, DEK 71] splits an FP number x into two FP numbers x_h and x_ℓ such that $x = x_h + x_\ell$ and x_h is the result of rounding x to a smaller precision. More precisely, let $\varrho - s$ be this smaller precision, with s being an integer such that $2 \leq s \leq \varrho - 2$ and ϱ the working precision.

Algorithm 5.4 computes x_h and x_ℓ so that x_h fits on $\varrho - s$ digits and x_ℓ fits on s digits. It only uses basic FP operations in the working precision. The constant C can be stored beforehand, so the algorithm consists of four operations.

Figure 5.1 gives an intuition on how this algorithm works. We compute $p \approx (\beta^s + 1) \cdot x$ and $q \approx x - p \approx -\beta^s \cdot x$. Therefore, p and $-q$ are very close to each other and x_h is computed exactly. Moreover, $q \approx x - p$ and the exponent of q is about s plus the exponent of x, so that all the digits of x lesser than this value are lost. So we only have in x_h the first $\varrho - s$ bits of x. Of course, a drawing is not a proof and especially not a formal proof. We have to deal with many cases, including the possible values of the exponents of p and x_h. An important point is the value of the exponent of q.

Algorithm 5.4 Veltkamp's splitting of an FP number

Input: FP number x, integer s such that $2 \le s \le \varrho - 2$

Output: FP numbers x_h and x_ℓ such that x_h is a rounding to nearest of x at precision $\varrho - s$ and $x_h + x_\ell = x$

Let $C = \beta^s + 1$.

$$p = \circ^\tau(x \cdot C)$$

$$q = \circ^\tau(x - p)$$

$$x_h = \circ^\tau(p + q)$$

$$x_\ell = \circ^\tau(x - x_h)$$

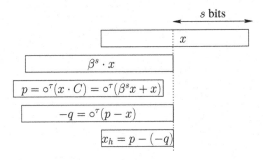

Figure 5.1. *Idea of Veltkamp's algorithm*

5.2.1.2. *Proof of Veltkamp's algorithm*

We first consider a rounding to nearest, without fixing the tie-breaking rule. The first lemma guarantees that an FP number is the rounding to nearest of any real value close enough to it. It will be used in order to prove that x_h is a rounding to nearest of x at a given precision, provided an error bound and some inequalities.

Lemma 5.10. Given an FP number f, if we have $\beta^{e+\varrho-1} \le z \le \beta^{e+\varrho}$ and $\beta^{e+\varrho-1} \le f$ and $|z - f| \le \frac{1}{2}\beta^e$, then f is a rounding to nearest of z.

In this lemma, e is an integer related to the magnitude of f. As both z and f are greater than $\beta^{e+\varrho-1}$, we do not fall into the pitfall of being just around a power of the radix (see section 1.1.1). Note that f is a rounding to nearest, but not necessarily with tie breaking to even.

Note that rather similar theorems for ensuring faithful rounding are available in section 6.3.1.

5.2.1.2.1. Sketch of the proof and radix discussion

Let us move to the proof of Veltkamp's algorithm in a generic radix β. We first assume that there is no underflow: all FP numbers are normal. We assume that the rounding is any rounding to nearest in precision ϱ. For an FP number x, we denote by e_x its canonical exponent $\text{cexp}(x) = \text{mag}(x) - \varrho$ as we assume all FP numbers are normal. We then denote $m_x = x\beta^{-e_x}$ the integer significand.

Here are the steps taken to prove algorithm 5.4:

1) $x_h = p + q$;

2) $e_q \leq e_p \leq s + 1 + e_x$;

3) $\beta = 2$ implies $e_q \leq s + e_x$;

4) either $e_q \leq s + e_x$ or we have both $q = -\beta^{\varrho+s+e_x}$ and $|x - x_h| \leq \frac{1}{2}\beta^{e_x+s}$;

5) $e_q \geq s + e_x$;

6) x_h is a rounding to nearest of x with precision $\varrho - s$, using lemma 5.10;

7) $x = x_h + x_\ell$;

8) $|x - x_h| \leq \frac{1}{2}\beta^{e_x+s}$;

9) $\beta = 2$ implies that x_ℓ fits on $s - 1$ bits.

Their detailed proofs can be found in [BOL 06a] and the most interesting parts can be found below. Assumption 3 is very useful as it exactly gives the exponent of q which is the key point of the proof in [DEK 71, SHE 97].

Here is a counter-example to assumption 3 when $\beta = 10$, $\varrho = 6$, and $s = 3$. For $x = 999,996$, we have $x \cdot 1001 = 1,000,995,996$ that is rounded into $p = 1,001,000,000$ and $x - p = -1,000,000,004$ therefore $q = \circ(x - p) = -1,000,000,000$ has for exponent $4 + e_x = 1 + s + e_x$.

As for assumption 9, it will be proved later (lemma 5.19) that x_ℓ can be represented with an integer significand smaller than or equal to $\frac{1}{2}\beta^s$. When $\beta = 2$, this significand is smaller than or equal to β^{s-1}, therefore x_ℓ can be represented on $s - 1$ bits.

This fact is well known in radix 2 but has counter-examples in greater radices, for example when $\beta = 10$, $\varrho = 6$, and $s = 3$. For $x = 123{,}456$, we have $x_h = 123{,}000$ and $x_\ell = 456$ cannot fit on 2 digits. In radix 2, the sign is enough to compensate one digit and x_ℓ can be shortened by one digit.

5.2.1.2.2. A proof with a generic radix but still no underflow

We still assume that there is no underflow. We want to prove that x_h is a rounding to nearest of x in precision $\varrho - s$. By symmetry of the rounding, we assume that $x >$

0. This is due to the fact that, whatever the tie-breaking rule τ and real x, we have $\circ^{\tau}(-x) = -\circ^{\tau'}(x)$ for a τ' computed from τ (the tie breaking rule definition can be found in section 3.3.1.1). This means that it is enough to prove the property with x positive and any tie-breaking rule to ensure it for all non-zero x and all tie-breaking rules. Therefore, we assume that $x > 0$, so we have $p > 0$, $q < 0$, and $x_h > 0$.

Lemma 5.11. $x_h = p + q$.

Proof. Let $\varepsilon = \frac{1}{2}\beta^{1-\varrho}$. We use Sterbenz' theorem (see section 5.1.1) to prove it. First, $q = \circ^{\tau}(x - p) \geq \circ^{\tau}(-p) = -p$ by monotony of the rounding. Second, if $f = \circ^{\tau}(z)$ and f is normal, then $|z| \leq |f|(1 + \varepsilon)$, so

$$p = (p - x) + x \leq |q|(1 + \varepsilon) + \frac{xC}{\beta^s + 1} \leq |q|(1 + \varepsilon) + |p|\frac{1 + \varepsilon}{\beta^s + 1}.$$

Therefore, $p \cdot \left(1 - \frac{1 + \varepsilon}{\beta^s + 1}\right) \leq |q|(1 + \varepsilon)$ leads to $p \leq 2|q|$.   ∎

Lemma 5.12. $e_q \leq e_p \leq s + 1 + e_x$.

Proof. As $|q| \leq |p|$, we have $e_q \leq e_p$. As $p = \circ^{\tau}((\beta^s + 1) \cdot x) \leq \circ^{\tau}(\beta^{s+1} \cdot x)$, we have $e_p \leq s + 1 + e_x$.   ∎

Lemma 5.13. Either $e_q \leq s + e_x$, or both $q = -\beta^{\varrho + s + e_x}$ and $|x - x_h| \leq \frac{1}{2}\beta^{e_x + s}$.

The result is a disjunction: either we are in the common case (like in radix 2), where the exponent of q will be known exactly, or we are in the exceptional case corresponding to the counter-example of the previous section: then q is slightly too large to have exponent $s + e_x$. Nevertheless, we can bound the difference between x and x_h. The proof is long and computational. It can be found in [BOL 06a].

Lemma 5.14. $e_q \geq s + e_x$.

Proof. We bound q using $p \geq |q|$:

$$|q| = |\circ^{\tau}(x - p)| \geq p - x - \tfrac{1}{2}\mathrm{ulp}(q) \geq \beta^s x - \tfrac{1}{2}\mathrm{ulp}(p) - \tfrac{1}{2}\mathrm{ulp}(q) \geq \beta^s - \mathrm{ulp}(p$$

Therefore, $|q| \geq m_x \beta^{s+e_x} - \beta^{s+1+e_x} \geq (m_x - \beta)\beta^{s+e_x}$ by lemma 5.12.

So if $m_x \geq \beta^{\varrho-1} + \beta$, we have $e_q \geq s + e_x$. If not, we may have $m_x = \beta^{\varrho-1}$. In this case, all the computations are exact and $-q = \beta^{s+\varrho-1+e_x}$ has its exponent equal to $s + e_x$.

The last possibility is that $\beta^{\varrho-1} + 1 \leq m_x \leq \beta^{\varrho-1} + \beta - 1$. Note that in radix 2, there is only one such FP number. Here, we precisely bound p and q:

$$x \cdot C \;\geq\; \beta^{e_x}\left(\beta^{\varrho-1}+1\right)\left(\beta^s+1\right) > \beta^{s+e_x}\left(\beta^{\varrho-1}+\beta^{\varrho-s-1}+1\right),$$
$$p \;\geq\; \beta^{s+e_x}\left(\beta^{\varrho-1}+\beta^{\varrho-s-1}+1\right),$$
$$p - x \;\geq\; \beta^{\varrho+s-1+e_x}+\beta^{\varrho-1+e_x}+\beta^{s+e_x}-\beta^{e_x}\left(\beta^{\varrho-1}+\beta-1\right),$$
$$p - x \;>\; \beta^{\varrho+s-1+e_x},$$
$$-q \;\geq\; \beta^{\varrho+s-1+e_x}.$$

In all cases, $e_q \geq s + e_x$. ∎

Lemma 5.15. $|x - x_h| \leq \frac{1}{2}\beta^{e_x+s}$.

Proof. We split between the two subcases of lemma 5.13. In the second case, it holds. In the first case, we have $e_q = s + e_x$ by lemma 5.14. This also implies that $|x - x_h| = |(x - p) - q| \leq \frac{1}{2}\mathrm{ulp}(q) = \frac{1}{2}\beta^{e_x+s}$. ∎

Lemma 5.16. $\beta^{\varrho-1+e_x} \leq x_h$.

Proof. If $m_x \geq \beta^{\varrho-1} + \frac{1}{2}\beta^s$, then $\beta^{\varrho-1+e_x} \leq x - \frac{1}{2}\beta^{s+e_x} \leq x - |x - x_h| \leq x_h$.

If not, let $i = m_x - \beta^{\varrho-1}$. We first have $i < \frac{1}{2}\beta^s$. Then, we can exactly give the values of p and q using i:

$$p = \circ^{\tau}(x\beta^s + x) = \beta^{s+e_x}\left(\beta^{\varrho-1}+\beta^{\varrho-1-s}+i\right),$$
$$q = \circ^{\tau}(x - p) = -\beta^{s+e_x}\left(\beta^{\varrho-1}+i\right).$$

So using lemma 5.11, we have $x_h = p + q = \beta^{\varrho-1+e_x}$. ∎

Lemma 5.17. x_h is a rounding to nearest of x at precision $\varrho - s$.

Proof. Let us apply lemma 5.10 with the value e being $e_x + s$.

We first have to prove that x_h can be represented on $\varrho - s$ bits. We know that $s + e_x \leq e_q \leq e_p \leq s + e_x + 1$, therefore $x_h = p + q$ can be represented with an integer significand and the exponent $s + e_x$. The corresponding integer significand m is then such that

$$|m| = |x_h|\beta^{-s-e_x} \leq (|x| + |x - x_h|)\beta^{-s-e_x} < \beta^{\varrho-s} + \frac{1}{2}.$$

So $|m| \leq \beta^{\varrho-s}$. This implies that there exists an FP number v bounded on $\varrho - s$ bits that is equal to x_h. It has exponent $s + e_x$ and integer significand m except when $|m| = \beta^{\varrho-s}$ where it has exponent $1 + s + e_x$.

We then have $\beta^{e_x+\varrho-1} \leq x < \beta^{e_x+\varrho}$ as x is normal. From lemma 5.16, we have $\beta^{e_x+\varrho-1} \leq x_h$. Finally, we have $|x - x_h| \leq \frac{1}{2}\beta^{e_x+s}$ by lemma 5.15, which ends the proof. ∎

5.2.1.2.3. Why underflow does not matter

The reason why underflow does not matter here is that we only deal with FP additions. Indeed, multiplying x by C is only adding x and $x\beta^s$ (which is an FP number in the format). Recalling theorem 5.5, underflow produces an exact addition. Note that the underflow threshold on precision $\varrho - s$ is the same as in precision ϱ.

Theorem 5.18 (Veltkamp). If x is an FP number and x_h is computed using algorithm 5.4, then x_h is a rounding to nearest of x in precision $\varrho - s$.

Theorem 5.19 (Veltkamp_tail). The value $x - x_h$ can be represented with exponent e_x and an integer significand smaller than or equal to $\frac{1}{2}\beta^s$. Moreover, $x_\ell = \circ^\tau(x - x_h) = x - x_h$ and fits on s digits.

Proof. When x is normal, as $x_h \geq \beta^{\varrho-1+e_x}$, we know that x_h has a canonical exponent greater than e_x. When x is subnormal, its exponent is minimal, therefore the exponent of x_h is greater. So in any case, $x - x_h$ can be represented with exponent e_x. Moreover, $|x - x_h| \leq \frac{1}{2}\beta^{e_x+s}$, so the integer significand is $|x - x_h|\beta^{-e_x} \leq \frac{1}{2}\beta^s < \beta^\varrho$ using lemma 5.15. So x_ℓ is representable, hence computed exactly. ∎

In the case $\beta = 2$, the integer significand is smaller than 2^{s-1}, hence fits on $s - 1$ bits.

5.2.1.3. *About the tie-breaking rule*

The question is: provided algorithm 5.4 is done in rounding to nearest with a given tie-breaking rule $\circ^\tau$, does it provide $x_h = \circ^\tau_{\varrho-s}(x)$ rounded with the same tie-breaking rule?

5.2.1.3.1. Rounding to nearest, tie breaking to even

When tie breaking is to even, the answer is yes. Assuming that all the computations are performed using $\circ$, the goal is to prove that $x_h = \circ_{\varrho-s}(x)$.

Lemma 5.20. Given an FP number f, if we have $\beta^{e+\varrho-1} \leq z \leq \beta^{e+\varrho}$ and $\beta^{e+\varrho-1} \leq f$ and $|z - f| < \frac{1}{2}\beta^e$, then $f = \circ^\tau(z)$ and f is the only possible rounding to nearest.

The only difference with lemma 5.10 is that the "less than or equal to" $\frac{1}{2}\beta^e$ has become a "less than" to guarantee that f is the only rounding to nearest.

Putting together lemmas 5.20 and 5.17, we easily prove that $x_h = \circ_{\varrho-s}(x)$ when $|x - x_h| < \frac{1}{2}\beta^{e_x+s}$.

The remaining case is the tie-breaking case when $|x - x_h| = \frac{1}{2}\beta^{e_x+s}$ and is more complex as it involves many subcases (odd or even radix, exponent of x_h, and exponent of p); see [BOL 06a] for more details.

Theorem 5.21 (Veltkamp_Even). If x is an FP number and x_h is computed using algorithm 5.4 and tie breaking to even, then x_h is the rounding to nearest with tie breaking to even of x in precision $\varrho - s$.

5.2.1.3.2. Rounding to nearest, tie breaking away from zero

For rounding to nearest, tie breaking away from zero, the answer is no: the result is not necessarily the rounding to nearest, tie breaking away from zero of the input. For example, let $\beta = 2$, $\varrho = 6$ and $s = 3$. If $x = 100,100_2$, then $p = 101,001,000_2$ and $q = -100,101,000_2$ so $x_h = 100,000_2$, which is not the expected result $101,000_2$.

We have proved that Veltkamp's algorithm is correct whatever the radix and the precision, with some results related to the tie-breaking rule. In particular, when a tie breaking rule τ is used, the computed x_h is a rounding to nearest of x in precision $\varrho - s$ but may not be the rounding to nearest with tie breaking rule τ of x. The main drawback of this algorithm is that it may overflow unnecessarily (see section 8.3.2).

5.2.2. Dekker's product

For two FP numbers x and y, algorithm 5.5 computes two FP numbers r_h and r_ℓ such that $r_h = \circ^\tau(x \cdot y)$ and $x \cdot y = r_h + r_\ell$ [DEK 71]. It uses only the working precision (ϱ bits). We rely on Veltkamp's algorithm (algorithm 5.4) with $s = \left\lceil \frac{\varrho}{2} \right\rceil$ to split x and y in two (near-)equal parts.

The idea of this algorithm is that there is no error in the computations of t_1, t_2, t_3, and r_ℓ. The multiplications are exact as the multiplicands are half-width numbers. The additions are exact since they mostly cancel. Indeed, we know r_ℓ will fit on ϱ digits. Figure 5.2 gives an intuition on the respective exponents and sizes of the various variables. In particular, $|t_1|$ is much smaller than $|r_h|$. But $|r_\ell|$ is much smaller, being smaller than or equal to half an ulp of r_h.

5.2.2.1. Proof of Dekker's product

It seems that Dekker believed the algorithm to be correct whatever the radix [DEK 71]. It was discovered later by Linnainmaa [LIN 81] that this was not the case. This algorithm always works in radix 2 but may fail for other radices when the precision is odd (see below).

To guarantee that a computation is exact, we prove that the infinitely precise value is in the format (lemma 3.30). Following the FLT format definition of section 3.1.2.1, we exhibit a possible exponent $e \geq e_{\min}$ (meaning that $f \cdot 2^{-e}$ is an integer) and prove that $|f| < \beta^{e+\varrho}$. The proofs are mostly inequalities, as possible exponents will be naively deduced at each step. As before, we denote by e_x the canonical exponent of an FP number x.

Algorithm 5.5 Dekker's product for computing the error of a multiplication

Input: FP numbers x and y

Output: FP numbers r_h and r_ℓ such that $r_h + r_\ell = x \cdot y$

$$(x_h, x_\ell) = \textit{Veltkamp}(x)$$

$$(y_h, y_\ell) = \textit{Veltkamp}(y)$$

$$r_h = \circ^\tau (x \cdot y)$$

$$t_1 = \circ^\tau [-r_h + x_h \cdot y_h]$$

$$t_2 = \circ^\tau [t_1 + x_h \cdot y_\ell]$$

$$t_3 = \circ^\tau [t_2 + x_\ell \cdot y_h]$$

$$r_\ell = \circ^\tau [t_3 + x_\ell \cdot y_\ell]$$

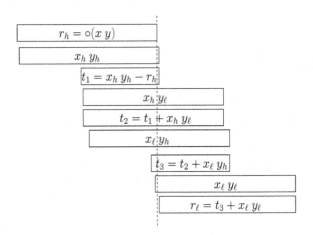

Figure 5.2. *Idea of Dekker's product*

From theorems 5.18 and 5.19 about Veltkamp's algorithm, we know that x_h can be represented with exponent $e_x + s$, that x_ℓ can be represented with exponent e_x, and that $|x_\ell| \leq \beta^{s+e_x}/2$. Similar properties hold for y, y_h, and y_ℓ. As explained above, we choose $s = \lceil \varrho/2 \rceil$ so that x_h and y_h will be on $\lfloor \varrho/2 \rfloor$ digits and x_ℓ and y_ℓ on $\lceil \varrho/2 \rceil$ digits.

From these properties, we deduce several inequalities about the various values $|x_\ell \cdot y_\ell|$, $|x_\ell \cdot y_h|$, $|x_h \cdot y_\ell|$, and $|x_h \cdot y_h|$. Moreover, as $|x \cdot y| \leq (\beta^\varrho - 1)^2 \beta^{e_x + e_y}$, we have $e_{r_h} \leq \varrho + e_x + e_y$. Thus, $|x \cdot y - r_h| \leq \beta^{\varrho + e_x + e_y}/2$. Furthermore, as x and y are normal, we have $|r_h| \geq \beta^{2\varrho - 2 + e_x + e_y}$, thus $e_{r_h} \geq \varrho - 1 + e_x + e_y$.

We aim at proving that all the FP computations are exact. This holds for the multiplications of halves:

– we can represent $x_h \cdot y_h$ with exponent $2s + e_x + e_y$ as $2\lfloor \varrho/2 \rfloor \leq \varrho$,

– we can represent $x_h \cdot y_\ell$ with exponent $s + e_x + e_y$ as $\lfloor \varrho/2 \rfloor + \lceil \varrho/2 \rceil \leq \varrho$,

– we can represent $x_\ell \cdot y_h$ with exponent $s + e_x + e_y$ as $\lfloor \varrho/2 \rfloor + \lceil \varrho/2 \rceil \leq \varrho$.

After the multiplications, let us consider the additions; the t_is are computed exactly:

– t_1 can be exactly represented with exponent $\varrho - 1 + e_x + e_y$. Moreover,

$$
|-r + x_h \cdot y_h| \leq |r - x \cdot y| + |x_h \cdot y_\ell| + |x_\ell \cdot y_h| + |x_\ell \cdot y_\ell|
$$
$$
< \tfrac{1}{2}\beta^{\varrho+e_x+e_y} + \beta^{\varrho+s+e_x+e_y} + \tfrac{3}{4}\beta^{2s+e_x+e_y} < \beta^{2\varrho-1+e_x+e_y}.
$$

– t_2 can be exactly represented with exponent $s + e_x + e_y$. Moreover,

$$
|t_1 + x_h \cdot y_\ell| = |-r + x_h \cdot y_h + x_h \cdot y_\ell| \leq |r - x \cdot y| + |x_\ell \cdot y_h| + |x_\ell \cdot y_\ell|
$$
$$
< \tfrac{1}{2}\beta^{\varrho+e_x+e_y} + \tfrac{1}{2}\beta^{\varrho+s+e_x+e_y} + \tfrac{1}{2}\beta^{2s+e_x+e_y} < \beta^{\varrho+s+e_x+e_y}.
$$

– t_3 can be exactly represented with exponent $s + e_x + e_y$. Moreover,

$$
|t_2 + x_\ell \cdot y_h| = |-r + x_h \cdot y_h + x_h \cdot y_\ell + x_\ell \cdot y_h| \leq |r - x \cdot y| + |x_\ell \cdot y_\ell|
$$
$$
\leq \tfrac{1}{2}\beta^{\varrho+e_x+e_y} + \tfrac{1}{4}\beta^{2s+e_x+e_y} < \beta^{\varrho+s+e_x+e_y}.
$$

What is left is the last computation: if $x_\ell \cdot y_\ell$ is computed exactly, then r_ℓ is computed exactly too, as it is equal to $x \cdot y - \circ^\tau(x \cdot y)$, which is representable when no underflow occurs.

Lemma 5.22. Provided that no underflow occurs, if $x_\ell \cdot y_\ell$ is representable, then $x \cdot y = r_h + r_\ell$.

The problem is that $x_\ell \cdot y_\ell$ fits on $2 \cdot \lceil \varrho/2 \rceil$ digits and this can exceed ϱ by 1 in some cases. There are various solutions:

– If some extra precision is available, then it can be used just for the computation of $x_\ell \cdot y_\ell$ to guarantee the exactness of the result.

– If the radix is 2, then x_ℓ and y_ℓ fit on $s-1$ bits, therefore $x_\ell \cdot y_\ell$ fits on $2 \cdot \lceil \varrho/2 \rceil - 2 \leq \varrho$ bits and the computation is exact.

– If the precision is even, then $2 \cdot \lceil \varrho/2 \rceil = \varrho$ and the computation is exact.

Lemma 5.23. Provided that $x \cdot y = 0 \lor |x \cdot y| \geq \beta^{e_{\min}+2\varrho-1}$, if $\beta = 2$ or ϱ is even, then $x \cdot y = r_h + r_\ell$.

This states the correctness of Dekker's product (algorithm 5.5). The "no underflow" assumption is explicit in this theorem: x or y may be subnormal, but the requirement is that of theorem 5.9 for the error term to be represented.

5.2.2.2. *When underflow happens*

Underflow may happen in this algorithm and its consequences are harmful: instead of all FP computations being exact, they may all become erroneous. Instead of a mathematical equality, the correctness statement is a bound on the error of the computations.

The problem is then twofold: if we are above the underflow threshold for one computation, then it is exact. If we are under the underflow threshold, then the error of the considered computation is less than $\frac{1}{2}\beta^{e_{\min}}$. Properties are simplified by the use of an extended format with the same precision but without underflow, that is, an FLX format.

This is formalized by the following Coq definition. For given real numbers a, a', and ε, we define $\mathrm{Underf\_Err}(a, a', \varepsilon)$ by both a is representable in the working format, a' is representable in the extended format, $|a - a'| \leq \varepsilon\beta^{e_{\min}}$ holds and, if the canonical exponent of a' in the extended format is greater than the underflow threshold of the working format, then $a = a'$. A similar property is defined in section 6.6.

We can then initialize the error computation by the following lemma, which can be applied to the multiplications:

Lemma 5.24. If a fits into the extended format, then we have $\mathrm{Underf\_Err}(\circ^\tau(a), a, 0.5)$.

The most used lemma is the following one, corresponding to the computations of the t_is. For a number u representable in the extended format, the value e_u denotes the canonical exponent in that format.

Lemma 5.25. Assume that both $\mathrm{Underf\_Err}(x, x', \varepsilon_x)$ and $\mathrm{Underf\_Err}(y, y', \varepsilon_y)$ hold. Let $z' = x' - y'$. If z' is representable in the extended format, if $e_{z'} \leq \min(e_{x'}, e_{y'})$, and if $\varepsilon_x + \varepsilon_y \leq \beta^{\varrho-1}$, then we have $\mathrm{Underf\_Err}(\square(x - y), x' - y', \varepsilon_x + \varepsilon_y)$.

Proof. If $e_{z'} \geq e_{\min}$, then we have both $e_{x'} \geq e_{\min}$ and $e_{y'} \geq e_{\min}$. So $x = x'$ and $y = y'$, by definition of $\mathrm{Underf\_Err}$. We deduce $z = z'$ and the result.

If $e_{z'} < e_{\min}$, then the hypothesis $\varepsilon_x + \varepsilon_y \leq \beta^{\varrho-1}$ guarantees

$$|x - y| \leq |x' - y'| + (\varepsilon_x + \varepsilon_y)\beta^{e_{\min}} < \beta^\varrho\beta^{e_{\min}-1} + \beta^{\varrho-1}\beta^{e_{\min}} \leq \beta^{e_{\min}+\varrho}.$$

Thus lemma 5.5 applies: $\square(x - y) = x - y$. So the distance between $\square(x - y)$ and $x' - y'$ is no larger than $\varepsilon_x + \varepsilon_y$. ∎

Note that the assumption $\varepsilon_x + \varepsilon_y \leq \beta^{\varrho-1}$ is easily fulfilled for all the operations of algorithm 5.5. By propagating the Underf_Err property, we get the following lemma that says that algorithm 5.5 either produces the correct result or a result correct within 3.5 times the smallest subnormal positive number. Even when underflow happens, the result is nearly correct.

Lemma 5.26. If $\beta = 2$ or ϱ is even, then $|x \cdot y - (r_h + r_\ell)| \leq \frac{7}{2}\beta^{e_{\min}}$.

The underflow problem has been tackled by Ogita, Rump, and Oishi [OGI 05], where they give a bound of $5\beta^{e_{\min}}$. There is no real proof of it and the authors confirm this bound is rough. The $\frac{7}{2}$ bound could probably be sharpened too. In particular, if the radix is 2, it may probably be reduced to 3.

Let us now state the final theorem. We assume the precision is greater than 3 and the underflow threshold $\beta^{e_{\min}}$ is smaller than 1.

Theorem 5.27 (Dekker). Assume that either β is 2 or ϱ is even. Let r_h and r_ℓ be computed by algorithm 5.5.

Then $x \cdot y = 0 \vee |x \cdot y| \geq \beta^{e_{\min}+2\varrho-1}$ implies that $x \cdot y = r_h + r_\ell$ and anyway, $|x \cdot y - (r_h + r_\ell)| \leq \frac{7}{2}\beta^{e_{\min}}$.

We have proved the correctness of Dekker's product under mild assumptions. Even if the hypothesis "the radix is 2 or the precision is even" is far from elegant, it is very easy to check. Moreover, it is indeed the case on the decimal basic formats of the IEEE-754 standard [IEE 08] on 64 bits ($\varrho = 16$) or 128 bits ($\varrho = 34$) but not on the decimal storage format on 32 bits ($\varrho = 7$). For instance, let $x = y = 1{,}234{,}567$, then $x_h = y_h = 1{,}230{,}000$, $x_\ell = y_\ell = 4{,}567$, and $r_h = 1{,}524{,}156 \cdot 10^6$ so r_ℓ should be equal to $-322{,}511$ which is in the format. But $\circ^\tau(x_\ell \cdot y_\ell) = \circ^\tau(20{,}857{,}489) = 20{,}857{,}490$ is not an exact computation. Therefore the computed r_ℓ is $-322{,}510$, hence incorrect. In this decimal storage format on 32 bits, the computation can be exact if the result of $x_\ell \cdot y_\ell$ can be stored with at least one more digit, therefore some extra-precision may save this algorithm. Of course, when an FMA is available, algorithm 5.3 should be used instead.

5.3. Remainder of the integer division

Another interesting lossless computation is the remainder of the integer division. More precisely, let us consider two FP numbers x and y, and n an integer close to x/y. We are interested in the remainder $x - n \cdot y$. An important point is the choice of n: let us choose the *binary64* numbers $x = 2^{-100}$, $y = 1$ and $n = \lceil x/y \rceil = \lceil 2^{-100} \rceil = 1$. Then, $x - n \cdot y = 2^{-100} - 1$ is not a *binary64* number.

The IEEE-754 standard only considers computing n to nearest, tie breaking to even. The x86 processor has two operations: FPREM for computing n toward 0 and FPREM1 for computing n to nearest, tie breaking to even. All these operations are lossless computations.

Theorem 5.28 (format_REM). Assume a generic FP format described by a monotone function φ. Consider a valid rounding *rnd* : $\mathbb{R} \to \mathbb{Z}$ (as defined in section 3.2.2) and two FP numbers x and y in the format such that

$$|x/y| < \tfrac{1}{2} \Rightarrow rnd(x/y) = 0.$$

Then $x - rnd\,(x/y) \cdot y$ is representable in the format.

This means that, in the particular case of small $|x/y|$ (less than 0.5), $rnd(x/y)$ should not be ± 1. If it is, we have the previous counter-example.

Proof. We assume without loss of generality that $x > 0$ and $y > 0$. The zero cases are indeed trivial and the negative cases can be deduced using the opposite rounding (possibly changing the tie breaking rule).

Let $n = rnd\,(x/y)$. The case $n = 0$ is trivial. Let us first assume the easiest case: $n \geq 2$. This implies $x > y$. As φ is monotone, it means that $\text{cexp}(y) \leq \text{cexp}(x)$. We denote by m_x (respectively m_y) the integer such that $x = m_x \beta^{\text{cexp}(x)}$ (respectively $y = m_y \beta^{\text{cexp}(y)}$), as both x and y are representable FP numbers. Therefore $x - n \cdot y = \left(m_x \beta^{\text{cexp}(x)-\text{cexp}(y)} - n m_y\right) \beta^{\text{cexp}(y)}$. This is a correct significand/exponent representation provided that $\text{cexp}(y)$ is larger than the canonical exponent of $x - n \cdot y$. This holds by the monotony of φ and the properties of n.

The most interesting case is $n = 1$. From the hypotheses, we deduce that $|x/y| = x/y \geq \tfrac{1}{2}$. So $\tfrac{y}{2} \leq x$. We then have to prove that $x - y$ is in the format and we rely on Sterbenz' theorem (theorem 5.1). It remains to prove that $x \leq 2y$. As $n = rnd\,(x/y) = 1$, we have that $|x/y - 1| < 1$ so $x/y < 2$, hence the result. ∎

Let us now simplify the statement of theorem 5.28 by considering specific functions.

Lemma 5.29 (format_REM_ZR). Assume a generic FP format described by a monotone function φ. Consider two FP numbers x and y in the format. Then $x - \lfloor x/y \rfloor \cdot y$ is representable in the format.

Lemma 5.30 (format_REM_N). Assume a generic FP format described by a monotone function φ and a tie-breaking rule τ. Consider two FP numbers x and y in the format. Then $x - $ Znearest $\tau\,(x/y) \cdot y$ is representable in the format.

5.4. Remainders of the FP division and square root

As for the integer division, other EFTs exist for FP division and square root [BOL 03, BOH 91]. An interesting point is that these remainders may be effectively computed using an FMA. The proofs are rather similar to that of the remainder of the integer division. In particular, they both rely on the following lemma.

Lemma 5.31 (generic_format_plus_prec). Assume an FP format described by a function φ and an integer ϱ such that $\forall e \in \mathbb{Z}, \varphi(e) \leq e - \varrho$. Let x and y be two unbounded FP numbers of exponents e_x and e_y such that

$$|x + y| < \beta^{\varrho + \min(e_x, e_y)}.$$

Then $x + y$ is representable in the format.

This lemma focuses on FLX-like formats but is otherwise similar to lemmas 5.2 and 5.5. Note also that the addends do not have to be representable.

5.4.1. Remainder of FP division

A difference with the integer division is that we now assume an FLX format, meaning we have an FP format with ϱ digits and an unbounded exponent range. It exactly fits the hypothesis on the format of lemma 5.31.

Theorem 5.32 (div_error_FLX). Let x and y be two FP numbers in the FLX format and let $\square$ be a rounding. Then

$$x - \square\,(x/y) \cdot y$$

is representable in the FLX format.

Note that the corresponding Coq statement does not require y to be nonzero. Indeed, the real division is a total function, therefore $x/0$ is a real (without any information about it) and so is $\square(x/0)$. Then this real multiplied by 0 is 0, so $x - \square\left(\frac{x}{0}\right) \cdot 0 = x$ is an FP number.

Proof. As explained, the result holds when $y = 0$. It also holds when $x = 0$ and when $\square\,(x/y) = 0$. Now let us apply lemma 5.31 to x and the FP exact multiplication of $\square\,(x/y)$ and y (see section 3.3.1.3). We have to bound $|x - \square\,(x/y) \cdot y|$. First, there exists a relative error δ such that $|\delta| < \beta^{1-\varrho}$ and $\square\,(x/y) = (x/y)(1 + \delta)$. Then $|x - \square\,(x/y) \cdot y| = |\delta x| \leq |x| < \beta^{\varrho + e_x}$.

Let e_r be the canonical exponent of $\square\,(x/y)$. It remains to prove the inequality $|x - \square\,(x/y) \cdot y| < \beta^{\varrho + e_r + e_y}$. Since $|x - \square\,(x/y) \cdot y| = |y| \cdot |x/y - \square\,(x/y)| < |y|\beta^{e_r}$, there is left to prove that $|y| < \beta^{\varrho + e_y}$ which holds as y is in the FLX format. ∎

5.4.2. Remainder of the square root

We now assume an FLX format with a precision greater than one and rounding to nearest (with any tie-breaking rule τ).

Theorem 5.33 (sqrt_error_FLX_N). Let x be an FP number in an FLX format with $\varrho > 1$. Then

$$x - \circ^{\tau}\left(\sqrt{x}\right)^2$$

is representable in the FLX format.

As before, we do not have to assume $x \geq 0$, as the Coq square root of a negative number is zero. If $x < 0$, we have $x - \circ^{\tau}\left(\sqrt{x}\right)^2 = x - \circ^{\tau}(0)^2 = x$, which is in the format.

Proof. As explained, this easily holds when $x < 0$. It also easily holds when $x = 0$. Let e_r and e_x be the canonical exponent of $\circ^{\tau}\left(\sqrt{x}\right)$ and x respectively. We apply lemma 5.31 to x and the FP exact multiplication of $\circ^{\tau}(\sqrt{x})$ by itself (see section 3.3.1.3).

First, we have to prove that $|x - \circ^{\tau}(\sqrt{x})^2| < \beta^{\varrho + e_x}$. It is sufficient to prove $|x - \circ^{\tau}(\sqrt{x})^2| \leq |x|$, as x is in the FLX format. As both x and $\circ^{\tau}(\sqrt{x})$ are non-negative, it is sufficient to prove that $\circ^{\tau}(\sqrt{x})^2 \leq 2x$, which is easy as $\circ^{\tau}(\sqrt{x}) \approx \sqrt{x}$ with a relative error smaller than $\frac{\beta^{1-\varrho}}{2}$.

Then, we have to prove $|x - \circ^{\tau}(\sqrt{x})^2| < \beta^{\varrho + 2e_r}$. This holds as

$$|x - \circ^{\tau}(\sqrt{x})^2| = |\sqrt{x} - \circ^{\tau}(\sqrt{x})| \cdot |\sqrt{x} + \circ^{\tau}(\sqrt{x})| \leq \tfrac{1}{2}\beta^{e_r} \cdot (\sqrt{x} + \circ^{\tau}(\sqrt{x})).$$

As $\circ^{\tau}(\sqrt{x})$ is in the FLX format with canonical exponent e_r, we have the inequalities $|\circ^{\tau}(\sqrt{x})| < \beta^{\varrho + e_r}$ and $\sqrt{x} < \beta^{\varrho + e_r}$. Thus, we deduce $|x - \circ^{\tau}(\sqrt{x})^2| < \beta^{\varrho + 2e_r}$. ∎

5.5. Taking the square root of the square

Another lossless computation is the FP square root of the FP square of an FP number. Note that the FP square of the FP square root is not a lossless computation: consider radix 2, $\varrho = 4$, and $x = 1010_2$, then $\circ^{\tau}[\sqrt{x}^2] = \circ^{\tau}(11.01_2^2) = 1011_2$.

In radix 2, the square root of the square is indeed lossless as $\circ^\tau \left[\sqrt{x^2}\right] = |x|$. Cody and Waite used this property as a test for the square root function [COD 80, p. 12 and p. 28]. It is also stated in one of Kahan's web papers [KAH 83, p. 29]. This holds thanks to the square root properties. It suffices to guarantee that the FP number nearest to $\sqrt{\circ^\tau (x^2)}$ is indeed $|x|$. Let us detail the proof and see what happens in other radices.

In an IEEE-754 standard format, the computation of $\sqrt{x^2}$ may underflow and overflow. For small positive x, the result will be 0 and for large finite x, it will be $+\infty$. For the proof, we assume an unbounded exponent range. In practice, the previous result is valid only for medium-range x, that is, when $|x|$ is about between 2^{-511} and 2^{511} in the binary64 format.

Another use of this property is when considering the following FP computation:

$$r = \circ^\tau \left[\frac{a}{\sqrt{a^2 + b^2}} \right].$$ [5.1]

If there were no rounding, we would have $-1 \leq r \leq 1$. But rounding may get in the way, and it would be reasonable to add a test before calling an arcsin function for instance. It is easy to see that the worst case corresponds to $b = 0$, where r reduces to $\circ^\tau \left[a/\sqrt{a^2}\right]$. It is then sufficient (but not necessary!) to prove that $\circ^\tau \left[\sqrt{a^2}\right] = |a|$, as this implies that r will be either 1 or -1.

The square root of the square is a lossless computation for small radices including 2, as explained in section 5.5.1. When it is not the case, we are nevertheless interested in the properties of this value [MAS 16] in order to remove tests, as explained in section 5.5.2.

5.5.1. *When the square root of the square is exact*

First, let us consider a small radix: 2, 3, or 4. In this case, it is lossless, whatever the tie-breaking rules.

Lemma 5.34. Assume $\beta \leq 4$, $\varrho > 2$, and an unbounded exponent range. Let τ_1 and τ_2 be tie-breaking rules. For any FP number x,

$$\circ^{\tau_1} \left(\sqrt{\circ^{\tau_2} (x^2)} \right) = |x|.$$

This is an extension of a published proof that holds for radix 2 [BOL 15b].

5.5.2. *When the square root of the square is not exact*

Then, let us consider a few counter-examples: with a radix $\beta = 10$ and a precision $\varrho = 4$, if $x = 31.66$, then we have $\circ[\sqrt{x^2}] = 31.65$. With $\beta = 1000$ and $\varrho = 2$, if $x = 31.662$, then we have $\circ[\sqrt{x^2}] = 31.654$. The last example shows that the difference may be of several ulps of the exact result x.

The computation is not lossless, but it does not mean that r of equation [5.1] can be larger than 1:

Lemma 5.35. Assume an unbounded exponent range with $\varrho > 2$. For i between 1 and 5, let τ_i be a tie-breaking rule. Then, for any FP number a and for any real number b,

$$-1 \leq \circ^{\tau_1} \left(\frac{a}{\circ^{\tau_2} \left(\sqrt{\circ^{\tau_3} \left(\circ^{\tau_4} (a^2) + \circ^{\tau_5} (b^2) \right)} \right)} \right) \leq 1.$$

Proof. This proof is rather long, with many subcases. We will only focus on the most interesting points. Let r be the value defined by equation [5.1]; we want to prove $-1 \leq r \leq 1$. It is sufficient to prove that

$$-1 \leq \circ^{\tau_1} \left(\frac{a}{\circ^{\tau_2} \left(\sqrt{\circ^{\tau_4} (a^2)} \right)} \right) \leq 1$$

by monotony of the rounding. We also assume $0 < a$ without loss of generality. Let $z = \circ^{\tau_2} \left(\sqrt{\circ^{\tau_4} (a^2)} \right)$.

The first intermediate result is that if $a \leq \sqrt{\beta} \cdot \beta^{\mathrm{mag}(a)-1}$, then $z = a$. In this case, the rounding error of the multiplication is small (as $a^2 \leq \beta^{2\mathrm{mag}(a)-1}$). Note that the above counter-examples correspond to FP values just above this threshold.

In the other cases, we may have $z \neq a$, and it may be off by several ulps. So we will exhibit some z' such that $z' \leq z \leq a$ and $\circ^{\tau}(a/z') = 1$. By monotony of rounding, this will ensure $r = 1$. We define $z' = a - n \cdot \mathrm{ulp}(a)$ with

$$n = \left\lceil \frac{a}{\mathrm{ulp}(a)} \frac{1}{2 + \beta^{1-\varrho}} \right\rceil - 1.$$

We then prove $0 \leq n \leq \left\lceil \frac{\beta}{2} \right\rceil - 1$ and thus that z' is an FP number. We also prove that

$$2n\beta^{\varrho-1}\mathrm{ulp}(a) \left(1 + \tfrac{1}{2}\beta^{1-\varrho} \right) < a \qquad [5.2]$$

holds, provided that either $\beta > 5$ or both $\beta = 5$ and $\varrho > 3$. We also prove that [5.2] implies $\circ^\tau(a/z') = 1$.

This means that lemma 5.35 is proved, except in the case $\beta = 5$ and $\varrho = 3$. As the exponent does not matter here, this case can be studied exhaustively. Indeed, there are only 125 possible integer significands. So, using the operators of section 3.3, Coq computes the value of z and verifies $z = a$ for each of the 125 possible values of a. ∎

5.6. Remainders for the fused-multiply-add (FMA)

We have seen that the error of addition and multiplication are representable, with underflow conditions for multiplication. But this is not the case for the FMA: two FP numbers may be needed to represent exactly the error in rounding to nearest. Consider for example in *binary64* the computation $\circ\left(3\left(1 + 2^{-52}\right) + 2^{100}\right)$. The FP result is 2^{100} and the error term is $3 + 3 \cdot 2^{-52}$, which does not fit in one FP number. When rounding to nearest, tie breaking to even, the error terms are indeed $3 + 2^{-50}$ and -2^{-52}. We will not only prove that the error can be represented using two FP numbers but we will also provide algorithms to compute them. This can be done either exactly (computing both error terms) as in section 5.6.1 or approximately to get a quick estimation of the first error term as in section 5.6.2.

5.6.1. *EFT for the FMA*

In an ideal world, we would like to compute three values: $r_h = \circ^\tau(ax + y)$, r_m and r_ℓ such that $ax + y = r_h + r_m + r_\ell$ exactly, $|r_m + r_\ell| \leq \frac{1}{2}\text{ulp}(r_h)$, and $|r_\ell| \leq \frac{1}{2}\text{ulp}(r_m)$. Algorithm 5.6 correctly computes these error terms, even if it is rather costly [BOL 05, BOL 11c].

Algorithm 5.6 Computation of the error of an FMA

Input: FP numbers a, x, and y

Output: FP numbers r_h, r_m, and r_ℓ such that $r_h + r_m + r_\ell = ax + y$

$$
\begin{aligned}
r_h &= \circ^\tau(ax + y) \\
u_h &= \circ^\tau(ax) \\
u_\ell &= \circ^\tau(ax - u_h) \\
(\alpha_h, \alpha_\ell) &= 2\text{Sum}(y, u_\ell) \\
(\beta_h, \beta_\ell) &= 2\text{Sum}(u_h, \alpha_h) \\
\gamma &= \circ^\tau(\circ^\tau(\beta_h - r_h) + \beta_\ell) \\
(r_m, r_\ell) &= \text{Fast2Sum}(\gamma, \alpha_\ell)
\end{aligned}
$$

Let us explain this algorithm. First r_h is computed thanks to the FMA. Then ax is decomposed as the sum of two FP numbers u_h and u_ℓ, using algorithm 5.3 (EFT for multiplication when an FMA is available). We then consider $u_h + u_\ell + y$ instead of $ax + y$. Even if it is usually difficult to get the correctly rounded sum of three numbers [BOL 08], it is easy here as it is r_h. We just have to get the two error terms, meaning $u_h + u_\ell + y - r_h$ as the sum of two FP numbers.

We first compute the "small error" α_ℓ, as the error term of $y + u_\ell$. Then all the other values will be larger than this one as explained in Figure 5.3. Then we combine the remaining larger values α_h, u_h, and $-r_h$, in the correct order to get a single FP number γ. We then just have to combine γ and α_ℓ to get our results r_m and r_ℓ. Note that this last computation can be done using a Fast2Sum for efficiency.

This explanation does not cover the cases where a value is zero but this is handled in the formal proof. Underflow is also complex: first, ax may not be representable as the sum of two FP numbers if too small. Then, a small relative error bound on the round-off error requires values to be normal and such an error bound is needed to bound α_ℓ compared to the other FP numbers.

Theorem 5.36 (ErrFMA_correct). We assume that $\varrho \geq 3$, $\beta \geq 2$, and that β is even. Let a, b, x be FP numbers (either normal or subnormal). Let r_h, u_h, u_ℓ, α_h, α_ℓ, β_h, β_ℓ, γ be computed as in algorithm 5.6. We have a few non-underflow hypotheses:

– either $ax = 0$ or $\beta^{e_{\min}+2\varrho-1} \leq |a \cdot x|$;

– either $\alpha_h = 0$ or $\beta^{e_{\min}+\varrho} \leq |\alpha_h|$;

– either $\beta_h = 0$ or $\beta^{e_{\min}+\varrho+1} \leq |\beta_h|$;

– either $r_h = 0$ or $\beta^{e_{\min}+\varrho-1} \leq |r_h|$ (that is, r_h is normal).

Then

$$a \cdot x + y = r_h + \gamma + \alpha_\ell.$$

More details about this proof can be found in [BOL 05]. The first non-underflow hypothesis corresponds to ax being large enough for its error to be representable. The other assumptions may seem useless, but they are used to ensure relative error bounds that guarantee that values are near enough one to another to cancel. The key point is indeed that the computation of γ is lossless. It is not straightforward that those results still hold when some FP operations underflow.

5.6.2. Approximate remainder of the FMA

Algorithm 5.6 returns the correct error of an FMA as the sum of two FP numbers at the cost of 20 operations. Algorithm 5.7 computes an approximation of this error of an

FMA with only 12 operations. The accuracy of this algorithm is given by theorem 5.37. Contrarily to the previous algorithm, we make no assumption on the radix β.

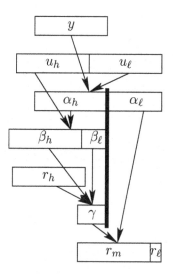

Figure 5.3. *Intermediate values of algorithm 5.6*

Algorithm 5.7 Computation of an approximation of the error of an FMA

Input: FP numbers a, x, and y
Output: FP numbers r_h and r_ℓ such that $r_h = \circ^\tau(ax + y)$ and $r_h + r_\ell \approx ax + y$

$$
\begin{aligned}
r_h &= \circ^\tau(ax + y) \\
u_h &= \circ^\tau(ax) \\
u_\ell &= \circ^\tau(ax - u_h) \\
(v_h, v_\ell) &= 2\mathrm{Sum}(y, u_h) \\
r_\ell &= \circ^\tau[(v_h - r_h) + (u_\ell + v_\ell)]
\end{aligned}
$$

Theorem 5.37 implies that $|r_h + r_\ell - (ax + y)| < 2 \cdot \beta^{3-2\varrho} \cdot |r_h|$, therefore we have at least $\varrho - 2$ correct digits following r_h as shown in the upper part of Figure 5.4. But this does not mean that r_ℓ is nearly correct. Indeed, it may be quite inaccurate but it implies $|r_\ell| \ll \mathrm{ulp}(r_h)$ as exemplified in the lower part of Figure 5.4. The error may be as much as $r_\ell/4$ in some cases. For instance, consider $\beta = 2$, $\varrho = 4$, $a = 10100_2$, $x = 1101_2$, and $y = 1101_2$. Then $r_h = 100100000_2$ and $r_\ell = -100_2 = -4$, while an exact r_ℓ should be $-101_2 = -5$.

Theorem 5.37 (ErrFmaAppr_correct). We assume that $\varrho \geq 4$. Let a, x, y be FP numbers (either normal or subnormal). Let r_h, p_h, p_ℓ, u_h, u_ℓ, t, v, r_ℓ be computed as in algorithm 5.7. We assume that r_h, r_ℓ, v_h, and $\circ^\tau(u_\ell + v_\ell)$ are either normal or zero. We also assume $ax = 0 \lor |ax| \geq \beta^{e_{\min}+2\varrho-1}$, so that p_ℓ is the error of $\circ^\tau(ax)$. Then

$$|r_h + r_\ell - (ax + y)| \leq \left(\frac{3\beta}{2} + \frac{1}{2}\right)\beta^{2-2\varrho}|r_h|.$$

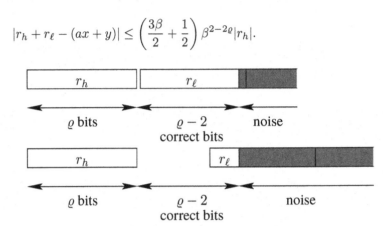

Figure 5.4. *Correct bits of Algorithm 5.7.*

Proof. This proof mostly relies on tight error bounding and a careful study of the numerous subcases. The main steps of the proofs are as follows:

– First, t is computed exactly. This property has two subcases depending on whether $|p_\ell| \leq \frac{1}{4}|p_h + y|$ or not. In the first case, Sterbenz' theorem applies to ensure $t = u_h - r_h$, and the magnitude of v is proved to be much smaller than that of t. In the second case, it means a large cancellation in the computation of u_h and a nonzero p_ℓ. Then $u_h - r_h$ is proved to be an FP number with exponent $e_a + e_x$.

– When $u_\ell = 0$, the theorem holds, as r_ℓ is the correct rounding of $r_h - (ax + y)$.

– We now assume that $u_\ell \neq 0$, therefore there was no cancellation in the computation of u_h. We then prove that $e_{p_h} \leq e_{u_h} + 1$ and that $e_{u_h} \leq e_{r_h} + 1$.

– We prove that $e_{p_h} \leq e_{u_h} + 1 = e_{r_h} + 2$ is not possible.

– At last, we prove $|u_h - r_h| \leq (\beta + 1) \cdot \text{ulp}(r_h)$ and the theorem.

More subcases exist, depending on the possible zero value of some intermediate FP computations or inputs. They are trivial, but must be handled in the formal proof nonetheless. More details about this proof can be found in [BOL 11c]. ∎

Note that the non-underflow assumptions are slightly more general than assuming an FLX format, as some variables such as p_ℓ and u_ℓ may be subnormal.

6

Example Proofs of Advanced Operators

Chapter 5 has covered error-free transformations (EFTs) and lossless computations. In this chapter, the considered operators get more complicated. We can no longer get rid of inaccuracies, so the goal is to reduce them or to control them. The purpose of this chapter is twofold: first, to give more guarantees on advanced operators of the literature, potentially useful to the reader; second, to cover a variety of techniques and concerns when verifying FP algorithms. This is why the proofs are not detailed much (compared to the first chapters) and only the prominent points (such as case distinctions, difficulties, and peculiar definitions) are given.

Here is the list of these operators. Section 6.1 presents an algorithm for computing the area of a triangle. Section 6.2 details the argument reduction used in Cody and Waite's approximation to the exponential. Section 6.3 presents some results about the faithfulness of Horner's scheme for polynomial evaluation. Section 6.4 shows an FP implementation of integer division. Section 6.5 considers various algorithms for computing the average of two FP numbers. Section 6.6 presents a robust algorithm for the orientation of three points in the plane. Robustness means here that the algorithm is either accurate or able to point to the fact that it is not accurate enough. Section 6.7 describes an accurate algorithm for computing the order-2 discriminant.

Only one operator provides the correct mathematical result, that is the integer division of section 6.4. All the other results are approximations of an ideal result. This approximation may be as good as possible: in the case of the computation of the average of section 6.5, the result may be the correct rounding of the real average. It may also be faithful, as in the polynomial evaluation of section 6.3. Good error bounds (a few ulps) are given in section 6.1 for computing the area of a triangle and in section 6.7 for computing the order-2 discriminant.

The previous error bounds are statically computed, but dynamic bounds may be useful too, as shown in section 6.6. To correctly compute the correct orientation of three points, the algorithm tests the potential result against a dynamic error bound. When that result strictly exceeds the error bound, it is guaranteed to have the correct sign. The problem of whether to trust an FP test is also present in section 6.7 when computing the order-2 discriminant.

Partly automated proofs are used in the example of argument reduction (section 6.2), integer division (section 6.4), and for the dynamic bounds of section 6.6 for the orientation test. They are unfortunately not enough to get the best error bounds in some cases. In particular, manual proofs can exhibit that some computations inside the operator are exact. This is the case for the triangle area (section 6.1) and for the order-2 discriminant (section 6.7). This latest example is the only one that also relies on EFTs for the multiplication.

A last point is the safety of these algorithms. Safety means that the corresponding programs will not fail. This is a common topic between this chapter and Chapter 8. Safety here includes preventing overflow, which is an important topic of section 6.5 for computing the average. Another safety property appears in section 6.1 for the computation of the area of a triangle: the input of a square root must be nonnegative.

6.1. Accurate computation of the area of a triangle

The following algorithm comes from Kahan. It computes the area of a triangle, given the side lengths as FP numbers. The common formula is two millenia old and attributed to Heron of Alexandria:

$$\sqrt{s\,(s-a)\,(s-b)\,(s-c)}, \quad \text{where } s = \frac{a+b+c}{2}.$$

This formula, however, is inaccurate for needle-like triangles like the one in Figure 6.1: either the result is inaccurate or the computation stops due to a negative square root created by round-off errors [KAH 86]. Kahan's formula is mathematically equivalent, but more accurate:

$$\frac{1}{4}\sqrt{(a+(b+c))\,(c-(a-b))\,(c+(a-b))\,(a+(b-c))}.$$

The parentheses are *not* to be removed to guarantee that the square root will carry out on a nonnegative number and that the result is accurate.

This formula was first published on the Internet [KAH 86], and then published in Goldberg's article [GOL 91] with a bound of 11ε, where ε stands for $\frac{1}{2}\beta^{1-p}$, that is, 2^{-53} in *binary64*. In this section, we prove both an improved error bound of 4.75ε and that the square root is always safe under some assumptions on the FP format [BOL 13a]. These assumptions are mild enough to cover all binary and decimal IEEE-754 formats.

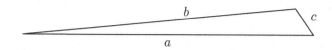

Figure 6.1. *A needle-like triangle.*

An interesting point of the proof is that it is done twice: once in the FLX format and then in the FLT format. See Table 3.2 for a comparison of some Flocq FP formats, including FLX and FLT. The idea is to first get a good error bound and then keep this error bound while handling underflow. As the result does not hold with subnormal numbers due to the large relative error bounds of underflowing multiplications, we will assume that the intermediate values are large enough not to underflow. This is done by requiring that the last value is large enough (see the end of the section). Only the proof with unbounded exponent range is described here.

Another interesting point is how much this example can be automated. An experiment can be done with the Gappa tool (see section 4.3.1) and the error bound automatically generated is quite good: 5.5ε, provided that we help the tool by providing it with a bound on b/a. This is a way to get a good error bound with nearly no human interaction. It proves that $a - b$ is computed exactly but is unable to guess that $t_4 = \circ[c - (a - b)]$ is computed exactly without further hint.

Theorem 6.1. Assume that $e_{\min} \leq -3 - p$, that $\frac{1}{4}$ fits in the format and that $\varepsilon = \frac{1}{2}\beta^{1-\varrho} \leq \frac{1}{100}$. Let a, b, and c be three FP numbers such that $0 \leq c \leq b \leq a \leq b + c$.

Let $\Delta = \frac{1}{4}\sqrt{(a + b + c)\,(a + b - c)\,(c + a - b)\,(c - a + b)}$ and let
$$\tilde{\Delta} = \circ\left[\frac{1}{4}\sqrt{(((a + (b + c))\,(a + (b - c)))\,(c + (a - b)))\,(c - (a - b))}\right].$$

If $\frac{1}{4}\beta^{\left\lceil \frac{e_{\min}+\varrho-1}{2}\right\rceil} < \tilde{\Delta}$, we have

$$\left|\tilde{\Delta} - \Delta\right| \leq \left(\frac{23}{4}\varepsilon + 38\varepsilon^2\right)|\Delta|.$$

What is assumed is the use of the FLT format with reasonable parameters: $e_{\min} \leq -3 - \varrho$ and $\varepsilon \leq \frac{1}{100}$. A stronger assumption is that $\frac{1}{4}$ fits in the format. The last two assumptions are proved to hold in radix 2 with at least 7 bits and in radix 10 with at least 3 digits. Note that the ordering of the multiplications has been precisely chosen (see below). Now let us sketch the proof of this theorem.

Let us fix some notations: $t_1 = \circ[a + (b + c)]$, $t_2 = \circ[a + (b - c)]$, $t_3 = \circ[c + (a - b)]$, $t_4 = \circ[c - (a - b)]$, $M = \circ[((t_1 \cdot t_2) \cdot t_3) \cdot t_4]$, and $\tilde{\Delta} = \circ\left[\frac{1}{4}\sqrt{M}\right]$.

We prove several properties on this algorithm. The first one is that it will not fail due to the square root: all the values, t_1, t_2, t_3, and t_4 are nonnegative. As for the rounding error, several lemmas allow us to improve over forward analysis.

Lemma 6.2. $a - b$ is computed exactly.

Proof. As $b \leq a \leq b + c \leq 2a$, Sterbenz' theorem (theorem 5.1) can be applied directly. ∎

Lemma 6.3. t_4 is computed exactly.

Proof. By using lemma 6.2, we have $t_4 = \circ[c - (a - b)] = \circ(c - (a - b))$. It remains to prove that $x = c - (a - b)$ is an FP number. It is sufficient to prove that x can be represented with an exponent greater than or equal to $e_{\min}$ and an integer significand smaller than β^p (see section 3.1.2.1). Let us take the canonical exponent of c denoted by e_c as exponent. As $0 \leq c \leq b \leq a$, the canonical exponents of a and b are greater than or equal to e_c. So $m_x = x \, 2^{-e_c} = (c - (a - b)) \, 2^{-e_c}$ is an integer. Let us now bound m_x. As $|x| = |c - (a - b)| = c - (a - b) \leq c = |c|$, the integer $|m_x|$ is smaller than the absolute value of the integer significand of c. As c is in the format, its integer significand has an absolute value smaller than β^p, so $|m_x| < \beta^p$. ∎

These exact computations give a better error bound than just forward analysis. Lemma 6.4 is useful to decrease the error on $t_2 = \circ[a + (b - c)]$. As all values are nonnegative and as $a \geq b \geq b - c$, the rounding error done in computing $b - c$ is much smaller than half an ulp of t_2.

Lemma 6.4. Given two FP numbers f and $\tilde{x}$ and real numbers x, $e \geq 0$, such that $|\tilde{x} - x| \leq e \cdot |x|$ and $0 \leq x \leq f$, we have

$$|\circ(f + \tilde{x}) - (f + x)| \leq \left(\varepsilon \cdot (1 + e) + \frac{e}{2} \right) \cdot |f + x|.$$

The proof is rather simple and relies on the $0 \leq x \leq f$ hypothesis for the sharper error bound. This lemma states that the relative error e is multiplied by 0.5, so the previous errors have less impact on the final relative error.

Except for these three lemmas, the proof of theorem 6.1 is a straightforward forward error analysis which gives a relative error bounded by $\frac{23}{4}\varepsilon$ plus some ε^2. This can be tightened in radix 2: in this case, multiplying by $\frac{1}{4}$ is exact and the error bound is $\frac{19}{4}\varepsilon + 33\varepsilon^2$, this is about $4.75\,\varepsilon$ instead of Goldberg's $11\,\varepsilon$.

As for underflow, the idea is to prove first that $t_4 \leq t_3 \leq t_2 \leq t_1$. Then, we choose to multiply these values in this order: $((t_1 \cdot t_2) \cdot t_3) \cdot t_4$. Hence, if an underflow occurs at any point, it also occurs in the last multiplication. Therefore, as long as the final result $\tilde{\Delta}$ is large enough, we are sure no operation did underflow.

The corresponding program using *binary64* numbers and operations is proved in section 8.3.3, taking overflow into account.

6.2. Argument reduction

Let us consider the process of computing an accurate FP approximation of an elementary function. We focus on the example of the exponential function for the *binary64* format. The interesting part of the input domain is $[-746; 710]$, as the values taken by exponential at inputs outside are either too small (and thus rounded to 0) or too large (and thus rounded to $+\infty$). The computation process usually involves three steps [MUL 16]:

1) compute a reduced argument $x' \approx x - k \cdot \log 2$ with k an integer;

2) perform an FP evaluation of a function f approximating exp: $y' \approx f(x')$;

3) reconstruct the final result $y \approx y' \cdot 2^k$.

It might seem that only the second step would be needed. But the domain of x is so wide that the degree of f (assuming it is a polynomial or a rational function) would have to be tremendously large to ensure a modicum of accuracy. To get suitable performances, we have to reduce the domain of approximation. This is the purpose of the first step: k is an integer close to $x/\log 2$ and x' lies in the interval $[-\frac{1}{2}\log 2; \frac{1}{2}\log 2]$ (a bit enlarged to account for approximate computations). The last step reverts the argument reduction. For the exponential function in radix 2, this last step happens to be an exact computation (unless it underflows), so we can assume that the relative error between y and $\exp x$ is the relative error between y' and $\exp(x - k \cdot \log 2)$. We want this relative error to be small enough for guaranteeing a given specification.

This error can be decomposed into three relative errors:

1) The error between y' and $f(x')$ is a round-off error; its bounds can be verified using the gappa tactic or results from section 6.3.

2) The error between $f(x')$ and $\exp x'$ is a method error; its bounds can be verified using the interval tactic (see section 4.2).

3) A bound on the relative error between $\exp x'$ and $\exp(x - k \cdot \log 2)$ can be trivially inferred from the absolute error between x' and $x - k \cdot \log 2$, due to the properties of the exponential.

Let us focus on bounding the absolute error between x' and $x - k \cdot \log 2$, which is the missing piece to bound the relative error between y and $\exp(x)$. Section 6.2.1 explains the algorithm for reducing x to x' in a way such that the error is small; given the range of x', a good argument reduction should give an absolute error bounded by about $\frac{1}{2}\mathrm{ulp}(\frac{1}{2}\log 2) = 2^{-55}$. Section 6.2.2 then details which hypotheses the Gappa tool needs in order to automatically prove this error bound. Finally section 6.2.3 summarizes the whole proof of the function, including the polynomial approximation and its round-off error.

6.2.1. Cody and Waite's argument reduction

Cody and Waite's algorithm is one of the first attempts at performing an accurate argument reduction [COD 80, p. 61]. Algorithm 6.1 shows what the C code would look like. The reduced argument x' is stored in x_reduced. The code first computes the integer k. Its value does not have to be equal to $\lfloor x/\log 2 \rfloor$; it can be one of the adjacent integers due to errors, as long as $|x - k \cdot \log 2|$ is not much larger than $\frac{1}{2}\log 2$.

To compute the reduced argument, the algorithm relies on a highly accurate approximation of $\log 2$: $|\mathrm{Log2h} + \mathrm{Log2l} - \log 2| \leq 2^{-102}$. Moreover, the Log2h constant is chosen so that its radix-2 significand ends with 11 zeros. As a consequence, its FP product with the integer k is performed exactly, since $|k| \leq 2^{11}$. The FP subtraction to x is also exact, due to Sterbenz' theorem (theorem 5.1). Thus the only round-off errors come from the product k * Log2l and its subtraction.

Algorithm 6.1 Cody and Waite's argument reduction when computing $\exp(x)$

```
double InvLog2 = 0x1.71547652b82fep0;
double Log2h = 0xb.17217f7d1cp-4; // 42 bits out of 53
double Log2l = 0xf.79abc9e3b398p-48;
double k = nearbyint(x * InvLog2);
double x_reduced = (x - k * Log2h) - k * Log2l;
```

Let us see how to formally prove that the absolute error between x' and $x - k \cdot \log 2$ is bounded by $(1+2^{-16}) \cdot 2^{-55}$. We will also prove $|x'| \leq 355 \cdot 2^{-10}$, as this property is needed for a full verification of an approximation of the exponential function. The Coq statement is as follows, with sub and mul denoting FP subtraction and multiplication in the *binary64* format.

```
Lemma argument_reduction :
  forall x : R,
  generic_format (FLT_exp (-1074) 53) x ->
  -746 <= x <= 710 ->
  let k := nearbyint (mul x InvLog2) in
  let x' := sub (sub x (mul k Log2h)) (mul k Log2l) in
  Rabs x' <= 355 / 1024 /\
  Rabs (x' - (x - k * ln 2)) <= 65537 * pow2 (-71).
```

We would like the whole formal proof to be automated. Unfortunately, the Gappa tool is unable to prove the goal when x is close to 0. More precisely, given the set of lemmas at its disposal, the tool fails to notice that the FP subtraction $\circ[x - k \cdot \mathtt{Log2h}]$ is actually exact. That is not much of an issue though. Indeed, x' is equal to x in that case ($k = 0$), so the correctness of the argument reduction trivially follows.

Therefore, the very first step of the proof is a case distinction depending on whether $|x| \leq T$ or $|x| \geq T$, for some threshold T. This threshold is chosen to be small enough so that $|x| \leq T$ implies $k = 0$ (trivial case), yet large enough so that Gappa can prove that the subtraction is exact. After some trials and errors, we find that $T = 5/16$ is a suitable threshold. When $|x| \leq 5/16$, the most intricate part of the proof is to deduce $k = 0$, but we can use the gappa tactic to discharge this equality. So we are left to prove that the argument reduction also behaves properly when $x \in [-746; -5/16] \cup [5/16; 710]$.

6.2.2. Bounding the error using Gappa

To ensure that the gappa tactic succeeds in formally proving the correctness on the domain $[-746; 710]$, we have to do a bit of forward reasoning so that the tool has enough hypotheses. To do so, let us guess the kind of reasoning the tool will apply and where it might fail to progress.

First of all, in order to verify $|x'| \leq 355 \cdot 2^{-10}$ in the above Coq statement, Gappa must be able to compute some tight bounds on the following subexpression of x' (ignoring the rounding operators, for clarity):

$$x - \lfloor x \cdot \mathtt{InvLog2} \rceil \cdot \mathtt{Log2h}.$$

This is a subtraction between values that are close to each other (especially when x is large), but Gappa cannot see it. Fortunately, it becomes obvious if one replaces the first occurrence of x with $x \cdot \mathtt{InvLog2} \cdot \mathtt{InvLog2}^{-1}$. The two subtracted sub-expressions now have the same structure (see section 4.3.7):

$$(x \cdot \mathtt{InvLog2}) \cdot \mathtt{InvLog2}^{-1} - \lfloor x \cdot \mathtt{InvLog2} \rceil \cdot \mathtt{Log2h},$$

so Gappa has no trouble giving some tight bounds on it. Therefore, the bound on x' will be automatically verified as long as we assert $x = x \cdot \mathtt{InvLog2} \cdot \mathtt{InvLog2}^{-1}$. This equality can be proved using the field tactic (see section 4.1.1).

Let us now look at the rightmost side of the above Coq statement: $|x' - (x - k \cdot \log 2)| \leq 65537 \cdot 2^{-71}$. This time, the tool has to compute some tight bounds on the following expression (again ignoring the rounding operators):

$$((x - k \cdot \mathtt{Log2h}) - k \cdot \mathtt{Log2l}) - (x - k \cdot \log 2).$$

As before, the two subtracted sub-expressions do not have the same structure, so Gappa will not be able to help us. We can recover this structure by replacing $\log 2$ with $\mathsf{Log2h} + \delta$. The expression then becomes

$$((x - k \cdot \mathsf{Log2h}) - k \cdot \mathsf{Log2l}) - ((x - k \cdot \mathsf{Log2h}) - k \cdot \delta).$$

Gappa has no trouble bounding such an expression, as long as we tell the tool how close $\mathsf{Log2l}$ is to $\delta = \log 2 - \mathsf{Log2h}$. Indeed, since Gappa does not know anything about the log function, it does not have any way to actually compute an enclosure of either δ or $\mathsf{Log2l} - \delta$. Fortunately, the interval tactic (see section 4.2) makes it straightforward to prove $\mathsf{Log2l} - \delta \in [-2^{-102}; 0]$.

To summarize, the gappa tactic succeeds in discharging

$$|x'| \leq 355 \cdot 2^{-10} \wedge |x' - (x - k \cdot \log 2)| \leq 65537 \cdot 2^{-71},$$

once the following properties have been proved by the user:

1) $|x| \geq 5/16$ by case distinction;

2) $x = x \cdot \mathsf{InvLog2} \cdot \mathsf{InvLog2}^{-1}$ using field;

3) $x - k \cdot \log 2 = x - k \cdot \mathsf{Log2h} - k \cdot (\log 2 - \mathsf{Log2h})$ using ring;

4) $\mathsf{Log2l} - (\log 2 - \mathsf{Log2h}) \in [-2^{-102}; 0]$ using interval.

6.2.3. Complete proof for the exponential function

We now have a Coq proof that gives some bounds on the reduced argument x' and on its absolute error with respect to $x - k \cdot \log 2$. Given the bounds on x', we can now bound the method error on a rational approximation of the exponential. This approximation has the same form as in section 4.2.2.3, except that p and q now have degree 2 and their *binary64* coefficients are given by algorithm 6.2.

$$f(x') = 2 \cdot \left(\frac{x' \cdot p(x'^2)}{q(x'^2) - x' \cdot p(x'^2)} + 0.5 \right).$$

Section 4.2.2.3 has shown how to use the interval tactic with Taylor models to prove a bound on the absolute error between the above approximation and the exponential. This proof script was for an approximation meant for a *binary32* implementation of exponential, but the tactic also succeeds for a larger and thus more accurate rational approximation. This gives the following lemma.

```
Lemma method_error :
  forall t : R,
  let t2 := t * t in
  let p := p0 + t2 * (p1 + t2 * p2) in
  let q := q0 + t2 * (q1 + t2 * q2) in
  let f := 2 * ((t * p) / (q - t * p) + 1/2) in
  Rabs t <= 355 / 1024 ->
  Rabs ((f - exp t) / exp t) <= 23 * pow2 (-62).
Proof.
intros t t2 p q f Ht.
unfold f, q, p, t2, p0, p1, p2, q0, q1, q2 ; simpl ;
interval with (i_bisect_taylor t 9, i_prec 70).
Qed.
```

Algorithm 6.2 Coefficients of the rational approximation of exp

---

```
Definition p0 := 1 * pow2 (-2).
Definition p1 := 4002712888408905 * pow2 (-59).
Definition p2 := 1218985200072455 * pow2 (-66).
Definition q0 := 1 * pow2 (-1).
Definition q1 := 8006155947364787 * pow2 (-57).
Definition q2 := 4573527866750985 * pow2 (-63).
```

Note that, while section 4.2.2.3 only used degree-5 Taylor models, here the much more accurate approximation forces us to use degree-9 models. Otherwise too many subdivisions of the input interval would be needed, which would prevent the proof from completing in a reasonable amount of time.

Now that we have a bound on the method error, we need to bound the round-off error between the floating-point result y' (before reconstruction) and the infinitely-precise value $f(x')$. Then we need to compose all the errors together. We also need to take into account the reconstruction used to produce the final result. Algorithm 6.3 shows both the implementation as a Coq function and its correctness property: the relative error between the final result and the exponential is bounded by 2^{-51}.

Section 4.3.1 has shown how to use Gappa to prove a bound on the relative error between y' and $P(x')$ when P is a *binary32* rational approximation of exponential. Again, there is nothing special about the *binary32* implementation, so Gappa succeeds just as well for *binary64*. In fact, it can even directly prove an error bound between y' and $\exp(x - k \cdot \log 2)$ as long as bounds on the method error and on the argument reduction error are available as hypotheses.

Algorithm 6.3 Definition and specification of Cody and Waite's exponential

---

Definition cw_exp (x : R) :=
  let k := nearbyint (mul x InvLog2) in
  let t := sub (sub x (mul k Log2h)) (mul k Log2l) in
  let t2 := mul t t in
  let p := add p0 (mul t2 (add p1 (mul t2 p2))) in
  let q := add q0 (mul t2 (add q1 (mul t2 q2))) in
  pow2 (Zfloor k + 1) *
  (add (div (mul t p) (sub q (mul t p))) (1/2)).

Theorem exp_correct :
  forall x : R,
  generic_format radix2 (FLT_exp (-1074) 53) x ->
  Rabs x <= 710 ->
  Rabs ((cw_exp x - exp x) / exp x) <= 1 * pow2 (-51).

Since we have such hypotheses at hand, the proof of the theorem is mostly automatized by the gappa tactic. One just has to manually handle the reconstruction of the result, which amounts to proving the following straightforward equality:

$$\frac{y' \cdot 2^k - \exp x}{\exp x} = \frac{y' - \exp(x - k \cdot \log 2)}{\exp(x - k \cdot \log 2)}.$$

This concludes the formal verification of algorithm 6.3. But we have cheated a bit here. Indeed, the theorem can only be applied to the actual result of an implementation when this result represents the real $y' \cdot 2^k$. This is obviously not the case when the reconstruction of the result overflows, which might happen since $\exp(710)$ exceeds 2^{1024}. This is not the case either when $y' \cdot 2^k$ falls in the underflow range. Note that this issue with underflows is a shortcoming of neither the correctness statement nor the algorithm. Indeed, no algorithm for the exponential function can ensure a small relative error in the underflow range; they can only ensure a small absolute error. Thus, to summarize, what we are able to state about this *binary64* implementation is that its relative error with respect to the exponential is less than 2^{-51}, provided that there is neither underflow nor overflow during the reconstruction of the result.

6.3. Faithful rounding of Horner evaluation

As explained in section 6.2, computing an elementary function typically begins with an argument reduction followed by a polynomial evaluation and a reconstruction. This section takes another look at the second step and its accuracy. As it has gone through argument reduction, the input to polynomial evaluation is a value x considered small. Given how the polynomial is designed, the low-degree terms are usually the more significant ones, while the high-degree terms can be seen as error-correcting terms [MUL 16]. Said otherwise, the terms are computed in a good order given their magnitudes. For instance, the elementary function $\exp(x)$ may be approximated by $1 + x + \frac{x^2}{2} = 1 + x(1 + \frac{x}{2})$ for a small x. For this kind of input and polynomial, Horner evaluation perfectly fits. At each step, the previous errors are multiplied by x, known to be small. This evaluation is therefore especially accurate.

The next question is how accurate the result can be. Usually, an error bound such as 1 ulp or 0.54 ulps is given, as done in sections 6.1 and 6.7. But the result has another property: it is faithfully rounded (see section 1.4.1), which means that the result is either the rounding down or up of the exact value. More generally, we will prove that, under certain conditions, the Horner evaluation of a polynomial is a faithful rounding of the expected mathematical value [BOL 04b]. We consider an FP format with underflow (the FLT format defined in section 3.1.2.1), precision $\varrho > 1$, minimal exponent e_{min}, and radix 2.

This section is organized as follows: section 6.3.1 gives some basic blocks to guarantee faithfulness, including the special cases near powers of the radix. Sections 6.3.2 and 6.3.3 show how to guarantee that an FP Horner evaluation is a faithful rounding of the expected mathematical value, with hypotheses either on the FP or the mathematical values.

6.3.1. How to guarantee faithfulness?

Faithfulness has been defined in section 1.4.1. It guarantees a one-ulp error (as it is either the rounding up or down). However, this is not equivalent: an error bounded by one ulp does not imply faithfulness, due to the discontinuity of the ulp function at powers of the radix. It is therefore not trivial to ensure [RUM 08, GRA 15a]. So we propose two other criteria that ensure faithfulness, even near powers of the radix.

More precisely, given a real x and an FP number f, we want to prove that f is a faithful rounding of x. We assume without loss of generality that $f \geq 0$. The possible cases for f are represented in Figures 6.2 and 6.3: the successor is always one ulp greater than f and the predecessor may be either one ulp or half an ulp smaller (when f is a power of the radix, subnormals excepted).

Lemma 6.5 corresponds to Figure 6.2. Let us denote f^- the predecessor of f (see sections 1.1.1 and 1.1.2). If $|f - x|$ is smaller than $\mathrm{ulp}(f^-)$, then faithfulness is guaranteed. Note that for positive f, the ulp of f^- is either equal to (in most cases) or smaller than the ulp of f.

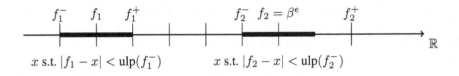

Figure 6.2. *How to prove faithfulness: first case*

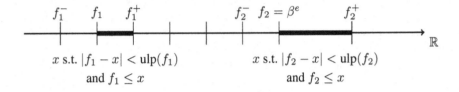

Figure 6.3. *How to prove faithfulness: second case*

Lemma 6.5. Let f be a positive FP number and let x be a real such that

$$|f - x| < \mathrm{ulp}\left(f^-\right).$$

Then f is a faithful rounding of x.

As seen in Figure 6.2, the condition of lemma 6.5 is adequate in most cases (away from a power of the radix, as for f_1). It is overly restrictive in some cases (near a power of the radix, as for f_2).

To handle the remaining cases, we prove lemma 6.6, which corresponds to Figure 6.3. If $|f - x|$ is smaller than $\mathrm{ulp}(f)$ and $f \leq x$, then faithfulness is guaranteed.

Lemma 6.6. Let f be a positive FP number and x a real such that

$$|f - x| < \mathrm{ulp}(f) \quad \text{and} \quad f \leq x.$$

Then f is a faithful rounding of x.

Note that rather similar theorems for ensuring correct rounding are available in section 5.2.1.2.

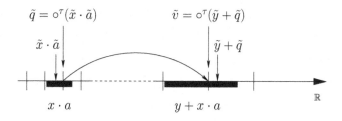

Figure 6.4. *Faithfulness of the last Horner step*

6.3.2. *Faithfulness for the last Horner iteration (hypotheses on FP values)*

Let us go back to Horner evaluation and especially its last step $y + x \cdot a$. We compute $\tilde{v} = \circ^\tau [f_0 + x \cdot (f_1 + x \cdot \ldots)]$ with rounding to nearest with any tie-breaking rule τ and we want to ensure that $\tilde{v}$ is a faithful rounding of some mathematical value near $f_0 + x \cdot (f_1 + x \cdot \ldots)$, assuming $|f_0| \gg |x \cdot (f_1 + x \cdot \ldots)|$. This last hypothesis needs to be more specific. It may also be expressed using either the computed FP values (theorem 6.8) or the mathematical values (theorem 6.9 in section 6.3.3).

The following theorem means that, if the final value is at least four times larger than the result of the last multiplication and if the accumulated errors (that include the previous rounding errors and possibly a method error) are small enough, then the final value is a faithful rounding of its ideal value. Intuitively, we assume $\tilde{y} \approx y$, $\tilde{x} \approx x$, $\tilde{a} \approx a$, and $4 |x \cdot a| \lesssim |y|$. As seen in Figure 6.4, the accumulated errors (both the roundings and the method error) are small enough so that the final $\tilde{v}$ is a faithful rounding of $y + x \cdot a$. Let us first consider the case of a positive $\tilde{v}$.

Theorem 6.7. Let y, x, and a be real numbers and let $\tilde{y}$, $\tilde{x}$, and $\tilde{a}$ be FP numbers. Let $\tilde{q} = \circ^\tau (\tilde{x} \cdot \tilde{a})$ and $\tilde{v} = \circ^\tau (\tilde{y} + \tilde{q})$. Let us assume that $\tilde{v} > 0$, and that

$$4|\tilde{q}| \leq |\tilde{v}| \quad \text{and} \quad |\tilde{y} - y| + |\tilde{x} \cdot \tilde{a} - x \cdot a| < \tfrac{1}{4}\text{ulp}\left(\tilde{v}^-\right).$$

Then $\tilde{v}$ is a faithful rounding of $y + x \cdot a$.

Proof. Several special cases have to be handled separately: when $\tilde{q}$ is subnormal and when $|\tilde{v}| \leq 2^{e_{\min} + \varrho}$ (in this case $\tilde{v} = \tilde{y} + \tilde{q}$ by theorem 5.5). We only focus on the general case.

Assume that $|\tilde{q}|$ is normal. As $4|\tilde{q}| \leq |\tilde{v}|$, we have $\text{ulp}(\tilde{q}) \leq \tfrac{1}{2}\text{ulp}\left(\tilde{v}^-\right)$. In this case, we then have $|\tilde{q} - \tilde{x} \cdot \tilde{a}| \leq \tfrac{1}{2}\text{ulp}(\tilde{q}) \leq \tfrac{1}{4}\text{ulp}\left(\tilde{v}^-\right)$.

As we wish to use either lemma 6.5 or lemma 6.6, we need to bound the value $|\tilde{v} - (y + x \cdot a)|$ by using both the previous result and the last theorem hypothesis:

$$
\begin{aligned}
|\tilde{v} - (y + x \cdot a)| &\leq |\tilde{v} - (\tilde{y} + \tilde{q})| + |\tilde{q} - \tilde{x} \cdot \tilde{a}| + |\tilde{y} - y| + |\tilde{x} \cdot \tilde{a} - x \cdot a| \\
&< |\tilde{v} - (\tilde{y} + \tilde{q})| + |\tilde{q} - \tilde{x} \cdot \tilde{a}| + \tfrac{1}{4}\mathrm{ulp}\left(\tilde{v}^-\right) \\
&\leq |\tilde{v} - (\tilde{y} + \tilde{q})| + \tfrac{1}{2}\mathrm{ulp}\left(\tilde{v}^-\right).
\end{aligned}
$$

Then, we have two cases. If $\tilde{y} + \tilde{q} \leq \tilde{v} = \circ^r(\tilde{y} + \tilde{q})$, the rounding of $\tilde{v}$ is upward and $\tilde{y} + \tilde{q}$ is nearer to $\tilde{v}$ than to $\tilde{v}^-$. Therefore, $|\tilde{v} - (\tilde{y} + \tilde{q})| \leq \tfrac{1}{2}\mathrm{ulp}\left(\tilde{v}^-\right)$. So the result holds by lemma 6.5.

Otherwise, $\tilde{y} + \tilde{q} \geq \tilde{v}$ and $|\tilde{v} - (y + x \cdot a)| < \tfrac{1}{2}\mathrm{ulp}(\tilde{v}) + \tfrac{1}{2}\mathrm{ulp}\left(\tilde{v}^-\right) \leq \mathrm{ulp}(\tilde{v})$. So the result holds by lemma 6.6. ∎

We now generalize the theorem to the cases when $\tilde{v}$ is nonpositive. We denote by f^* the "predecessor toward zero" of f. It is defined as f^- when $x \geq 0$ and f^+ otherwise. Note that $\mathrm{ulp}(0)$ has a meaningful value (see section 3.1.3.3), it is equal to $\beta^{e_{\min}}$ here in an FLT format. The predecessor of 0 is therefore $-\beta^{e_{\min}}$.

Theorem 6.8. Let y, x, and a be real numbers and $\tilde{y}$, $\tilde{x}$, and $\tilde{a}$ be FP numbers. Let $\tilde{q} = \circ^r(\tilde{x} \cdot \tilde{a})$ and $\tilde{v} = \circ^r(\tilde{y} + \tilde{q})$. Let us assume that

$$
4|\tilde{q}| \leq |\tilde{v}| \quad \text{and} \quad |\tilde{y} - y| + |\tilde{x} \cdot \tilde{a} - x \cdot a| < \tfrac{1}{4}\mathrm{ulp}\left(\tilde{v}^*\right).
$$

Then $\tilde{v}$ is a faithful rounding of $y + x \cdot a$.

Proof. When $\tilde{v} > 0$, this is exactly theorem 6.7. When $\tilde{v} < 0$, various symmetries are used to get back to theorem 6.7. So we only have to consider the case $\tilde{v} = 0$. As $4|\tilde{q}| \leq |\tilde{v}|$, we have $\tilde{q} = 0$ and then deduce that $\tilde{y} = 0$. The addition is exact, but the multiplication may lead to a round-off error. Thus, we have

$$
\begin{aligned}
|\tilde{v} - (y + x \cdot a)| &= |\tilde{q} - \tilde{x} \cdot \tilde{a} + \tilde{y} - y + \tilde{x} \cdot \tilde{a} - x \cdot a| \\
&< |\tilde{q} - \tilde{x} \cdot \tilde{a}| + \tfrac{1}{4}\mathrm{ulp}\left(\tilde{v}^-\right) \\
&< 2^{e_{\min}-1} + 2^{e_{\min}-1} = \mathrm{ulp}(\tilde{v}).
\end{aligned}
$$

As $\tilde{v}$ is not a power of the radix, this error bound guarantees the faithfulness. ∎

6.3.3. Faithfulness for the last Horner iteration (hypotheses on mathematical values)

The previous theorem is valuable as it links a computed value with an ideal value, but it requires knowledge on the outputs $\tilde{q}$ and $\tilde{v}$. We would prefer hypotheses only on the inputs $\tilde{y}$, $\tilde{x}$, and $\tilde{a}$.

Theorem 6.9. Let y, x, and a be real numbers and $\tilde{y}$, $\tilde{x}$, and $\tilde{a}$ be FP numbers. Let $\tilde{v} = \circ^\tau[\tilde{y} + \tilde{x} \cdot \tilde{a}]$. Let us assume that

$$\frac{5 + 4 \cdot 2^{-\varrho}}{1 - 2^{-\varrho}} \left(|\tilde{x} \cdot \tilde{a}| + 2^{e_{\min} - 1} \right) \leq |\tilde{y}| \quad \text{and}$$

$$|\tilde{y} - y| + |\tilde{x} \cdot \tilde{a} - x \cdot a| \leq 2^{-2-\varrho} \cdot \left(1 - 2^{1-\varrho} \right) \cdot |\tilde{y}| - 2^{-2-\varrho} \cdot |\tilde{x} \cdot \tilde{a}| - 2^{e_{\min} - 2}.$$

Then $\tilde{v}$ is a faithful rounding of $y + a \cdot x$.

The proof is cumbersome and long as it involves heavy manual computations but not difficult. It relies on naive error analysis (see section 1.2), so that $\tilde{v}$ and $\tilde{q}$ are bounded by values depending only on $\tilde{y}$, $\tilde{x}$, and $\tilde{a}$ (and the precision ϱ).

Note that the hypotheses can be simplified for the sake of readability as a corollary.

Lemma 6.10. Let y, x, and a be real numbers and $\tilde{y}$, $\tilde{x}$, and $\tilde{a}$ be FP numbers. Let $\tilde{v} = \circ^\tau[\tilde{y} + \tilde{x} \cdot \tilde{a}]$. Let us assume $\varrho \geq 4$,

$$6 \left(|\tilde{x} \cdot \tilde{a}| + 2^{e_{\min} - 1} \right) \leq |\tilde{y}|, \quad \text{and}$$

$$|\tilde{y} - y| + |\tilde{x} \cdot \tilde{a} - x \cdot a| \leq \tfrac{2}{3} 2^{-2-\varrho} \cdot |\tilde{y}| - 2^{e_{\min} - 2}.$$

Then $\tilde{v}$ is a faithful rounding of $y + x \cdot a$.

The last inequality may be difficult to ensure when $\tilde{y} \neq y$. Even if $\tilde{a} = a$ and $\tilde{x} = x$, the last inequality indeed requires $|\tilde{y} - y| \leq \tfrac{2}{3} 2^{-2-\varrho} \cdot |\tilde{y}|$ which is an error much smaller than half an ulp (about $\frac{1}{12}\mathrm{ulp}(y)$). Thus, a correct rounding of y is generally not accurate enough. A solution may be to use Gal's method [GAL 86, STE 05].

When an FMA is available, Horner's scheme becomes both faster and more accurate than using an FP multiplication followed by an FP addition . Instead of $\circ^\tau[f_0 + x \cdot (f_1 + x \cdot \ldots)]$, it becomes $\tilde{v} = \circ^\tau(f_0 + x \cdot \circ^\tau(f_1 + x \cdot \ldots))$. The condition $|f_0| \gg |f_1 \cdot x|$ then becomes useless, and a similar result holds, with hypotheses on the mathematical values [BOL 04b]. Note that [BOL 04b] also provides a Maple program to check the theorems hypotheses.

6.4. Integer division computed using FMA

Intel Itanium processors have no dedicated hardware for divisions, be they integer divisions or FP divisions. So it is up to libraries to implement such functions in software. This section shows how one can specify and prove some code for dividing two unsigned 16-bit integers.

Section 6.4.1 shows what the assembly code looks like and section 6.4.2 shows how its correctness can be specified in Coq. Section 6.4.3 explains which parts of this correctness can be handled automatically and how to prove the other parts manually. Finally, section 6.4.4 is a digression on the topic of the frcpa instruction used in the algorithm for computing the inverse approximate of the divisor.

6.4.1. Assembly code

Algorithm 6.4 proposes an Itanium assembly snippet for dividing two unsigned 16-bit integers a and b [COR 00]. Each line can be decomposed as follows. The "(p6)" part indicates that the instruction is conditionally executed, depending on the value of predicate p6. When no predicate is present, the instruction is always executed. Then comes the actual instruction and its operands; destination registers and predicates are on the left-hand side of the equal sign, source registers are on the right-hand side. Registers f6 to f9 store 82-bit FP numbers (64 bits of precision without hidden bit, 17 bits of exponents, 1 bit of sign); register f0 is wired so as to always contain $+0$. Finally, two semicolons at the end of a line mark the end of a bundle, that is, a set of instructions that can be executed in parallel, as for instance fma.s1 and fnma.s1 at lines 2–3.

Algorithm 6.4 Itanium assembly for dividing two unsigned 16-bit integers

```
1  // Inputs: dividend a in f6, divisor b in f7, 1 + 2⁻¹⁷ in f9
2      frcpa.s1   f8,p6=f6,f7  ;;  // f8 ← f7⁻¹ approximately
3  (p6) fma.s1    f6=f6,f8,f0      // f6 ← ∘(f6 × f8)
4  (p6) fnma.s1   f7=f7,f8,f9  ;;  // f7 ← ∘(−f7 × f8 + f9)
5  (p6) fma.s1    f8=f7,f6,f6  ;;  // f8 ← ∘(f7 × f6 + f6)
6      fcvt.fx.trunc.s1 f8=f8      // f8 ← ⌊f8⌋
7  // Output: ⌊a/b⌋ in f8
```

The integer inputs a and b are initially stored in the FP registers f6 and f7. The first instruction (frcpa) at line 2 computes an approximation of $1/b$ and stores it in f8 if neither a nor b are equal to zero. Otherwise, if a or b is equal to zero, the result of the division is stored in f8 and predicate p6 is cleared, so the next three instructions are not executed. We will ignore that degenerate case in the following.

The approximation stored in f8 has two properties. First, its relative error is less than $2^{-8.886}$. This bound is documented [INT 06] but it can also be derived, as shown in section 6.4.4. Second, its precision is 11, that is, all the bits except for the 11 most significant ones are set to zero. The second instruction at line 3 multiplies this number by a so as to get a crude approximation of a/b, which is stored in f6. Due to the low accuracy of frcpa, this approximation cannot yet be truncated to an integer, as the result would be potentially different from $\lfloor a/b \rfloor$.

The purpose of the next two instructions at lines 4–5 is to refine the approximation of a/b in f6 using FMAs; the refined approximation is stored in f8. The last instruction at line 6 truncates this FP number in order to get $\lfloor a/b \rfloor$, hopefully. Algorithm 6.5 expresses the code as a set of equations. Note that $\circ(\cdot)$ denotes a rounding to nearest, with tie breaking to even, a precision of 64 bits, and a minimal exponent of -65597.

Algorithm 6.5 Computation of $q = \lfloor a/b \rfloor$ for two integers $a, b \in [1; 65535]$

$$y_0 \approx 1/b$$
$$q_0 = \circ(a \cdot y_0)$$
$$e_0 = \circ(1 + 2^{-17} - b \cdot y_0)$$
$$q_1 = \circ(e_0 \cdot q_0 + q_0)$$
$$q = \lfloor q_1 \rfloor$$

6.4.2. Coq specification

In order to prove the correctness of this algorithm, we just have to prove that, for any integers $a, b \in [1; 65535]$, we have $q = \lfloor a/b \rfloor$. The Coq implementation of the algorithm is as follows.

```
Definition div_u16 (a b : Z) : Z :=
  let y0 := frcpa b in
  let q0 := fma a y0 0 in
  let e0 := fnma b y0 (1 + pow2 (-17)) in
  let q1 := fma e0 q0 q0 in
  Zfloor q1.
```

The fma (resp. fnma) function is easily defined by combining the round function (see section 3.2.2) with $(x, y, z) \in \mathbb{R}^3 \mapsto x \cdot y + z$ (respectively $-x \cdot y + z$). The definition of frcpa is slightly more complicated. It will be detailed in section 6.4.4 but for now, let us just axiomatize the properties we are interested in:

```
Axiom frcpa : R -> R.
Axiom frcpa_spec : forall x : R,
  1 <= Rabs x <= 65535 ->
  generic_format radix2 (FLT_exp (-65597) 11) (frcpa x) /\
  Rabs (frcpa x - 1/x) <= 4433*pow2 (-21) * Rabs(1/x).
```

Finally, the correctness lemma for the algorithm is stated as follows.

```
Lemma div_u16_spec : forall a b : Z,
  (1 <= a <= 65535)%Z ->
  (1 <= b <= 65535)%Z ->
  div_u16 a b = (a / b)%Z.
```

6.4.3. *Formal proof using Gappa*

Let us explain why the algorithm works. The main idea is that it behaves like Newton's iteration. Indeed, if ε_0 denotes the relative error between y_0 and b^{-1} (and thus also the relative error between $q_0 = a \cdot y_0$ and a/b), we get the following approximates:

$$
\begin{aligned}
q_1 &\approx (1 + e_0) \cdot q_0 \\
&\approx (2 - b \cdot y_0 + 2^{-17}) \cdot (a \cdot y_0) \\
&\approx (2 - b \cdot b^{-1} \cdot (1 + \varepsilon_0) + 2^{-17}) \cdot a \cdot b^{-1} \cdot (1 + \varepsilon_0) \\
&\approx (1 - \varepsilon_0 + 2^{-17}) \cdot (a/b) \cdot (1 + \varepsilon_0) \\
&\approx (a/b) \cdot (1 - \varepsilon_0^2 + 2^{-17} \cdot (1 + \varepsilon_0)).
\end{aligned}
$$

If not for the 2^{-17} addend, we would have the traditional quadratic convergence: from a relative error ε_0 for q_0, we get a relative error ε_0^2 for the refined approximation q_1. In other words, from the 8-bit accuracy of frcpa, we get a 16-bit accurate result. But this accuracy would be pointless if q_1 ends up being less than $\lfloor a/b \rfloor$, since the final truncation would thus be off by one. That is why the seemingly superfluous perturbation 2^{-17} is added. This constant causes an overestimation large enough to force $q_1 \geq \lfloor a/b \rfloor$, yet small enough to preserve the quadratic convergence [COR 00].

The Gappa tool (see section 4.3.1) cannot recognize a quadratic convergence, and even if it were able to, it might not be able to recognize this overestimated one. So we have to tell the tool about it. We do not have to take into account the rounding operators when doing so, since Gappa can relate rounded and non-rounded expressions using absolute or relative round-off errors. So we start by removing the rounding operators from q_0, e_0, and q_1 of algorithm 6.5 and by giving names to these new expressions:

$$
\begin{aligned}
q_0' &= a \cdot y_0, \\
e_0' &= 1 + 2^{-17} - b \cdot y_0, \\
q_1' &= e_0' \cdot q_0' + q_0'.
\end{aligned}
$$

The convergence can now be expressed as a simple equality relating q_1' and ε_0:

$$
q_1' = (a/b) \cdot (1 - \varepsilon_0^2 + 2^{-17} \cdot (1 + \varepsilon_0)). \tag{6.1}
$$

Given the definition of ε_0, the field tactic (see section 4.1.1) has no trouble proving equation [6.1], since neither a nor b is equal to zero. Given that equation, the bounds on a and b, the fact that they are integers, and the specification of frcpa, Gappa succeeds in computing tight bounds on the relative error ε between the rounded result q_1 and a/b. In particular, it succeeds in proving that it is both nonnegative and less than 2^{-16}:

$$
\varepsilon = \frac{q_1 - a/b}{a/b} \in [2^{-18.28}; 2^{-16.99}].
$$

Now we have to relate this relative error with the integer parts $\lfloor q_1 \rfloor$ and $\lfloor a/b \rfloor$. This part of the proof has to be done manually. More precisely, we need the integer parts of q_1 and a/b to be equal, that is, we want to prove the following inequalities:

$$
\lfloor a/b \rfloor \le q_1 < \lfloor a/b \rfloor + 1. \tag{6.2}
$$

Since b is positive, we have

$$
b \cdot \lfloor a/b \rfloor = a - (a \bmod b).
$$

By definition of ε, we have

$$
b \cdot q_1 - a = a \cdot (q_1/(a/b) - 1) = a \cdot \varepsilon.
$$

So the inequalities in equation [6.2] are equivalent to

$$
-(a \bmod b) \le a \cdot \varepsilon < b - (a \bmod b).
$$

Since $a \bmod b \in [0; b-1]$, we just have to prove $a \cdot \varepsilon \in [0; 1)$ to deduce equation [6.2] and thus that $\lfloor q_1 \rfloor$ and $\lfloor a/b \rfloor$ are equal.

To summarize, there are three steps in the correctness proof:

1) Prove that the correctness is implied by $a \cdot \varepsilon \in [0; 1)$ with ε the relative error between q_1 and a/b using the above properties of the integer part and the modulo.

2) Exhibit a pseudo-quadratic convergence for the relative error between q_1' (the non-rounded version of q_1) and a/b using the field tactic.

3) Verify that $a \cdot \varepsilon \in [0; 1)$ holds using the gappa tactic.

Note that the correctness of this algorithm for the unsigned division of 16-bit integers was originally proved in HOL Light, as well as all the other division algorithms for Itanium [HAR 00]. The major difference here is that the tedious process of formally proving $a \cdot \varepsilon \in [0; 1)$ is automated using Gappa.

6.4.4. *Specification of* frcpa

In section 6.4.2, the frcpa function was specified using axioms. In particular, we assumed that, for any real x such that $|x| \in [1; 65535]$, the relative error between frcpa(x) and $1/x$ was bounded by $4433 \cdot 2^{-21} \simeq 2^{-8.886}$. The bound $2^{-8.886}$ comes from the description of the instruction but let us see how we can actually derive it from its pseudocode [INT 06, p. 3:116]. This algorithm takes an FP number x and returns an FP number y approximating x^{-1}. Note that significands are 64-bit wide with no hidden bit and exponents are biased.

```
i = x.significand{62:55};
y.significand = (1 << 63) | (recip_table[i] << 53);
y.exponent = FP_REG_EXP_ONES - 2 - x.exponent;
```

The pseudocode works as follows. It first extracts the eight most significant bits (not counting the most significant bit, which is 1) of the significand of x. This 8-bit value i is used to index the recip_table array containing 10-bit values. The 256 integers ranging from 1020 down to 1 contained in the array are documented along the pseudocode. The significand of y is then obtained by prepending 1 to the 10-bit value v_i read from the array. The exponent of y is chosen from the exponent of x so that the sum of their unbiased values is -1. Thus, x, y, and i satisfy the following properties for some exponent e:

$$\left(1 + \frac{i}{256}\right) \cdot 2^e \leq x < \left(1 + \frac{i+1}{256}\right) \cdot 2^e,$$

$$y = \left(1 + \frac{v_i}{1024}\right) \cdot 2^{-1-e}.$$

To express the statement on the error bound in Coq, we start by representing the array as a list of numbers.

Definition recip_table : list R :=
  1020 :: 1012 :: 1004 :: 996 ::
  (* ... 248 values ... *)
  7 :: 5 :: 3 :: 1 :: nil.

Since the relative error between y and $1/x$ does not depend on the exponent e, we set $e = 0$ without loss of generality. The statement about the relative error can thus be expressed as follows, where (S i) denotes the successor of a natural number i, INR coerces a natural number into a real number, and nth extracts an element of a list.

Lemma frcpa_spec :
  forall (i : nat) (x : R),
  (0 <= i < 256)%nat ->
  INR (256 + i) / 256 <= x < INR (256 + S i) / 256 ->
  let y := (1024 + nth i recip_table 0) / 2048 in
  Rabs (y - 1 / x) <= 4433 * pow2 (-21) * Rabs (1 / x).

This lemma looks like it could be discharged by the interval or gappa tactics if i was a constant. Indeed, x would then be enclosed between literal rational numbers, while y would not contain any mention of a list anymore. This leads to a rather uncommon kind of formal proof, since it enumerates all the possible values for i (see also section 5.5).

The proof scheme is as follows. Let us suppose that we have proved the bound for all the values of i up to $j - 1$. So the hypothesis on i now looks like $j \leq i < 256$. It can be transformed into $(i = j) \lor (j + 1 \leq i < 256)$. The case $i = j$ is handled by rewriting i into j (which is a literal constant, e.g. $j = 0$ at first), performing some simplifications, and then calling interval (see section 4.2). Then we repeat the whole procedure to handle the other case $j + 1 \leq i < 256$. The proof loop terminates once the hypothesis becomes $256 \leq i < 256$.

So the proof script ends up calling the interval tactic 256 times, once for each possible value of i. That way, the bound on the relative error on the result of the frcpa instruction is automatically and formally proved. As for proving that all the bits of the result are cleared except for the 11 most significant ones, this is obvious from the content of recip_table and could be formally proved computationally.

6.5. Average of two FP numbers

Let us now consider that we want to compute the average of two FP numbers. This example tries to show how to deal with overflow problems. Indeed, the naive formula (x+y)/2 is highly accurate, but may fail due to overflow, even if the correct result would be in the range. For example, if we consider the largest FP number Ω, then

$\square[(\Omega+\Omega)/2]$ overflows while the correct result is Ω. This problem has been known for decades and has been thoroughly studied by Sterbenz, among other "carefully written programs" [STE 74, chap 9].

Like Sterbenz, we consider an FP format with underflow (the FLT format defined in section 3.1.2.1) with a precision ϱ, a minimal exponent e_{min}, and a radix equal to 2. The following algorithms are known to fail in radix 10 [STE 74, p. 244]. The corresponding rounding to nearest, tie breaking to even, is denoted by round_flt.

Sterbenz' study is especially interesting as he does not give a complete program: he gives hints about how to circumvent overflow and he specifies which requirements the algorithm is supposed to satisfy. All are achievable and they cover both the safety and the accuracy of the wanted program. As they are used through the section, the requirements are denoted with equation numbering:

– average(x, y) is within a few ulps of $(x + y)/2$ [6.3]
(this will be quantified later as either 0.5 or 1.5);

– min$(x, y) \le$ average$(x, y) \le$ max(x, y);   [6.4]

– average$(x, y) =$ average(y, x);   [6.5]

– average$(-x, -y) = -$average(x, y);   [6.6]

– average(x, y) has the same sign as $(x + y)/2$.   [6.7]

Sterbenz also requires one property related to underflow:

– average$(x, y) = 0$ if and only if $y = -x$. This should hold except in case of underflow.   [6.8]

We are able to prove a stronger requirement that partly handles subnormals. Remember that $2^{e_{min}}$ is the smallest positive FP number.

– $|(x + y)/2| \ge 2^{e_{min}}$ implies average$(x, y) \ne 0$.   [6.9]

Sterbenz also requires the program not to fail:

– the program never overflows.   [6.10]

Note that correct rounding easily fulfills requirements [6.3]–[6.9] but was considered hard to achieve in [STE 74].

Sterbenz suggested several algorithms to compute the average:

– $\circ[(x + y)/2]$ is accurate, but may overflow when x and y have the same sign;

– $\circ[x + (y - x)/2]$ is less accurate but it does not overflow if x and y have the same sign;

– $\circ[x/2 + y/2]$ cannot overflow; it is not as accurate as the first algorithm in case of underflow.

These three ways of computing the average are studied in the next three sections 6.5.1, 6.5.2, and 6.5.3. Then the algorithm hinted by Sterbenz is given in section 6.5.4 and a more accurate algorithm is given in section 6.5.5.

6.5.1. *The* avg_naive *function*

The avg_naive function is the simplest one to compute the average.

Definition avg_naive (x y : R) := round_flt (round_flt (x + y) / 2).

That is to say, avg_naive$(x, y) = \circ[(x + y)/2]$.

In our algorithmic model without overflow, this function is highly accurate, as it computes the correctly-rounded average.

Theorem 6.11. For all FP numbers x and y,

$$\circ[(x + y)/2] = \circ\left((x + y)/2\right).$$

Proof. Let us compare $|x + y|$ to $2^{p+e_{\min}}$. If $|x + y| \le 2^{p+e_{\min}}$, then $\circ(x + y)$ has the minimal exponent, meaning a subnormal number or belonging to the first normal binade. It is therefore computed without error as explained in section 5.1.2.2. Thus $\circ[(x + y)/2] = \circ\left((x + y)/2\right)$.

If $|x + y| > 2^{p+e_{\min}}$, then $|\circ(x + y)| \ge 2^{p+e_{\min}}$ by monotony of the rounding. In this case, the division by 2 is exact. Thus $\circ[(x+y)/2] = \circ(x+y)/2 = \circ\left((x + y)/2\right)$. ∎

As written above, correct rounding easily fulfills requirements [6.3]–[6.9]. The difficult point is requirement [6.10] (no overflow) that can be fulfilled by requiring x and y to be of opposite signs.

6.5.2. *The* avg_half_sub *function*

The avg_half_sub function is more complex, as it is designed to prevent overflow when x and y have the same sign (which will be assumed only for some of the requirements).

Definition avg_half_sub (x y : R) :=
  round_flt (x + round_flt (round_flt (y - x) / 2)).

That is to say, avg_half_sub$(x, y) = \circ[x + (y - x)/2]$.

Some of the basic requirements are easy to prove:

- Requirement [6.6]: avg_half_sub$(-x, -y) = -$avg_half_sub(x, y).
- Requirement [6.8]: $(x + y)/2 = 0$ implies avg_half_sub$(x, y) = 0$.

Proving requirement [6.7] (avg_half_sub(x, y) has the same sign as $(x + y)/2$) is also not difficult.

Proving requirement [6.4] (min$(x, y) \leq$ avg_half_sub$(x, y) \leq$ max(x, y)) is harder as several rounded operations are involved, including potential underflow. Let us assume first that $x \leq y$ holds. We have to prove that $x \leq \circ[x + (y - x)/2] \leq y$.

The left-hand inequality is based on the monotony of the rounding operator and is quite simple: as $y \geq x$, we have $\circ(y - x) \geq 0$, so $\circ[(y - x)/2] \geq 0$. By monotony of the rounding operator, $\circ[x + (y - x)/2] \geq \circ(x)$. Since x is in the FP format, we thus have $\circ[x + (y - x)/2] \geq \circ(x) = x$.

The right-hand inequality is more difficult: $\circ[x + (y - x)/2] \leq y$. Since we have assumed $x \leq y$, the rounding down $\triangledown(y - x)$ is either zero or positive and we distinguish these two cases. If $\triangledown(y - x) = 0$, then $x = y$ (see section 5.1.2.1) and the result holds. Let us now assume $\triangledown(y - x) > 0$ and prove that $\circ[(y - x)/2] \leq y - x$ holds, which implies the inequality we are interested in by monotony of the rounding operator. If $y - x$ is an FP number hence exactly computed, this is trivial. Otherwise we prove that $\circ[(y - x)/2] \leq \triangledown(y - x)$ by manipulation of real number inequalities and by the study of whether $y - x$ is rounded up or down. Then we have left to prove that $\triangledown(y - x) \leq y - x$, which holds by definition.

When $y < x$, we have to prove that $y \leq$ avg_half_sub$(x, y) \leq x$. This is equivalent to proving $-x \leq -$avg_half_sub$(x, y) \leq -y$. Since $-$avg_half_sub$(x, y) =$ avg_half_sub$(-x, -y)$ holds, the previous proof can be instantiated with the values $-x$ and $-y$, as $-x \leq -y$.

For the remaining requirements, we now assume that x and y have the same sign. More precisely, it means that either x and y are both nonnegative or both nonpositive. A difficult one to prove is requirement [6.9], which states that $2^{e_{\min}} \leq |(x + y)/2|$ implies avg_half_sub$(x, y) \neq 0$. This relies on the following intermediate fact.

Lemma 6.12. If f is a positive FP number, then even when underflow occurs, we have $\circ(f/2) < f$.

Proof. This proof is a case distinction. If $|f| \geq 2^{\varrho + e_{min}}$, then $\circ(f/2) = f/2 < f$. Otherwise $f = m \cdot 2^{e_{min}}$ with $|m| < 2^{\varrho}$, so we have left to prove that the integer rounding to nearest even of $m/2$ is strictly smaller than m. This is done by studying m: as $f > 0$, then $m \geq 1$. When $m = 1$, the result holds as the tie-breaking rule is to even and $0 < 1$. For larger m, we prove that this integer rounding is smaller than $m/2 + 1/2$ which is smaller than m. ∎

Let us prove that avg_half_sub$(x, y) \neq 0$ holds, assuming x and y have the same sign and $2^{e_{min}} \leq |(x + y)/2|$. Without loss of generality, we can assume $x \leq y$ by using the same trick as in the proof of requirement [6.4]. We prove that $\circ[x + (y - x)/2] \neq 0$ by contradiction. If the result of an FP addition is zero, the addends are opposite, as explained in section 5.1.2.1. Therefore $x = -\circ[(y - x)/2] \leq 0$. If x is zero, we can deduce both $2^{e_{min}} \leq |y/2|$ and $\circ(y/2) = 0$, which is impossible since $2^{e_{min}}$ is representable. Thus x is negative and so is y. As a result, $y - x \leq -x$ holds, which implies $\circ[(y - x)/2] \leq \circ(-x/2)$. Thus, by lemma 6.12, we deduce $\circ[(y - x)/2] < -x$, which is a contradiction.

An important requirement is requirement [6.3] which bounds the rounding error. The first subcase is when $|(x + y)/2|$ is exactly $2^{e_{min} - 1}$. As x and y have the same sign, the only possible values for (x, y) are $(0, \pm 2^{e_{min}})$ or $(\pm 2^{e_{min}}, 0)$. This special case is not difficult, but must be studied differently from the general case. The general case corresponds to avg_half_sub(x, y) being nonzero. Then, following the same reasoning as with avg_naive, either $\circ(y - x)$ is computed exactly or the division in $\circ[(y - x)/2]$ is. The final rounding error is therefore small and bounded as follows:

$$\left| \text{avg\_half\_sub}(x, y) - \frac{x + y}{2} \right| \leq \tfrac{3}{2}\text{ulp}\left(\frac{x + y}{2} \right),$$

provided that x and y have the same sign.

The last missing property is requirement [6.5]: it does not hold since the values of avg_half_sub(x, y) and avg_half_sub(y, x) may be different, contrary to what happens with avg_naive. For example, if $y = 1$ and $x = -2^{-53}$, then avg_half_sub$(x, y) = 0.5 - 2^{-53}$ while avg_half_sub$(y, x) = 0.5$.

6.5.3. *The* avg_sum_half *function*

The avg_sum_half function is rather simple, even if it contains two multiplications by 0.5. This is not a problem on recent architectures as the cost of addition and multiplication is nearly the same.

```
Definition avg_sum_half (x y : R) :=
  round_flt (round_flt (x/2) + round_flt (y/2)).
```

That is to say, avg_sum_half$(x, y) = \circ[x/2 + y/2]$.

Provided that $|x|$ is not too small, this function computes the correctly-rounded average. Note that we could have chosen y instead of x, the important point being testing a single value.

Theorem 6.13. For all FP numbers x and y such that $2^{e_{\min}+2\varrho+1} \leq |x|$,

$$\circ[x/2 + y/2] = \circ\left((x+y)/2\right).$$

This might not hold when both $|x|$ and $|y|$ are too small. For example, if $x = y = 2^{e_{\min}}$, then the average is also $2^{e_{\min}}$, while the algorithm returns 0. Note that the assumption $2^{e_{\min}+2\varrho+1} \leq |x|$ can be replaced by $2^{e_{\min}+2\varrho+1} \leq |y|$ by symmetry.

Proof. As x is large enough, we have $\circ(x/2) = x/2$. There are two subcases, depending on the magnitude of y. If $|y| \geq 2^{p+e_{\min}}$, then we also have $\circ(y/2) = y/2$. Thus $\circ[x/2 + y/2] = \circ(x/2 + y/2) = \circ((x+y)/2)$.

If $|y| < 2^{\varrho+e_{\min}}$, then it is subnormal and the division may be inexact. However, x is large enough that this error is too small to impact the result. More precisely, we prove that $\circ[x/2 + y/2] = \circ(x/2 + \circ(y/2)) = x/2 = \circ((x+y)/2)$, by using twice the following result: Given an FP number f and a real h such that $2^{\varrho+e_{\min}} \leq |f|$ and $|h| \leq \frac{1}{4}\text{ulp}(f)$, then $\circ(f + h) = f$. ∎

As with the avg_naive function, correct rounding implies all the previous requirements and gives a half-ulp error bound, provided that either $|x|$ or $|y|$ is large enough.

We can now combine the three presented functions to define comprehensive algorithms to compute an accurate or correctly-rounded approximation of the average.

6.5.4. Sterbenz' accurate algorithm

Following Sterbenz' ideas and the previous definitions, algorithm 6.6 defines an accurate average function. Its properties are derived from the properties of avg_naive and avg_half_sub. The proofs of requirements [6.3]–[6.9] are sometimes long as many subcases have to be studied but they are straightforward. The longest proof is that of requirement [6.6], average$(-x, -y)$ = $-$average(x, y), as all sign possibilities (positive, negative, and zero) have to be considered. What is left to prove is that no overflow occurs (requirement [6.10]), this will be detailed in section 8.3.4 when describing the corresponding C program.

Algorithm 6.6 Accurate computation of the average (Sterbenz)

if x and y do not have the same sign

    return $\circ[(x + y)/2]$

else

    if $|x| \leq |y|$

        return $\circ[x + (y - x)/2]$

    else

        return $\circ[y + (x - y)/2]$

6.5.5. Correctly-rounded algorithm

From the previous properties of avg_naive and avg_sum_half, algorithm 6.7 can be defined, which returns the correctly-rounded average. This program returns $\circ((x + y)/2)$, therefore fulfilling requirements [6.3]–[6.9]. An interesting point is the value of C. Correct rounding holds whenever C is greater than or equal to $2^{e_{min}+2\varrho+1}$ in our model without overflow. We may therefore increase this value as long as overflows are prevented.

Algorithm 6.7 Correct computation of the average

let $C := 2^{e_{min}+2\varrho+1}$

if $C \leq |x|$

    return $\circ[x/2 + y/2]$

else

    return $\circ[(x + y)/2]$

To summarize, we have two different algorithms, the original one from Sterbenz and a more accurate one, both formally proved [BOL 15a]. The more accurate algorithm needs to load the constant C in an FP register but it requires less tests than Sterbenz' algorithm. The corresponding program verifications for both versions are in section 8.3.4. They include all the overflow checking, which are key results in these examples.

6.6. Orientation of three points

In the context of computational geometry, it is common to manipulate geometric predicates that qualify the relative positioning of points. For instance, some predicates might characterize whether a point is inside the sphere that goes through four other points. While one would like to implement such a predicate using FP arithmetic for efficiency reasons, there is a major difference with most other FP algorithms. The result of the predicate does not approximate a real number; it approximates a discrete value:

either the point is inside the sphere, or it is on the surface, or it is outside. There is no straightforward notion of closeness for the result of such a predicate. If the predicate gets it wrong, this might invalidate basic geometric theorems on which algorithms from computational geometry rely [KET 04]. So the goal is to devise implementations that always return the correct result [YAP 95]. More precisely, we will focus on devising FP implementations that either compute the exact result or answer that they do not know what the exact result is. Such an implementation can then act as the first stage of an exact predicate as it is done in the CGAL library [MEL 07].

Let us consider the simplest of the geometric predicates, orient2d, which characterizes on which side of a line formed by two points a third point of the 2D plane is located, as shown in Figure 6.5. Using the coordinates of these points, the predicate can be expressed as follows:

$$\text{orient2d}(p_1, p_2, p_3) = \text{sign} \begin{vmatrix} x_2 - x_1 & x_3 - x_1 \\ y_2 - y_1 & y_3 - y_1 \end{vmatrix}.$$

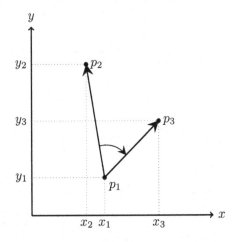

Figure 6.5. *Clockwise orientation of three points p_1, p_2, and p_3.*

When the three points are aligned, the determinant is zero. When p_1, p_2, and p_3 are oriented clockwise, its sign is negative. When they are counterclockwise, it is positive. This sign can be computed using four exact subtractions to get the two vectors, then two exact multiplications and one exact subtraction to compute the 2×2 determinant, and finally a comparison to zero. While an implementation using an exact arithmetic is possible, it is slow and should only be used as a last resort. So we will see if we can make use of FP arithmetic to perform a fast computation of the sign and to properly detect when the computed sign is possibly incorrect.

Section 6.6.1 shows how a naive implementation of the orient2d predicate using FP operations behaves and how to make use of its geometric properties to turn it into a more robust version. This new version depends on various "magical" constants; section 6.6.2 shows how to compute them and how to prove their correctness. Then section 6.6.3 shows what that robust algorithm looks like. Finally, section 6.6.4 gives the example of a slightly more complicated geometric predicate: deciding where a point is located with respect to a circle defined by three other points.

6.6.1. *Naive FP implementation*

Let us assume that the input coordinates are representable as *binary64* FP numbers. If FP operations were exact, one would get a fast implementation by using a straightforward translation to C code:

```
int orient2d(double x1, double y1, double x2,
             double y2, double x3, double y3)
{
  double det = (x2 - x1) * (y3 - y1)
             - (x3 - x1) * (y2 - y1);
  if (det < 0) return NEGATIVE;
  if (det > 0) return POSITIVE;
  return ZERO;
}
```

Since FP operations are not usually exact, round-off errors can cause det to have a sign different from the exact determinant that will be denoted by *det*. Note that the round-off error of the last FP subtraction does not matter, since we are only interested in the sign of the result and FP subtraction preserves the sign of the exact computation, as both 0 is in the format and the rounding is monotone. The round-off errors of the six first operations, however, are propagated and can cause the final sign to be incorrectly computed. Not only might det be computed as zero even if the points are not aligned but its sign might even be the opposite of the exact one.

Figure 6.6 illustrates the situation. Points p_2 and p_3 are fixed at positions $p_2 = (3.125, 3.125 - 2^{-51})$ and $p_3 = (3.25 + 2^{-51}, 3.25)$. Point p_1 visits all the points with *binary64* coordinates around $(1.1, 1.1)$, distance between two neighboring points is 2^{-52}. Each of the three pictures shows the sign of the determinant depending on the position of p_1, which is located in the lower-right part of every picture. When p_1 is on a diagonal line in that part, the three points are aligned, so the sign should be zero. In the upper left corner of every picture, the three points are oriented counterclockwise, while in the lower right corner, they are oriented clockwise.

Figure 6.6a shows the ideal situation: the arithmetic operations are performed exactly and the computed sign matches the orientation of the three points. Figure 6.6b shows the situation with *binary64* arithmetic: some orientations outside the diagonal line are incorrectly computed as aligned (white points); there are even some cases where the sign is computed as positive while it should be negative and conversely.

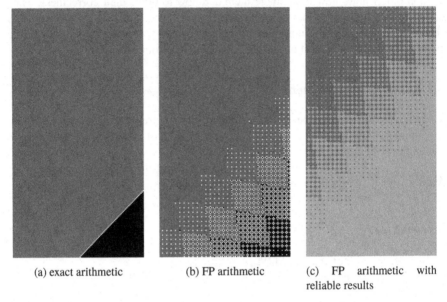

(a) exact arithmetic	(b) FP arithmetic	(c) FP arithmetic with reliable results

Figure 6.6. *Results of the* orient2d *algorithm depending on the arithmetic. White points mean that the points are computed as aligned, black points as oriented clockwise, and dark gray points as counterclockwise. Light gray points in Figure 6.6c mean a possibly incorrect result.*

Rather than always using a correct but slower code, we can try to devise some mechanism that detects whether the sign of det is possibly incorrect. In that case only, the slower code (e.g. exact arithmetic) would be used to compute the proper sign. Figure 6.6c illustrates this situation: some mechanism (described later in algorithm 6.8) detects that the computed result is possibly incorrect (light gray area). All the incorrect orientations in Figure 6.6b are now properly detected but there are also some orientations that, while correctly computed, would cause the slower code to be called to recompute them.

For the detection mechanism to be useful, it has to satisfy a few properties:

1) It will never say that the computed sign is reliable when it is not.

2) The mechanism should be fast, so as to not defeat the point of using FP arithmetic to compute the determinant. For instance, it should be cheaper than running the whole algorithm using interval arithmetic.

3) False negatives should not occur too often, as that causes some slower code to be run while the sign was already correctly computed with FP arithmetic.

If we knew *a priori* a tight bound Δ on the absolute error between the computed FP value det and the ideal value *det*, we could easily turn the naive algorithm into a robust one: we would just compare the magnitude of det with this bound Δ; if it is larger, the computed result has the same sign as the exact result. This would satisfy the three properties we expect: soundness (if the bound is correct), speed (hardly more FP operations), usefulness (if the bound is tight). The C code would look as follows:

```
int orient2d(double x1, double y1, double x2,
             double y2, double x3, double y3)
{
  double det = (x2 - x1) * (y3 - y1)
             - (x3 - x1) * (y2 - y1);
  double delta = ...;
  if (det < -delta) return NEGATIVE;
  if (det > +delta) return POSITIVE;
  return UNKNOWN;
}
```

Unfortunately, we cannot compute a tight bound *a priori*. Indeed, a bound on the absolute error would not be tight since it would have to work for large inputs and thus would be useless for small inputs. A bound on the relative error would not make any sense either, since we are interested in the sign of det. Indeed, if its sign can be wrong, it means that the relative error of the algorithm is unbounded.

So we want to compute a bound on the absolute error between det and *det* at execution time. Evaluating det using interval arithmetic would provide us with such a bound, but we want some faster mechanism that does not involve directed rounding modes. First, let us consider a more specific problem. We ask Gappa (see section 4.3.1) what is the bound Δ_1 on the absolute error when all the differences $|x_2 - x_1|$, $|y_2 - y_1|$, $|x_3 - x_1|$, and $|y_3 - y_1|$ are bounded by 1. The coordinates do not have to be bounded. For the *binary64* format, the tool answers that $\Delta_1 = 2^{-51}$.

Now, let us suppose that the differences $|x_2 - x_1|$ and $|x_3 - x_1|$ are no longer bounded by 1, but by 2. Gappa then computes an error bound twice as large. Similarly, if we were to double the range of the differences on the y inputs, we would get a doubled error bound. This behavior can be explained by the form of the formulas used for forward error analysis (see section 1.2). As a result, we could expect the following value Δ to be a general bound on the absolute rounding error:

$$\Delta = \Delta_1 \cdot M \quad \text{with } M = \left\| \begin{array}{c} x_2 - x_1 \\ x_3 - x_1 \end{array} \right\|_\infty \cdot \left\| \begin{array}{c} y_2 - y_1 \\ y_3 - y_1 \end{array} \right\|_\infty. \qquad [6.11]$$

The vector norm used in this formula is just the maximum of the absolute values, so the computation of Δ would not be too expensive. Indeed, the absolute value is a cheap operation, so the overhead with respect to the straightforward implementation of the predicate would be two maxima and two multiplications. While Δ is defined using exact operations on real numbers, we will later see how to replace them with FP operations. So, with respect to speed, the above definition of Δ is suitable.

Unfortunately, there are several issues on the correctness side. First of all, when the FP products of det underflow, the value Δ is no longer an upper bound on the absolute rounding error. Indeed, Δ_1 is a value obtained assuming that no underflow occurs. But even for the normal case, one has to be careful when reusing the value Δ_1 computed by Gappa. While this value is correct, it is only meaningful when the input differences are bounded by a power of 2. This is easily seen by asking Gappa the error bound for some other ranges. For a *binary64* format rounded to nearest, Gappa answers $\Delta_1 = 2^{-51}$ when the input difference ranges are $[-1; 1]$. If we extend the ranges to $[-1.01; 1.01]$, Gappa answers $\Delta_{1.01} = 2^{-49.98}$. So a slight modification of the inputs causes the error bound computed by Gappa to be twice as large. In a way, the bound Δ_1 computed by Gappa is too tight for our purpose. The next section will show how to compute a constant factor that is suitable for equation [6.11].

6.6.2. *Homogeneous error bound*

Choosing a suitable value of Δ_1 is not sufficient; we also have to find under which condition equation [6.11] gives a proper bound on the absolute rounding error. We would like to obtain the following property with M as in equation [6.11]:

$$2^{-1022} \leq M \Rightarrow |\text{det} - det| \leq 2^{-49.99\cdots} \cdot M \qquad [6.12]$$

Our approach should also be generic enough so that it can be applied to more complicated geometric predicates. We will follow the structure of the expressions, inductively computing for every sub-expression a bound on its rounding error. This bound will be computed as the product between some expression (ultimately M) and a constant factor (ultimately Δ_1). The property satisfied by this bound will be expressed as a relation $\mathcal{R}$, from which we would deduce equation [6.12].

This relation $\mathcal{R}$ is quite complicated, since it should support various operations as well as underflows. So we start with a simpler relation $\mathcal{R}_1$, which will be progressively enriched until we get $\mathcal{R}$.

$$\mathcal{R}_1(\tilde{x}, x, m, \delta) \overset{\text{def}}{=} |\tilde{x} - x| \leq \delta \cdot m.$$

Let us start with the case of an exact addition. One can easily prove the following formula:

$$\forall \tilde{u}, \tilde{v}, u, v, m, \delta_u, \delta_v \in \mathbb{R},$$

$$\mathcal{R}_1(\tilde{u}, u, m, \delta_u) \wedge \mathcal{R}_1(\tilde{v}, v, m, \delta_v) \quad \Rightarrow \quad \mathcal{R}_1(\tilde{u} + \tilde{v}, u + v, m, \delta_u + \delta_v). \quad [6.13]$$

The formula would be similar for an exact subtraction. Notice how m appears in all three instances of $\mathcal{R}_1$. So given two properties, one should first ensure that they use the same value of m. It means that we need a formula such as the following one:

$$\forall \tilde{x}, x, m_1, m_2, \delta \in \mathbb{R}, \quad m_1 \leq m_2 \quad \Rightarrow$$

$$\mathcal{R}_1(\tilde{x}, x, m_1, \delta) \quad \Rightarrow \quad \mathcal{R}_1(\tilde{x}, x, m_2, \delta). \quad [6.14]$$

Unfortunately, there are some trivial counterexamples, e.g. $m_1 = 0$, $m_2 = 1$, and $\delta = -1$. They can be avoided by enriching $\mathcal{R}_1$ with the property $\delta \geq 0$.

Let us look at an exact multiplication now. The way δ is propagated is slightly more complicated so we take some inspiration from forward error analysis (see section 4.3.2).

$$|\tilde{u} \cdot \tilde{v} - u \cdot v| \leq |\tilde{u} - u| \cdot |v| + |u| \cdot |\tilde{v} - v| + |\tilde{u} - u| \cdot |\tilde{v} - v|.$$

Unfortunately, we do not have any way to compute bounds on $|u|$ and $|v|$ yet. So we modify $\mathcal{R}_1$ to keep enough information around. As with the bound $\delta \cdot m$ on the error $|\tilde{x} - x|$, the bound on the exact value will be relative to m, as follows: $|x| \leq b \cdot m$. This leads to the second iteration of $\mathcal{R}$:

$$\mathcal{R}_2(\tilde{x}, x, m, b, \delta) \stackrel{\text{def}}{=} \begin{cases} \delta \geq 0, \\ |x| \leq b \cdot m, \\ |\tilde{x} - x| \leq \delta \cdot m. \end{cases}$$

Formula [6.13] about addition is easily adapted: not only should the bounds δ on the errors be added but also the bounds b on the values. As for formula [6.14], we encounter the same problem as with δ: the bound b has to be nonnegative. The definition of $\mathcal{R}_2$ is sufficient to prove the following formula about the exact multiplication:

$$\forall \tilde{u}, \tilde{v}, u, v, m_u, m_v, b_u, b_v, \delta_u, \delta_v \in \mathbb{R},$$

$$\mathcal{R}_2(\tilde{u}, u, m_u, b_u, \delta_u) \wedge \mathcal{R}_2(\tilde{v}, v, m_v, b_v, \delta_v) \quad \Rightarrow$$

$$\mathcal{R}_2(\tilde{u} \cdot \tilde{v}, u \cdot v, m_u \cdot m_v, b_u \cdot b_v, \delta_u \cdot b_v + b_u \cdot \delta_v + \delta_u \cdot \delta_v). \quad [6.15]$$

Now that we have formulas to handle exact arithmetic operations, we need to devise a way to deal with rounding operators. Forward error analysis tells us that

$$|\circ(\tilde{x}) - x| \leq |\circ(\tilde{x}) - \tilde{x}| + |\tilde{x} - x| \leq \tfrac{1}{2}\mathrm{ulp}(\tilde{x}) + \delta \cdot m$$

when rounding to nearest. Since $\tilde{x}$ is bounded by $(b+\delta)\cdot m$, we would like the following formula to hold:

$$\forall \tilde{x}, x, m, b, \delta \in \mathbb{R},$$

$$\mathcal{R}_2(\tilde{x}, x, m, b, \delta) \quad \Rightarrow \quad \mathcal{R}_2(\circ(\tilde{x}), x, m, b, \delta + (b + \delta) \cdot \varepsilon). \quad [6.16]$$

The ε constant depends on the rounding direction and the format; it is $\varepsilon = 2^{-53}$ for *binary64* and rounding to nearest.

There is a small subtlety in the proof since $\tilde{x}$ might be subnormal. So a property such as $\mathrm{ulp}(\tilde{x}) \leq \varepsilon \cdot |\tilde{x}|$ is not guaranteed to hold. This difficulty can be avoided by relying on the monotony of the ulp function. So, when rounding to nearest, we have

$$|\circ(\tilde{x}) - x| \leq \tfrac{1}{2}\mathrm{ulp}(\tilde{x}) + \delta \cdot m \leq \tfrac{1}{2}\mathrm{ulp}((b + \delta) \cdot m) + \delta \cdot m.$$

As a result, formula [6.16] holds as long as we have

$$\tfrac{1}{2}\mathrm{ulp}((b + \delta) \cdot m) \leq \varepsilon \cdot (b + \delta) \cdot m.$$

The property above holds only when $(b + \delta) \cdot m$ is large enough. More precisely, it holds as soon as $\varepsilon \cdot (b + \delta) \cdot m$ is larger than the constant

$$\mu = \tfrac{1}{2}\mathrm{ulp}(0) = 2^{-1075}$$

for *binary64* with rounding to nearest. Unfortunately, we do not have enough information to ensure $\mu \leq \varepsilon \cdot (b + \delta) \cdot m$, so we enrich the definition of $\mathcal{R}$ again.

Before modifying $\mathcal{R}$, let us notice that proving $\mu \leq \varepsilon \cdot b \cdot m$ is sufficient, since δ is nonnegative. Moreover, if we knew that b is larger than 1, we would just need to prove $\mu \cdot \varepsilon^{-1} \leq m$. So we could try to define $\mathcal{R}$ as $\mu \cdot \varepsilon^{-1} \leq m \Rightarrow \mathcal{R}_2(\ldots)$. Unfortunately, such a definition would cause both formulas [6.14] and [6.15] to become incorrect. So rather than using m in the precondition, we add yet another argument to the relation. This is the final modification needed for defining $\mathcal{R}$. It is now rich enough to be propagated along FP computations.

$$\mathcal{R}(c, \tilde{x}, x, m, b, \delta) \quad \overset{\mathrm{def}}{=} \quad \mu \cdot \varepsilon^{-1} \leq c \Rightarrow \begin{cases} b \geq 1, \\ \delta \geq 0, \\ |x| \leq b \cdot m, \\ |\tilde{x} - x| \leq \delta \cdot m. \end{cases}$$

We can now prove the following formula about rounding operators:

$$\forall c, \tilde{x}, x, m, b, \delta \in \mathbb{R},$$

$$\mathcal{R}(c, \tilde{x}, x, m, b, \delta) \quad \Rightarrow \quad \mathcal{R}(\min(c, m), \circ(\tilde{x}), x, m, b, \delta + (b + \delta) \cdot \varepsilon).$$

The previous formulas are changed as follows:

$$\forall c, \tilde{u}, u, \tilde{v}, v, m, b_u, b_v, \delta_u, \delta_v \in \mathbb{R},$$

$$\mathcal{R}(c, \tilde{u}, u, m, b_u, \delta_u) \wedge \mathcal{R}(c, \tilde{v}, v, m, b_v, \delta_v) \quad \Rightarrow$$

$$\mathcal{R}(c, \tilde{u} + \tilde{v}, u + v, m, b_u + b_v, \delta_u + \delta_v).$$

$$\forall c, \tilde{u}, u, \tilde{v}, v, m_u, m_v, b_u, b_v, \delta_u, \delta_v \in \mathbb{R},$$

$$\mathcal{R}(c, \tilde{u}, u, m_u, b_u, \delta_u) \wedge \mathcal{R}(c, \tilde{v}, v, m_v, b_v, \delta_v) \quad \Rightarrow$$

$$\mathcal{R}(c, \tilde{u} \cdot \tilde{v}, u \cdot v, m_u \cdot m_v, b_u \cdot b_v, \delta_u \cdot b_v + b_u \cdot \delta_v + \delta_u \cdot \delta_v).$$

$$\forall c, \tilde{x}, x, m_1, m_2, b, \delta \in \mathbb{R}, \quad m_1 \leq m_2 \quad \Rightarrow$$

$$\mathcal{R}(c, \tilde{x}, x, m_1, b, \delta) \quad \Rightarrow \quad \mathcal{R}(c, \tilde{x}, x, m_2, b, \delta).$$

We have almost all the formulas needed for verifying orient2d. We just need one last formula for the base case of the induction. Indeed, to get formula [6.11], we do not want to start the induction from the coordinates but from their differences. To express the relation between the rounded differences and their exact values, we take advantage of the fact that the relative error of the FP sum is always bounded, even if the result is in the subnormal range:

$$\forall u, v \in F, \quad \mathcal{R}(\mu \cdot \varepsilon^{-1}, \circ(u - v), u - v, |u - v|, 1, \varepsilon).$$

Since the first argument is set to $\mu \cdot \varepsilon^{-1}$, the precondition of $\mathcal{R}$ always holds, so the property above implies

$$|\circ(u - v) - (u - v)| \leq \varepsilon \cdot |u - v|.$$

All the formulas above use real numbers to represent the bounds b and δ. To simplify the proof process in Coq, it is better to formalize and prove all these formulas using unbounded FP numbers (or rational numbers), so that the proof assistant will automatically compute the values of b and δ of the conclusion of a formula from its hypotheses. Note that this requires a forward-reasoning style of proof. For instance, the statement of the formula for the sum is expressed as follows, with the real addition to sum expressions and an exact addition Fplus of FP numbers to sum the bounds.

```
Rel c u1 v1 m b1 d1 -> Rel c u2 v2 m b2 d2 ->
Rel c (u1+u2) (v1+v2) m (Fplus b1 b2) (Fplus d1 d2).
```

When applied to the orient2d predicate using *binary64* arithmetic with rounding to nearest, we get the following relation (with an approximated δ for the sake of readability):

$$\mathcal{R}(M, \det, \det, M, 2, 2^{-49.99\cdots})$$

$$\text{with } M = \left\| \begin{matrix} x_2 - x_1 \\ x_3 - x_1 \end{matrix} \right\|_{\infty} \cdot \left\| \begin{matrix} y_2 - y_1 \\ y_3 - y_1 \end{matrix} \right\|_{\infty}, \quad [6.17]$$

from which we deduce equation [6.12].

Note that M appearing as the first argument of $\mathcal{R}$ (hence the requirement $2^{-1022} = \mu \cdot \varepsilon^{-1} \leq M$ in equation [6.12]) is specific to the orient2d predicate. The first argument will be more complicated for larger predicates, as can be seen in section 6.6.4.

6.6.3. *Writing a robust algorithm*

In equations [6.17] and [6.12], the values M and $2^{-49.99\cdots} \cdot M$ are expressed using exact arithmetic. But in the actual implementation of algorithm 6.8, FP computations are used, which means that the constant 2^{-1022} has to be increased to take into account any underestimation while approximating M. Similarly, the constant $\delta = 2^{-49.99\cdots}$ has to be increased to take into account any underestimation while approximating $\delta \cdot M$. Let us see how the final constants are obtained.

The relative error between the FP value m of algorithm 6.8 and the infinitely-precise value M is bounded by $\theta_3 = (1 + 2^{-53})^3 - 1$ (two subtractions and one multiplication), assuming the last multiplication does not underflow. Thus we have $M \cdot (1 - \theta_3) \leq m \leq M \cdot (1 + \theta_3)$. By choosing $C_1 = \triangle(2^{-1022} \cdot (1 + \theta_3))$, the following implications hold:

$$\begin{aligned} m \geq C_1 \quad &\Rightarrow \quad m \geq 2^{-1022} \cdot (1 + \theta_3) \\ &\Rightarrow \quad M \cdot (1 + \theta_3) \geq 2^{-1022} \cdot (1 + \theta_3) \\ &\Rightarrow \quad M \geq 2^{-1022}. \end{aligned}$$

By choosing $C_2 = \triangle(\delta \cdot (1 - \theta_3)^{-1})$, the following implications hold:

$$|\det| > \circ(C_2 \cdot m) \;\;\Rightarrow\;\; |\det| > C_2 \cdot m,$$

$$\Rightarrow\;\; |\det| > C_2 \cdot M \cdot (1 - \theta_3),$$

$$\Rightarrow\;\; |\det| > \delta \cdot M.$$

Note that the inequality $m \geq C_1$ also implies that the last multiplication performed when computing m did not underflow, so we do not need any special check to exclude that case.

Algorithm 6.8 Robust implementation of orient2d

```
int orient2d(double x1, double y1, double x2,
             double y2, double x3, double y3)
{
  double dx1 = x2 - x1;
  double dx2 = x3 - x1;
  double dy1 = y2 - y1;
  double dy2 = y3 - y1;
  double nx = fmax(fabs(dx1), fabs(dx2));
  double ny = fmax(fabs(dy1), fabs(dy2));
  double m = nx * ny;
  if (m < 0x1.0000000000002p-1022) return UNKNOWN;
  double det = dx1 * dy2 - dx2 * dy1;
  double delta_m = 0x1.0000000000003p-50 * m;
  if (det < -delta_m) return NEGATIVE;
  if (det > +delta_m) return POSITIVE;
  return UNKNOWN;
}
```

There could still be an issue with det ending up either infinite or NaN. The NaN case is simple: both final comparisons will fail, so the end result will be *unknown*. The infinite case is a bit more contrived. Let us see where the infinity comes from. First, if neither FP products $dx_1 \cdot dy_2$ nor $dx_2 \cdot dy_1$ are infinite yet their difference is infinite, they have to be of opposite signs. So the sign of the computed determinant is necessarily correct in that case. Second, if either FP product is infinite, then m is also infinite since it is larger than both products in absolute value. Therefore delta_m is $+\infty$ which means that the two final comparisons will fail, leading to an *unknown* result. So algorithm 6.8 is robust not only to round-off errors but also to overflow issues.

6.6.4. *Generalization to other predicates*

On the example of orient2d, the value of c ends up being equal to M after simplification in the final relation given by equation [6.17]. On larger examples, this might not be the case. Moreover, during the induction, the expression c could end up being different on both sides of a sum or product. In that case, the following formula can be used to unify both values of c to their minimum:

$$\forall c_1, c_2, \tilde{x}, x, m, b, \delta \in \mathbb{R}, \quad c_2 \leq c_1 \quad \Rightarrow$$

$$\mathcal{R}(c_1, \tilde{x}, x, m, b, \delta) \quad \Rightarrow \quad \mathcal{R}(c_2, \tilde{x}, x, m, b, \delta).$$

An example requiring several applications of this formula is the predicate for deciding the location of a point of the 2D plane with respect to the circle defined by three other points:

$$\text{incircle2d}(p_1, p_2, p_3, p_4) = \text{sign} \begin{vmatrix} x_2 - x_1 & y_2 - y_1 & (x_2 - x_1)^2 + (y_2 - y_1)^2 \\ x_3 - x_1 & y_3 - y_1 & (x_3 - x_1)^2 + (y_3 - y_1)^2 \\ x_4 - x_1 & y_4 - y_1 & (x_4 - x_1)^2 + (y_4 - y_1)^2 \end{vmatrix}.$$

The approach described in this section produces the following property for *binary64*:

$$2^{-1022} \leq \min(X^2, Y^2, M) \Rightarrow |\text{det} - det| \leq 2^{-45.99\cdots} \cdot M$$

$$\text{with } X = \begin{Vmatrix} x_2 - x_1 \\ x_3 - x_1 \\ x_4 - x_1 \end{Vmatrix}_\infty, \quad Y = \begin{Vmatrix} y_2 - y_1 \\ y_3 - y_1 \\ y_4 - y_1 \end{Vmatrix}_\infty, \quad M = X \cdot Y \cdot \max(X^2, Y^2).$$

As with the orient2d predicate, we could easily deduce a robust implementation of incircle2d from the above property.

This section has presented an approach for implementing geometric predicates using FP operations in such a way that they are fast yet always correct, though the result might be useless when the precision of the computations is not sufficient. This is done by dynamically computing a bound on the absolute round-off error of the computations and using it to filter out potentially incorrect results.

6.7. Order-2 discriminant

The next example is the accurate computation of the discriminant $b^2 - ac$ for three given FP numbers a, b, and c in radix 2 with a precision greater than 1. An example of use is a simple linear system such as a mechanical system with a mass, a spring, and a drag or an electrical system with an inductance, a capacitance, and a resistance [KAH 04]. After being disturbed by an impulse, this system may or may not pass through its equilibrium state before returning to it. This can be decided by determining whether the roots of a degree-2 polynomial are real or complex, whilst the roots tell how long the system takes to return to equilibrium.

The naive algorithm may give very inaccurate results when $b^2 \approx ac$, so Kahan provided an accurate algorithm. It is an interesting application for formal verification as the pen-and-paper proofs provided are described as "far longer and trickier" than the algorithms and programs and Kahan deferred their publication [KAH 04].

The value to be computed is quite similar to the one needed for the orientation of three points of section 6.6. But we do not look forward to robustness anymore, only to reasonable accuracy. As the naive algorithm is not accurate at all for canceling values, we have to somehow detect if a cancellation is possible and then use EFTs to recover enough accuracy (see section 5.2 for the EFT for multiplication).

Algorithm 6.9 gives the original C code. The fabs function computes the absolute value of an FP number. The Dekker function computes the error of the FP multiplication. It is algorithm 5.5 described in section 5.2 and algorithm 8.2 of section 8.3.2 for the program.

Algorithm 6.9 Accurate computation of the order-2 discriminant

```
double Kahan_discr (double a, double b, double c) {
  double p, q, d, dp, dq;
  p = b * b;
  q = a * c;
  if (p + q <= 3 * fabs(p - q))
    d = p - q;
  else {
    dp = Dekker(b, b, p); // p + dp = b * b
    dq = Dekker(a, c, q); // q + dq = a * c
    d = (p - q) + (dp - dq);
  }
  return d;
}
```

Our goal is to formally prove that this algorithm is accurate. In the pen-and-paper proof, the result is claimed to be within two ulps of the exact mathematical result [KAH 04, BOL 06b]. However, the test was mistakenly considered by the authors to be correct with respect to the real values. This does not hold as p + q <= 3 * fabs(p - q) may give a Boolean answer which is the opposite of an ideal test $p + q \leq 3 \cdot |p - q|$ computed using infinitely-precise operations [BOL 09b]. In section 6.7.1, we consider the case of the idealized algorithm with an infinitely-precise test. In section 6.7.2, we study the discrepancy cases in order to prove that the 2-ulps error bound holds nonetheless.

6.7.1. *Proof of the ideal algorithm*

Let us now sketch the proof. For the sake of simplicity, we assume that there is neither underflow or overflow. As written above, the idealized algorithm assumes that the test is made on real values: the comparison is $p + q \leq 3 \cdot |p - q|$ instead of the actual test $\circ(p + q) \leq \circ [3 \cdot |p - q|]$. The limits of this approach are discussed in section 6.7.2.

Let us prove that $b^2 - ac$ is within 2 ulps of d computed as explained. Let $\delta = |d - (b^2 - ac)|$, we will prove that $\delta \leq 2 \cdot \mathrm{ulp}(d)$. The full formal proof can be found in [BOL 09b]. Only the most interesting points are given here.

Here is a summary of the proof structure:

1) when $p + q \leq 3 \cdot |p - q|$;

2) when $p + q > 3 \cdot |p - q|$, then lemmas 6.14, 6.15, and 6.16;

    a) when $\mathrm{ulp}(p) = \mathrm{ulp}(q)$;

    b) when $|p - q| \geq 3 \min(\mathrm{ulp}(p), \mathrm{ulp}(q))$;

    c) when $|p - q|$ is either $2 \min(\mathrm{ulp}(p), \mathrm{ulp}(q))$ or $\min(\mathrm{ulp}(p), \mathrm{ulp}(q))$.

When $p + q \leq 3 \cdot |p - q|$, the proof is straightforward. Each operation may create a small error bounded by half an ulp and the hypothesis guarantees that there is no cancellation. We then easily have $\delta \leq 2 \cdot \mathrm{ulp}(d)$. Let us now assume that $p + q > 3 \cdot |p - q|$. This corresponds to the else branch, with intermediate values for the multiplication errors: $p + dp = b^2$ and $q + dq = ac$. First, if $p = q$ then d is correct within half an ulp as $d = \circ(b^2 - ac)$.

Lemma 6.14. *When $p + q > 3 \cdot |p - q|$, the computation of $p - q$ is exact.*

Proof. If $q \leq 0$, then we have $3 \cdot |p - q| < p + q \leq |p| - |q| \leq |p - q|$, which is absurd. Therefore $q > 0$. Similarly, we have $p > 0$. Indeed, we already have that $p = \circ(b^2) \geq 0$, but $p = 0$ is forbidden as it would mean $| - q| < q$.

So $p = |p| \leq |p - q| + |q| \leq \frac{1}{3}(p + q) + q$, which implies $p \leq 2q$. Similarly, $q \leq 2p$. Therefore the subtraction is exact by Sterbenz' theorem (see section 5.1.1). ∎

Lemma 6.15. *When $p + q > 3 \cdot |p - q|$ and $p \neq q$, we have $|\circ(dp - dq)| \leq 2|p - q|$.*

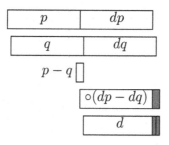

Figure 6.7. *Order-2 discriminant: worst case. The gray parts corresponds to the possibly incorrect bits.*

Proof. As $p \neq q$, we know that $|p - q| \geq \min(\mathrm{ulp}(p), \mathrm{ulp}(q))$. We also know that $\max(\mathrm{ulp}(p), \mathrm{ulp}(q)) \leq 2\min(\mathrm{ulp}(p), \mathrm{ulp}(q))$ as $\frac{q}{2} \leq p \leq 2q$. Thus, $|dp - dq| \leq |dp| + |dq| \leq \frac{1}{2}\mathrm{ulp}(p) + \frac{1}{2}\mathrm{ulp}(q) \leq \frac{3}{2}\min(\mathrm{ulp}(p), \mathrm{ulp}(q)) \leq \frac{3}{2}|p - q|$.

Therefore, $|\circ(dp - dq)| \leq 2 \cdot |p - q|$.   ∎

The bound might not seem tight enough but we are in the few cases where $p \approx q$. Therefore, $p - q$ may be very small, so this bound $(2 \cdot |p-q|)$ is not far from the lowest possible bound (just under $1.5 \cdot |p - q|$).

Lemma 6.16. Assume $p + q > 3 \cdot |p - q|$. If the computation of $dp - dq$ is exact, then $\delta \leq \frac{1}{2} \cdot \mathrm{ulp}(d)$.

Proof. As $p+q > 3\cdot|p-q|$, the computation of $p-q$ is exact by lemma 6.14. Moreover, if the computation of $dp - dq$ is exact, then we may simplify δ by $\delta = |\circ(b^2 - ac) - (b^2 - ac)| \leq \frac{1}{2}\mathrm{ulp}(d)$.   ∎

We have finished the intermediate lemmas and we bound δ in all possible cases:

– When $\mathrm{ulp}(p) = \mathrm{ulp}(q)$, the result $\delta \leq \frac{1}{2} \cdot \mathrm{ulp}(d)$ holds by lemma 6.16.

– When $|p - q| \geq 3\min(\mathrm{ulp}(p), \mathrm{ulp}(q))$, the result holds using lemma 6.14, lemma 6.15, and the monotony of ulp and $\circ$.

– The remaining case is when the ulps of p and q are different but $|p - q|$ is either $2\min(\mathrm{ulp}(p), \mathrm{ulp}(q))$ or $\min(\mathrm{ulp}(p), \mathrm{ulp}(q))$. It means that p and q are very near a power of 2. This is equivalent to considering $p = 1$ and q is either 1^- or 1^{--}. We just have to manually check these cases.

This very last case, corresponding to $p + q > 3 \cdot |p - q|$ and $|p - q|$ is either $2\min(\mathrm{ulp}(p), \mathrm{ulp}(q))$ or $\min(\mathrm{ulp}(p), \mathrm{ulp}(q))$, gives the error bound of two ulps.

Lemma 6.17. Provided that the test is correct and that no underflow or overflow occur, algorithm 6.9 returns a value within two ulps of the correct result $b^2 - ac$.

We refer the reader to [BOL 09b] for more details on the proof and in particular the handling of subnormal numbers.

6.7.2. *Proof when the test goes wrong*

While the idealized algorithm covers most possible inputs, there are some cases where the test may go wrong [BOL 09b], so it cannot serve as a basis for a complete formal proof.

Let us use a toy FP format with radix-2 and a 5-bit significand. Let $p = 27 = 11011_2$ and $q = 14 = 1110_2$, then $p + q = 41 > 39 = 3 \cdot |p - q|$. But $\circ(p + q) = \circ(41) = 40$ and $\circ(p - q) = 13$ and $\circ[3 \cdot |p - q|] = \circ(39) = 40$. Therefore

$$\circ(p + q) = \circ[3 \cdot |p - q|] \quad \text{and} \quad p + q > 3 \cdot |p - q|,$$

so the FP test and the real test disagree.

Nevertheless, in most cases, the FP and real tests agree, so the previous proofs may be reused. We just have to look into the potential disagreements. Here is a summary of the proof structure:

1) Assume that $\circ(p + q) \leq \circ[3 \cdot |p - q|]$ so that $d = \circ(p - q)$.

   a) If $p + q \leq 3 \cdot |p - q|$, we apply lemma 6.17.

   b) If $p + q > 3 \cdot |p - q|$, see section 6.7.2.2.

2) Assume that $\circ(p + q) > \circ[3 \cdot |p - q|]$ so that $d = \circ[(p - q) + (dp - dq)]$.

   a) If $p + q > 3 \cdot |p - q|$, we apply lemma 6.17.

   b) If $p + q \leq 3 \cdot |p - q|$, see section 6.7.2.1.

6.7.2.1. *The execution should have taken the most accurate path*

As explained, we assume that $p + q > 3 \cdot |p - q|$ holds, yet $\circ(p + q) \leq \circ[3 \cdot |p - q|]$. We split into two cases.

The first case is when $\text{ulp}(d) \geq \text{ulp}(q)$. By lemma 6.14, we have that $p - q$ is computed exactly, so $d = p - q$. We easily deduce that $\delta \leq 2 \cdot \text{ulp}(d)$.

The second case is when $\text{ulp}(d) < \text{ulp}(q)$. We then prove that q is very close to $p/2$ using the following lemma.

Lemma 6.18. Assume that both $p + q > 3 \cdot |p - q|$ and $\circ(p + q) \leq \circ[3 \cdot |p - q|]$. Assume that $\mathrm{ulp}(d) < \mathrm{ulp}(q)$. Then,

$$\frac{p}{2} \leq q \leq p\frac{1 + 2^{1-\varrho}}{2 + 2^{-\varrho}}.$$

The rest of the proof is less interesting and can be found in [BOL 09b]. It is based on rounding error manipulation and a case distinction on the respective values of $\mathrm{ulp}(p)$ and $\mathrm{ulp}(q)$.

6.7.2.2. The execution should have taken the short path

We now assume that $p + q \leq 3 \cdot |p - q|$ holds, yet $\circ(p + q) > \circ[3 \cdot |p - q|]$. We may assume without loss of generality that $p \geq q$.

Lemma 6.19. Assume that both $p + q \leq 3 \cdot |p - q|$ and $\circ(p + q) > \circ[3 \cdot |p - q|]$. Then

$$p - q \leq \frac{1}{3}(p + |q|)\frac{(1 + 2^{-\varrho})^2}{1 - 2^{-\varrho}}.$$

Proof. By straightforward error analysis.   ■

Lemma 6.20. Assume that both $p + q \leq 3 \cdot |p - q|$ and $\circ(p + q) > \circ[3 \cdot |p - q|]$. Then $0 < q$.

Proof. If $q \leq 0$, the inequality of lemma 6.19 becomes

$$p - q \leq \frac{1}{3}(p - q)\frac{(1 + 2^{-\varrho})^2}{1 - 2^{-\varrho}}.$$

This implies $1 \leq \frac{1}{3}(1 + \varepsilon)$ with $0 < \varepsilon$ small, which is absurd. Therefore $0 < q$.   ■

The case of the second disagreement is then solved by proving the following bound: $\circ(|dp - dq|) \leq 3 \cdot \mathrm{ulp}(\circ(p - q))$. Therefore, the error bound holds whatever the answers of the FP and real tests.

Theorem 6.21. Provided that neither underflow nor overflow occur, algorithm 6.9 returns a value within two ulps of the exact result $b^2 - ac$.

Underflow is complex to handle here, as the multiplication error may not be representable. Instead of assuming that there is no underflow, we require the inputs to be large enough so that the multiplication error is an FP number (see sections 5.2 and 8.3.2). This is the only requirement, as the remainder of the proof holds, even when subnormal numbers are involved. Overflow can be taken into account at the program level by assuming the input values are small enough.

7

Compilation of FP Programs

The IEEE-754 standard has improved the *portability* and *reproducibility* of the results of FP programs by mandating that every basic FP operation be performed as if the result was first computed with infinite precision and then rounded to the target format (see section 1.1.4). In particular, most hardware components providing FP arithmetic are nowadays compliant with this standard. Unfortunately, this still does not guarantee that an FP program will behave in a reproducible way, as both the programming language and the compiler can interfere with the proper execution of FP arithmetic.

Algorithm 7.1 Opposite of Fast2Sum of 1 and $2^{-53} + 2^{-78}$

```
int main() {
  double x, y, z;
  x = 1.;
  y = 0x1p-53 + 0x1p-78;    // y = 2⁻⁵³ + 2⁻⁷⁸
  z = x + y - x - y;        // parsed as ((x + y) - x) - y
  printf("%a\n", z);
  return 0;
}
```

In the code comment: `// ` $y = 2^{-53} + 2^{-78}$

Compilation options	Program result
-O0 (x86-32)	-0x1p-78
-O0 (x86-64)	0x1.ffffffp-54
-O1, -O2, -O3	0x1.ffffffp-54
-Ofast	0x0p+0

Table 7.1. *Results of algorithm 7.1 depending on the processor and compiler options.*

Algorithm 7.1 is a small example in C illustrating how the compilation process might impact the code written by the user. The example implements the opposite of the Fast2Sum algorithm (see section 5.1.3.1) and applies it to the inputs $x = 1$ and $y = 2^{-53} + 2^{-78}$. Since $|x| \geq |y|$, the program is expected to output the opposite z of the rounding error of the FP sum of x and y in *binary64* with rounding to nearest, tie breaking to even. Unfortunately as seen in Table 7.1, this very simple program compiled with GCC 6.2.0 gives three different answers on an x86 architecture depending on the instruction set and the chosen level of optimization. Of these three, only one corresponds to the expected output: 0x1.ffffffp-54.

The root cause for the discrepancies in the first three rows of the result table lies in the x86 architecture. Indeed, nowadays this architecture provides two FP units: one performs *binary32* and *binary64* computations, while the other one performs FP computations using 80-bit numbers (64 bits of precision), as detailed in section 1.4.2. So the result will depend on which FP unit and which format the compiler selects for performing each FP operation. The compiler may thus choose to round the infinitely-precise result either to extended precision, or to double precision, or first to extended and then to double precision. Note that, in all cases, y is computed exactly, so $y = 2^{-53} + 2^{-78}$ and the compilation choices only impact the computation of z. With the -O0 optimization level for the 32-bit instruction set, all the computations are performed with extended precision and rounded in double precision only once at the end. With the -O0 optimization level for the 64-bit instruction set, all the computations are performed with double precision. With level -O1 and higher, the intermediate value $\circ[(x + y) - x]$ is precomputed by the compiler as if performed with double precision; the program effectively computes only the last subtraction and the result does not depend on the instruction set.

With level -Ofast, the architecture no longer matters, since the program performs no FP computation; the executable code only outputs the constant 0. Indeed, this optimization level turns -funsafe-math-optimizations on, which allows for the reorganization of FP operations. Hence, GCC dutifully reorganizes $\circ[((x + y) - x) - y]$ into $\circ[(x + y) - (x + y)]$, then further simplifies it to 0. It is explicitly stated in GCC documentation that this option "can result in incorrect output for programs which depend on an exact implementation of IEEE or ISO rules/specifications for math functions".

If the code was using not only additions but also multiplications, we could experience yet another source of nonreproducibility, due to the potential use of the *fused multiply-add* operator. As an example, let us see how a C compiler may implement the FP evaluation of a * b + c * d. When an FMA is available, the various compilation schemes may lead to three different results: $\circ(\circ(a \cdot b) + \circ(c \cdot d))$, $\circ(a \cdot b + \circ(c \cdot d))$, and $\circ(c \cdot d + \circ(a \cdot b))$. Many more examples of strange FP behaviors can be found [MON 08, MUL 10].

As surprising as it may seem, discrepancies caused by the choice of the target format for each operation or by the use of the FMA operation are allowed by the ISO C standard [ISO 11], which leaves much freedom to the compiler in the way it implements FP computations. section 7.1 presents the semantics of FP arithmetic in various programming languages. These semantics happen to be vague and obscure, so that vastly incompatible architectures could be supported at the time these programming languages were first designed.

Sometimes, optimizing compilers take additional liberties with the source programs, generating executable code that exhibits behaviors not allowed by the specification of the source language, as was shown when using the -Ofast optimization level in the above example. If the user had not explicitly passed this option to the compiler, it could even be considered a case of *miscompilation*.

Algorithm 7.2 Example from GCC bug #323

```
void test(double x, double y) {
  const double y2 = x + 1.0;
  if (y != y2) printf("error\n");
}

int main() {
  const double x = .012;
  const double y = x + 1.0;
  test(x, y);
  return 0;
}
```

Consider algorithm 7.2 as another example of miscompilation. It is adapted from GCC's infamous bug 323.[1] For an x86 32-bit target at optimization level -O1, all versions of GCC prior to 4.5 miscompile this code as follows: the expression x + 1.0 in function test is computed in extended precision, as allowed by C, but the compiler omits to round it back to double precision when assigning to y_2, as prescribed by the C standard. Consequently, y and y_2 compare different, while they must be equal according to the C standard.

As the compiler gives so few guarantees on how it implements FP arithmetic, it therefore seems impossible to guarantee the result of a program, since most analyses of FP programs assume correct compilation and a strict application of the IEEE-754 standard where neither extended registers nor FMA are used. To take into account all the modifications the compiler might have performed on a code, a possible solution is

---

1 "Optimized code gives strange FP results", http://gcc.gnu.org/bugzilla/show_bug.cgi?id=323

to work at the level of the generated assembly code rather than the higher-level source language [NGU 11]. Another approach is to cover all the ways a compiler may have compiled each FP operation and to compute an error bound that stands correct whatever the compiler choices (see section 8.4.1); this however defeats the point of most clever uses of FP arithmetic presented in Chapters 5 and 6.

This chapter presents the CompCert C compiler as a way to ensure that any FP analysis or formal proof performed at the source level is still meaningful once the program has been compiled. Indeed, not only has this compiler been formally verified in Coq, so as to offer strong guarantees on the behavior of the generated assembly code with respect to the original C code [LER 09], but it also preserves most properties of an FP program in the resulting executable file by giving a much more precise and useful semantics to FP operations than mandated by C [BOL 15c]. Section 7.2 presents the compiler and focuses on a few aspects of the formalization and handling of FP arithmetic in CompCert such as parsing and output of FP constants, and constant propagation. Most of these aspects are just an application of the concepts presented in Chapter 3. For conversions from integers to FP numbers and conversely, the situation is slightly more complicated as processors do not always offer primitive operations to perform them, so the compiler has to emit code that emulates them, as shown in section 7.3.

7.1. Semantics of languages and FP arithmetic

Assume that a programmer has written an FP algorithm with IEEE-754-compliant operations in mind, the goal being to get portable code and reproducible results. The programmer then chooses a high-level programming language, since assembly languages would defeat the point of portability. Unfortunately, high-level language semantics are often rather vague with respect to FP operations, so as to account for as many execution environments as possible, even non-IEEE-754-compliant ones. So the programmer has to make some assumptions on how compilers will interpret the program. Unfortunately, different compilers may take different paths while still being compliant with the language standard, or they might depart from the standard for the sake of execution speed (possibly controlled by a compilation flag). Finally, the operating system and various libraries have a role to play too, as they might modify the default behavior of FP units or emulate features not supported in hardware, e.g. subnormal numbers.

This section gives an overview of some of the possible semantics of the resulting program through the lens of three major programming languages: Java (section 7.1.1), C (7.1.2), and Fortran (7.1.3).

7.1.1. *Java*

Java, being a relatively recent language, started with the most specified description of FP arithmetic. It proposes two data types that match the *binary32* and *binary64* formats of IEEE-754. Moreover, arithmetic operators are mapped to the corresponding operators from IEEE-754, but rounding modes other than default are not supported, and neither are the override of exceptional behaviors (e.g. terminating a program on FP overflow rather than producing an infinity in rounding to nearest). The latter is hardly ever supported by languages so we will not focus on it in the remainder of this chapter.

Unfortunately, a non-negligible part of the architectures the Java language was targeting only had access to x87-like FP units, which make it possible to set the precision of computation but not the allowed range of exponents. Thus these FP units behave as if they were working with exotic FP formats that have the usual IEEE-754 precision but an extended exponent range. On such architectures, complying with the Java semantics was therefore highly inefficient. As a consequence, the language later evolved and the FP semantics were relaxed to account for a potential extended exponent range:

> Within an expression that is not FP-strict, some leeway is granted for an implementation to use an extended exponent range to represent intermediate results. (15.4 FP-strict expressions, Java SE 7)

For programs that were dependent on the early IEEE-754-compliant behavior, the Java language specification introduced a `strictfp` keyword that can be used to annotate classes and methods.

7.1.2. *C*

The C language comes from a time where FP units were more exotic, so the wording of the standard leaves much more liberty to the compiler. Intermediate results can not only be computed with an extended range, they can also have an extended precision.

> The values of operations with floating operands [...] are evaluated to a format whose range and precision may be greater than required by the type. (5.2.4.2.2 Characteristics of floating types, C11)

In fact, most compilers interpret the standard in an even more relaxed way: values of local variables that are not spilled to memory might preserve their extended range and precision.

Note that this optimization opportunity also applies to the use of an FMA operator for computing the expression $a \cdot b + c$, as the intermediate product is then performed with a much greater precision.

While Annex F of the C standard allows a compiler to advertise compliance with IEEE-754 FP arithmetic if it supports a specified set of features, none of these features reduces the leeway compilers have in choosing intermediate formats. Section 8.4 gives some ways of defending against such compilers.

7.1.3. Fortran

The Fortran language gives even more leeway to compilers, allowing them to rewrite expressions as long as they do not change the value that would be obtained if the computations were performed with an infinitely-precise arithmetic.

Two expressions of a numeric type are mathematically equivalent if, for all possible values of their primaries, their mathematical values are equal. (7.1.5.2.4 Evaluation of numeric intrinsic operations, Fortran 2008)

The standard, however, forbids such transformations when they would violate the "integrity of parentheses". For instance, (a + b) - a - b can be rewritten as 0, but ((a + b) - a) - b cannot, since it would break the integrity of the outer parentheses.

This allowance for assuming FP operations to be associative and distributive has unfortunately contaminated compilers for other languages, which do not even have the provision of preserving parentheses. For instance, the seemingly innocuous -Ofast option of GCC will enable this optimization for the sake of speed, at the expense of the conformance with the C standard, as can be seen in Table 7.1.

7.2. Verified compilation

Ordinary compilers sometimes *miscompile* source programs: starting with a correct source, they can produce executable machine code that crashes or computes wrong results. Formally verified compilers come with a mathematical proof of *semantic preservation* that rules out all possibilities of miscompilation. Intuitively, the semantic preservation theorem says that the executable code produced by the compiler always executes as prescribed by the semantics of the source program. In the following, we study one such formally verified compiler: the CompCert C compiler [LER 09], with special emphasis on its handling of FP computations. Section 7.2.1 gives an overview of what the preservation theorem looks like.

Part of the context of this theorem covers how FP arithmetic is supported by CompCert. Section 7.2.2 gives an overview of this specification. The following properties can be inferred from the specification [BOL 15c]:

– The float and double types are mapped to the *binary32* and *binary64* formats, respectively. Extended-precision FP numbers are not supported: the long double type is either unsupported or mapped to *binary64*, depending on a compiler option;

– Conversions to an FP type, either explicit ("type casts") or implicit (at assignment, parameter passing, and function return), always round the FP value to the given FP type, discarding excess precision;

– Reassociation of FP operations, or "contraction" of several operations into one (e.g. a multiplication and an addition being contracted into a fused multiply-add) are prohibited. On target platforms that support FMA instructions, CompCert makes them available as compiler built-in functions, so they must be explicitly used by the programmer;

– All intermediate FP results in expressions are computed with, the precision that corresponds to their static C types, that is, the largest of the precisions of their arguments;

– All FP computations round to nearest, tie breaking to even, except conversions from FP numbers to integers, which round toward zero. The CompCert formal semantics make no provisions for programmers to change rounding modes at run-time;

– FP literal constants are also rounded to nearest, tie breaking to even.

Once FP arithmetic has been specified and the semantic preservation theorem has been stated, one has to actually write (and formally verify) a compiler that satisfies this theorem. Section 7.2.3 indicates how parsing and generation of FP literals is handled. Section 7.2.4 gives some examples of code generation for handling the comparison of FP numbers. Finally, section 7.2.5 shows how some code optimizations related to FP numbers are handled by the compiler.

7.2.1. A verified C compiler: CompCert

Before proving a semantic preservation theorem, we must make its statement mathematically precise. This entails

1) specifying precisely the program transformations (compiler passes) performed by the compiler;

2) giving mathematical semantics to the source and target languages of the compiler (in the case of CompCert, the CompCert subset of ISO C99 and ARM/PowerPC/x86 assembly languages, respectively).

The semantics used in CompCert associate *observable behaviors* with every program. Observable behaviors include normal termination, divergence (the program runs forever), and encountering an undefined behavior (such as an out-of-bounds array access). They also include traces of all input/output operations performed by the program: calls to I/O library functions (such as printf) and accesses to volatile memory locations. For example, the Fast2Sum program shown at the start of the chapter is predicted, by CompCert's C semantics, to have a single behavior, namely, normal termination with exit code 0 and a trace consisting of a single observable event: a call to `printf` with argument `0x1.ffffffp-54`.

Equipped with these formal semantics, one can state precisely the desired semantic preservation results. Here is one such result that is proved in CompCert:

Theorem 7.1 (Semantic preservation). Let S be a source C program. Assume that S is free of undefined behaviors. Further assume that the CompCert compiler, invoked on S, does not report a compile-time error, but instead produces executable code E. Then any observable behavior of E is one of the possible observable behaviors of S.

The statement of the theorem leaves two important degrees of freedom to the compiler. First, a C program can have several legal behaviors, owing to underspecification in expression evaluation order, and the compiler is allowed to pick any one of them. Second, undefined behaviors need not be preserved during compilation, as the compiler can optimize them away. This is not the only possible statement of semantic preservation: indeed, CompCert proves additional, stronger statements that imply the theorem above [LER 09]. The bottom line, however, is that the correctness of a compiler can be characterized in a mathematically-precise, yet intuitively understandable way, as soon as the semantics of the source and target languages are specified.

7.2.2. Formalization of FP arithmetic

With respect to FP arithmetic, three pieces need to be formalized in order to specify and verify the behavior of the compiler. First, the FP formats have to be formalized as well as some basic arithmetic operations. Second, a semantics of the assembly opcodes of the target processors has to be specified, using these formalized formats and operators. Third, a semantics of the C language has to be specified, using the same formats and operators. Note that the semantics of the C language used for the compiler does not have to be as underspecified as the C standard; some *undefined* and *implementation-defined* behaviors (according to the standard) might be fully specified by a compiler. To illustrate these three pieces of CompCert, we will be using the sum of two values of type `float` as a running example.

The Coq formalization used for the binary FP formats and operations is the one described in section 3.4, that is, an inductive type that supports signed zeros, signed infinities, finite numbers, and NaNs with a payload. The `float32` type is an instance of this inductive type for representing *binary32* FP numbers, while `float` represents *binary64* numbers. These types are then embedded into the set of values supported by the compiler through two dedicated constructors `Vsingle` and `Vfloat`:

```
Inductive val : Type :=
  | Vundef : val    (* result of an invalid operation *)
```

```
| Vint : int -> val
| Vlong : int64 -> val
| Vfloat : float -> val    (* binary64 number *)
| Vsingle : float32 -> val  (* binary32 number *)
| Vptr : block -> int -> val.
```

The FP addition of two float32 values f_1 and f_2 is denoted Float32.add f1 f2 in CompCert specifications. Note that, with respect to the addition defined in section 3.4.3, two arguments are missing: the rounding mode and the behavior in case of NaN. The mode is implicitly set to rounding to nearest, tie breaking to even, as this is the default rounding mode in C and CompCert offers no way to modify it on the fly. The case of NaNs is slightly more complicated, as their handling depends on the architecture: not all processors behave in the same way. Indeed, the IEEE-754 standard is underspecified when it comes to NaNs. So each architecture supported by CompCert has a customized specification to cover the following cases:

1) When an operation produces a NaN (e.g. $\infty - \infty$), what is its payload?

2) How are signaling and quiet NaNs encoded and how does an operation transform an input signaling NaN into an output quiet NaN?

3) When both operands are NaNs, which one is returned by the operation.

The behavior of a *binary32* addition over values of type val can then be defined as follows. This operation is meant to be used only when the two input operands are *binary32* numbers, so the Vundef value is returned in all the other cases.

```
Definition addfs (v1 v2 : val) : val :=
  match v1, v2 with
  | Vsingle f1, Vsingle f2 => Vsingle (Float32.add f1 f2)
  | _, _ => Vundef
  end.
```

The instruction set of the target processor can now be specified. First, instructions corresponding to FP operations are added to the set of opcodes. For instance, on the PowerPC architecture supported by CompCert, the fadds instruction takes three FP registers; the first one is the destination register while the last two are the source registers. All these registers are part of the FP bank of registers, which is denoted by the type freg. Thus the set of opcodes is defined as the following inductive type, which contains a constructor for denoting fadds:

```
Inductive instruction : Type :=
| Pfadds : freg -> freg -> freg -> instruction
| ...
```

Then the CompCert specification of the PowerPC states how such an instruction is executed by the processor; that is, reads the content of the input registers, it performs a *binary32* addition as previously formalized (rounding to nearest, tie breaking to even, with PowerPC-specific handling of NaNs), writes the value to the target register, and moves to the next instruction.

```
Definition exec_instr (f : function) (i : instruction)
    (rs : regset) (m : mem) : outcome :=
  match i with
  | Pfadds rd r1 r2 =>
    Next (nextinstr (rs#rd <- (addfs rs#r1 rs#r2))) m
  | ...
  end.
```

Finally, it remains to specify how the C addition should be performed. More generally, given two C expressions e_1 and e_2 of respective types ty_1 and ty_2, CompCert first classifies the binary operator to know whether it is an integer operation or an FP operation or a pointer operation. For instance, the bin_case_s constructor denotes an FP operation to be performed using *binary32* arithmetic. Then both input operands are converted via casts to a common type ty, which in this case is float. Finally, the corresponding version of the operator (*sop* for a *binary32* operator) is executed. In the case of a *binary32* addition, a function that behaves as addfs is passed as the *sop* argument.

```
Definition make_binarith (... sop : binary_operation)
    (e1 e2 : expr) (ty1 ty2 : type) :=
  let c := classify_binarith ty1 ty2 in
  let ty := binarith_type c in
  do e1' <- make_cast ty1 ty e1;
  do e2' <- make_cast ty2 ty e2;
  match c with
  | bin_case_s => OK (Ebinop sop e1' e2')
  | ...
  end.
```

Note that, while both the specification of the C language and that of the PowerPC instruction set use the same definition of what an addition of two *binary32* numbers is, this does not mean that an FP addition in the source code will necessarily be translated into an fadds PowerPC opcode. The theorem of semantic preservation only implies that, whichever sequence of opcodes the compiler chooses, its result is indistinguishable from the result of the fadds opcode. In particular, the compiler might have elided such an addition in case of optimization (section 7.2.5).

7.2.3. Parsing and output of FP constants

The evaluation of FP literals is delicate: literals are often written in decimal, requiring nontrivial conversion to an IEEE-754 binary format. For example, the strtod and strtof functions of the GNU C standard library incorrectly rounded the result in some corner cases.[2] To avoid these pitfalls, CompCert uses a simple but correct Flocq-based algorithm for evaluating these literals [BOL 15c]. In C, an FP literal consists of an integral part, a fractional part, an exponent part, and a precision suffix which indicates at which precision the literal should be evaluated. Each of these parts can be omitted, in which case 0 is used as default value for the first three parts. (This operation is done in an early stage of parsing in CompCert.) The integral and fractional parts may be written in either decimal or hexadecimal notation; the use of hexadecimal (in both parts) is indicated if the integral part begins with the prefix "0x". The exponent is given as a power of 2 if hexadecimal is used, and as a power of 10 if decimal is used.

CompCert's parsing algorithm first shifts the separator between the integer part and the fractional part to the right, while modifying the exponent, so that only an integer part is left. At this stage, the number is thus of the form $m \cdot \beta^e$ with m and e being arbitrary-precision integers. The second step consists of actually evaluating the FP number, using Flocq with the target format specified by the precision suffix. When $e \geq 0$, the algorithm computes the integer $m \times \beta^e$, then rounds the result to the nearest representable FP number, tie breaking to even. When $e < 0$, the algorithm first computes the integer β^{-e}, then performs the FP division $\circ(m/\beta^{-e})$ using Flocq (see sections 3.3.2 and 3.4.3). Notice that, since Flocq can formalize a multi-precision arithmetic, the numbers m and β^{-e} do not have to fit into the target format; the division can cope with arbitrarily large numbers. CompCert's algorithm also detects whether e is so large that the literal would be rounded to infinity, or whether e is so small that the literal would be rounded to zero, so as to avoid computing uselessly large powers of β.

Once the compilation process is over, CompCert has to perform the converse operation: converting FP constants to bit-patterns suitable for emission in the generated assembly code, that is, integers. For this purpose, CompCert uses the function bits_of_binary_float described in section 3.4.2. This function takes a binary_float number and returns the Z integer that encodes it, which can then be output as any other integer. Note that if the functions provided by Flocq had not been effective, they could not have been used by CompCert to parse and output FP constants.

---

2 "Incorrect rounding in strtod", http://sourceware.org/bugzilla/show_bug.cgi?id= 3479

7.2.4. *Code generation*

Most FP operations of the C language map directly to hardware-implemented instructions of the target platforms. Some operations, however, are not directly supported by some target platforms, forcing the compiler to implement these operations using sometimes convoluted combinations of other instructions [BOL 15c]. The correctness of these code generation strategies depends on the validity of algebraic identities over FP operations. Comparisons between FP numbers will serve as an illustration.

The PowerPC architecture provides an fcmp instruction that produces 4 bits of output: "less than", "equal", "greater", and "uncomparable", and conditional branch instructions that test any one of these bits. To compile a non-strict inequality test such as "less than or equal", CompCert produces code that performs the logical "or" of the "less than" and "equal" bits, then conditionally branches on the resulting bit. Semantically, this is justified by the identity $(x \leq y) \equiv (x < y \lor x = y)$, which holds for any two FP numbers x and y. Note that, even if two NaNs have the same sign/payload and thus are equal from the mathematical point of view of Coq, the comparison operators defined by the compiler still make them unordered.

On the x86 architecture, the comisd SSE2 instruction sets the ZF, CF, and PF condition flags in such a way that only the following relations (and their negations) between the operands x and y can be tested by a single conditional branch instruction:

– x and y are unordered or x == y (instruction je; and jne for the negated test);

– x >= y (jae, jb);

– x > y (ja, jbe);

– x and y are unordered (jp, jnp).

Therefore, on x86, a branch on equality x == y must be compiled as a comparison comisd followed by two conditional branches:

$$\text{if } (x == y) \text{ goto } L; \quad \rightarrow \quad \begin{array}{l} \text{comisd x, y} \\ \text{jp } L' \\ \text{je } L \end{array}$$
$$L' :$$

The first jp conditional branch handles the case where x and y are unordered. In this case, they compare as not equal, so the code branches to a label L' following the code. Once the unordered case is taken care of, the x86 condition e, meaning "equal or unordered", coincides with "equal", thus the je conditional branch to L implements the correct behavior.

The disequality is handled similarly, with the unordered case first and then the actual comparison:

```
if (x != y) goto L;     →      comisd x, y
                               jp L
                               jne L
```

Likewise, branches on x < y or x <= y can be compiled as comisd x y followed by two conditional branches (jp-jb and jp-jbe, respectively). One conditional branch however suffices if the compiler reverses the operands of the comisd instruction and tests y > x or y >= x instead. Again, the soundness of these code generation tricks follows from semantic properties of FP comparisons that are easily verified in Coq, namely, $(x < y) \equiv (y > x)$, $(x \leq y) \equiv (y \geq x)$, and the fact that the four outcomes of an FP comparison (less, equal, greater, unordered) are mutually exclusive.

In CompCert, the most convoluted code generation schemes for FP operations are found in the conversions between integers and FP numbers. This will be the topic of section 7.3.

7.2.5. Code optimization

While the CompCert compiler is in no way as optimizing as compilers that have not been formally verified, it nonetheless tries to produce non-naive assembly code. When it comes to FP code, improvements come from performing some constant propagation and some algebraic simplifications [BOL 15c].

7.2.5.1. Constant propagation

Constant propagation is a basic but important optimization in compilers. It consists of evaluating, at compile time, arithmetic and logical operations whose arguments can be statically determined. For instance, the Fast2Sum example of the introduction is reduced to the printing of a single constant; no FP operations are performed by the executable code. For another example, consider the following C code fragment:

```
inline double f(double x) {
  if (x < 1.0) return 1.0;
  else return 1.0 / x;
}
double g(void) {
  return f(3.0);
}
```

Combining constant propagation with function inlining, the body of function g is optimized into return 0x1.5555555555555p-2. Not only the division 1.0 / x but also the conditional statement x < 1.0 have been evaluated at compile time. These computations are performed using the functions described in section 3.4.3, making them independent of the FP arithmetic of the host platform running the compiler, and guaranteeing that the constant propagation optimization preserves the semantics of the source program. Note that, as in section 7.2.3, if the operations from section 3.4.3 had been described in an abstract non-algorithmic way, they would have been unusable for writing this compiler pass.

The CompCert C semantics does not provide a way for programmers to change the FP rounding mode during program execution, therefore the compiler can assume that all the FP arithmetic operations round to nearest, tie breaking to even. Otherwise constant propagation, as described above, would not satisfy the semantic preservation theorem on architectures that have dynamic rounding modes, since constant propagation might give a different result to actual execution. Therefore, programs that need other rounding modes fall outside the perimeter of CompCert's semantic preservation results. They can, however, be supported via the compiler option -ffloat-const-prop 0, which turns FP constant propagation off. A more general solution would be to extend CompCert so that it supports a dynamic mode, e.g. by representing it as a pseudo global variable. Constant propagation would then only happen if either the rounding mode is statically known, or if the result would be the same whatever the mode.

7.2.5.2. Algebraic simplifications

For integer computations, compilers routinely apply algebraic identities to generate shorter instruction sequences and use cheaper instructions. Examples include reassociation and distribution of constants (e.g. $(n - 1) \times 8 + 4$ becomes $n \times 8 - 4$); multiplication by certain constants being transformed into shifts, additions, and subtractions (e.g. $n \times 7$ becomes $(n \ll 3) - n$); and divisions by constants being replaced by multiplications and shifts [GRA 94].

For FP computations, there are much fewer opportunities for compile-time simplifications. The reason is that very few algebraic identities hold over FP arithmetic operations for all possible values of their FP arguments. CompCert implements two modest FP optimizations based on such identities. Note that, in the following description of the optimizations, $\oplus$, $\ominus$, $\otimes$, and $\oslash$ denote fully-defined FP arithmetic operations, including the way in which they handle exceptional inputs.

The first optimization is the replacement of *binary64* divisions with multiplications if the divisor is an exact power of 2:

$$x \oslash 2^n \quad \rightarrow \quad x \otimes 2^{-n} \quad \text{if } |n| < 1023.$$

This optimization is valuable, since FP multiplication is usually much faster than FP division. Such a replacement works for a handful of divisors other than powers of two [BRI 08], but they are not supported by CompCert.

A second optimization replaces FP multiplications by 2 with FP additions:

$$x \otimes 2 \quad \rightarrow \quad x \oplus x,$$
$$2 \otimes x \quad \rightarrow \quad x \oplus x.$$

FP multiplication and addition take about the same time on modern processors, but the optimized form avoids the cost of loading the constant 2 into an FP register.

As simple as the optimizations above are, their correctness proof in the case where x is a NaN already requires additional hypotheses about the payloads produced by FP operations. These hypotheses are, fortunately, satisfied on the three architectures targeted by CompCert.

Several other plausible FP optimizations are regrettably unsound for certain values of their arguments:

$$x \oplus (+0) \quad \nrightarrow \quad x, \qquad\qquad\qquad [7.1]$$
$$x \oplus (-0) \quad \nrightarrow \quad x, \qquad\qquad\qquad [7.2]$$
$$x \ominus (+0) \quad \nrightarrow \quad x, \qquad\qquad\qquad [7.3]$$
$$x \ominus (-0) \quad \nrightarrow \quad x, \qquad\qquad\qquad [7.4]$$
$$-(-x) \quad \nrightarrow \quad x, \qquad\qquad\qquad [7.5]$$
$$(-x) \oplus y \quad \nrightarrow \quad y \ominus x, \qquad\qquad\qquad [7.6]$$
$$y \oplus (-x) \quad \nrightarrow \quad y \ominus x, \qquad\qquad\qquad [7.7]$$
$$y \ominus (-x) \quad \nrightarrow \quad y \oplus x. \qquad\qquad\qquad [7.8]$$

Viewed as algebraic identities, equations [7.1] and [7.4] do not hold for $x = -0$; [7.2], [7.3], and [7.5] do not hold if x is a signaling NaN; and the two sides of [7.6], [7.7], and [7.8] produce NaNs of different signs if x (the negated argument) is NaN and y (the other argument) is not NaN.

The only way to exploit simplifications such as [7.1]–[7.8] above while preserving semantics is to apply them conditionally, based on the results of a static analysis that can exclude the problematic cases. As a trivial example of static analysis, in the then branch of a conditional statement if (x >= 1.0), we know that x is neither NaN nor

−0 nor subnormal, therefore optimizations [7.1]–[7.8] are sound. Another possibility would be for CompCert to support options similar to GCC's -ffinite-math-only, -fno-signed-zeros, and so on.

7.3. Conversions between integers and FP numbers

In the C language, conversions between FP and integer numbers occur either explicitly (*type casts*) or implicitly (during assignments and function parameter passing). There are many such conversions to implement. In the case of CompCert 2.0, there are four integer types and two FP types to consider, for a total of 8 integer-to-FP conversions and 8 FP-to-integer conversions. We write *dst_src* for the conversion from type *src* to type *dst*. The types of interest are, on one side, the FP formats f32 and f64, and, on the other side, the integer types s32, u32, s64, and u64 (signed or unsigned, 32 or 64 bits). For example, f64_u32 is the conversion from 32-bit unsigned integers to *binary64* FP numbers.

from \ to	f32	f64	f80
s32	AS	AS	X
u32	A	A	−
s64	−	−	X
u64	−	−	−

from \ to	s32	u32	s64	u64
f32	AS	A	−	−
f64	APS	A	−	−
f80	X	−	X	−

Table 7.2. *Conversions between integers and FP numbers that are natively supported on three processor architectures. A stands for ARM with VFPv2 or higher; P for PowerPC 32 bits; S for the SSE2 instructions of x86 in 32-bit mode; and X for the x87 extended-precision instructions of x86.*

Table 7.2 summarizes which of these conversions are directly supported by hardware instructions for each of CompCert's three target architectures. For completeness, we also list the conversion instructions that operate over the f80 extended-precision format of the x86 architecture. As shown by this table, none of CompCert's target architectures provides instructions for all 16 conversions. The PowerPC 32-bit architecture is especially unhelpful in this respect, providing only one FP-to-integer conversion (s32_f64) and no integer-to-FP conversions.

Any conversions not provided by a processor instruction must, therefore, be synthesized by the compiler as sequences of other instructions. These instruction sequences are often nonobvious, and their correctness proofs are sometimes delicate. This section lists some implementation schemes used by CompCert, plus several schemes observed in the code generated by GCC version 4. All of these schemes have been formally proved in Coq [BOL 15c].

Note that the C language leaves as undefined the behavior of converting a NaN or an infinity or an out-of-range value to an integer. Moreover, when the input values are in range, they are small enough that no overflow can occur during conversion. As a consequence, none of the following algorithms for converting an FP number to an integer can experience an exceptional behavior, unless we are in the case of an undefined behavior according to the C language. Similarly, when converting from an integer to an FP value, no overflow can occur either. Thus, while the formal proofs do take into account all the issues related to exceptional values, these issues will be ignored when describing how the various algorithms behave.

7.3.1. *From 32-bit integers to FP numbers*

Let us first focus on the conversion of 32-bit integers to FP numbers. Among CompCert's target architectures, only ARM VFP provides instructions for all four conversions f64_s32, f64_u32, f32_s32, and f32_u32, as can be seen from Table 7.2. The x86 architecture provides one SSE2 instruction for the f64_s32 conversion. Its unsigned counterpart, f64_u32, can be synthesized from f64_s32 by a case analysis that reduces the integer argument to an integer in the range $[0; 2^{31})$:

$$f64\_u32(n) = \text{if } n < 2^{31} \qquad\qquad [7.9]$$
$$\text{then } f64\_s32(n)$$
$$\text{else } f64\_s32(n - 2^{31}) \oplus 2^{31}.$$

Both the f64_s32 conversion and the FP addition in the else branch are exact, the latter because it is performed using *binary64* arithmetic.

Conversions from 32-bit integers to *binary32* are trivially implemented by first converting to *binary64*, then rounding to *binary32*:

$$f32\_s32(n) \quad = \quad f32\_f64(f64\_s32(n)),$$
$$f32\_u32(n) \quad = \quad f32\_f64(f64\_u32(n)).$$

The inner conversion is exact and the outer f32_f64 conversion rounds the result according to the current rounding mode, as prescribed by the IEEE-754 standard and the ISO C standards, appendix F.

The x87 extended-precision FP instructions provide the following alternative implementations, where the inner conversion is also exact:

$$f64\_s32(n) \quad = \quad f64\_f80(f80\_s32(n)),$$
$$f64\_u32(n) \quad = \quad f64\_f80(f80\_s64(n)),$$
$$f32\_s32(n) \quad = \quad f32\_f80(f80\_s32(n)),$$
$$f32\_u32(n) \quad = \quad f32\_f80(f80\_s64(n)).$$

These alternative instruction sequences however can involve more data transfers through memory than the SSE2 implementations described above.

The PowerPC 32-bit architecture offers a bigger challenge, since it fails to provide any integer-to-FP conversion instruction. The PowerPC compiler writer's guide [IBM 96] describes the following software implementation, based on bit-level manipulations over the *binary64* format combined with a regular FP subtraction:

$$\mathsf{f64\_u32}(n) \quad = \quad \mathsf{f64make}(\mathsf{0x43300000}, n) \ominus 2^{52}, \qquad\qquad [7.10]$$

$$\mathsf{f64\_s32}(n) \quad = \quad \mathsf{f64make}(\mathsf{0x43300000}, n + 2^{31}) \ominus (2^{52} + 2^{31}). \qquad [7.11]$$

We write $\mathsf{f64make}(h, \ell)$, where h and ℓ are 32-bit integers, for the *binary64* FP number whose in-memory representation is the 64-bit word obtained by concatenating h with ℓ. This $\mathsf{f64make}$ operation can easily be implemented by storing h and ℓ in two consecutive 32-bit memory words, then loading a *binary64* FP number from the address of the first word.

The reason why this clever implementation produces correct results is that the *binary64* number $A = \mathsf{f64make}(\mathsf{0x43300000}, n)$ is equal to $2^{52} + n$ for any integer $n \in [0; 2^{32})$. Thus,

$$\begin{aligned}
A \ominus 2^{52} \quad &= \quad \circ\big((2^{52} + n) - 2^{52}\big) && \text{(no exceptional values)} \\
&= \quad \circ(n) \; = \; n. && (n \text{ is representable})
\end{aligned}$$

Likewise, the *binary64* number $B = \mathsf{f64make}(\mathsf{0x43300000}, n + 2^{31})$ is equal to $2^{52} + n + 2^{31}$ for any integer $n \in [-2^{31}; 2^{31})$. Thus,

$$B \ominus \big(2^{52} + 2^{31}\big) = \circ\big((2^{52} + n + 2^{31}) - \big(2^{52} + 2^{31}\big)\big) = \circ(n) = n.$$

7.3.2. *From 64-bit integers to FP numbers*

None of CompCert's target architectures provide instructions for performing the conversions from 64-bit integers to FP numbers: $\mathsf{f64\_s64}$, $\mathsf{f64\_u64}$, $\mathsf{f32\_s64}$, and $\mathsf{f32\_u64}$. The closest equivalent is the $\mathsf{f80\_s64}$ conversion instruction found in the x87 subset of the x86 32-bit architecture, which gives the following implementations:

$$\begin{aligned}
\mathsf{f64\_s64}(n) \quad &= \quad \mathsf{f64\_f80}(\mathsf{f80\_s64}(n)), \\
\mathsf{f64\_u64}(n) \quad &= \quad \mathsf{f64\_f80}(\text{if } n < 2^{63} \\
&\qquad\qquad\quad \text{then } \mathsf{f64\_s64}(n) \\
&\qquad\qquad\quad \text{else } \mathsf{f64\_s64}(n - 2^{63}) \oplus 2^{63}),
\end{aligned}$$

and similarly for f32_s64 and f32_u64, replacing the final f64_f80 rounding by f32_f80. Since the 80-bit extended-precision FP format of the x87 has a 64-bit significand, it can exactly represent any integer in the range $(-2^{64}; 2^{64})$. Hence, all the FP computations in the formulas above are exact, except the final f64_f80 or f32_f80 conversions, which perform the correct rounding.

If the target architecture provides only conversions from 32-bit integers, it is always possible to convert a 64-bit integer by splitting it in two 32-bit halves, converting them, and combining the results. Writing $n = 2^{32} \cdot h + \ell$, where h and ℓ are 32-bit integers and ℓ is unsigned, we have

$$\text{f64\_s64}(n) \quad = \quad \text{f64\_s32}(h) \otimes 2^{32} \oplus \text{f64\_u32}(\ell), \qquad [7.12]$$

$$\text{f64\_u64}(n) \quad = \quad \text{f64\_u32}(h) \otimes 2^{32} \oplus \text{f64\_u32}(\ell). \qquad [7.13]$$

All operations are exact except the final FP addition, which performs the correct rounding. For the same reason, a fused multiply-add instruction can be used if available, without changing the result.

For the PowerPC 32-bit architecture, combining implementations [7.10] and [7.13] gives

$$\text{f64\_u64}(n) = \left(\text{f64make}(0\text{x43300000}, h) \ominus 2^{52}\right) \otimes 2^{32}$$
$$\oplus \left(\text{f64make}(0\text{x43300000}, \ell) \ominus 2^{52}\right).$$

A first improvement is to fold the multiplication by 2^{32} with the first f64make FP construction:

$$\text{f64\_u64}(n) = \left(\text{f64make}(0\text{x45300000}, h) \ominus 2^{84}\right)$$
$$\oplus \left(\text{f64make}(0\text{x43300000}, \ell) \ominus 2^{52}\right).$$

Indeed, just like $\text{f64make}(0\text{x43300000}, n) = 2^{52} + n$ for all 32-bit unsigned integers n, it is also the case that $\text{f64make}(0\text{x45300000}, n) = 2^{84} + n \cdot 2^{32}$.

One further improvement is possible:

$$\text{f64\_u64}(n) = \left(\text{f64make}(0\text{x45300000}, h) \ominus (2^{84} + 2^{52})\right)$$
$$\oplus \text{f64make}(0\text{x43300000}, \ell).$$

Indeed, $\text{f64make}(0\text{x45300000}, h)$ lies in the range $[2^{84}; 2^{84} + 2^{64})$, so it is within a factor of two of $2^{84} + 2^{52}$, which makes the FP subtraction exact (see section 5.1.1).

This leaves only the outer FP addition, which correctly rounds to the final result. A similar analysis for the signed integer case gives:

$$\texttt{f64\_s64}(n) = (\texttt{f64make}(\texttt{0x45300000}, h + 2^{31}) \ominus (2^{84} + 2^{63} + 2^{52}))$$
$$\oplus \texttt{f64make}(\texttt{0x43300000}, \ell).$$

Many of the implementation schemes for converting 32-bit integers to FP numbers listed in section 7.3.1 do not extend straightforwardly to the 64-bit case, because double rounding can occur (see section 1.4.2). For instance, assuming that the $\texttt{f64\_s64}$ conversion is available, it is not correct to define its unsigned counterpart $\texttt{f64\_u64}$ in the style of equation [7.9]:

$$\texttt{f64\_u64}(n) \neq \texttt{if } n < 2^{63}$$
$$\texttt{then f64\_s64}(n)$$
$$\texttt{else f64\_s64}(n - 2^{63}) \oplus 2^{63}.$$

Indeed, for some values of $n > 2^{63} + 2^{52}$, both the conversion $A = \texttt{f64\_s64}(n - 2^{63})$ and the FP addition $A \oplus 2^{63}$ are inexact, and the two consecutive rounding produce a result different from the correct single rounding of n to *binary64* format. For example, $n = 2^{63} + 2^{62} + 2^{10} + 1$ leads to $A = \circ(n - 2^{63}) = 2^{62} + 2^{10}$, which leads to $\circ(A + 2^{63}) = 2^{63} + 2^{62} \neq 2^{63} + 2^{62} + 2^{11} = \circ(n)$.

Looking at the assembly code generated by GCC for the PowerPC 64-bit architecture, we observe an elegant workaround for this issue:

$$\texttt{f64\_u64}(n) = \texttt{if } n < 2^{63} \qquad\qquad\qquad [7.14]$$
$$\texttt{then f64\_s64}(n)$$
$$\texttt{else } 2 \otimes \texttt{f64\_s64}((n \gg 1) \mid (n \,\&\, 1)).$$

This is an instance of the *round-to-odd* technique presented in section 1.4.3, since the computation $n' = (n \gg 1) \mid (n \,\&\, 1)$ has the effect of rounding $n/2$ to odd. Indeed, looking at the two low-order bits of n, we have

n	n'	
$4k$	$2k$	(even, but an exact quotient)
$4k + 1$	$2k + 1$	(quotient rounded up)
$4k + 2$	$2k + 1$	(exact quotient)
$4k + 3$	$2k + 1$	(quotient rounded down)

Therefore, in the case where $n \in [2^{63}; 2^{64})$, n' is the rounding to odd of $n/2$ to a precision of 63 bits, which is denoted $\square_{63}^{\text{odd}}(n/2)$. Thus, by applying theorem 3.47 with

$p = 53$ and $p + k = 63$, we can show that equation [7.14] is actually computing the following value:

$$\begin{aligned} 2 \otimes \text{f64\_s64}(n') &= \circ(2 \cdot \circ(\square_{63}^{\text{odd}}(n/2))) \\ &= \circ(2 \cdot \circ(n/2)) \\ &= \circ(n). \end{aligned}$$

Double rounding is also an issue when converting 64-bit integers to *binary32* FP numbers. Again, it is not correct to proceed as in the 32-bit case, simply converting to *binary64* then rounding to *binary32*:

$$\begin{aligned} \text{f32\_s64}(n) &\neq \text{f32\_f64}(\text{f64\_s64}(n)), \\ \text{f32\_u64}(n) &\neq \text{f32\_f64}(\text{f64\_u64}(n)). \end{aligned}$$

Indeed, for large enough values of n, the conversion to *binary64* is inexact, causing a double rounding error in conjunction with the subsequent rounding to *binary32*. For example, $n = 2^{54}+2^{30}+1$ leads to $\circ_{24}(\circ_{53}(n)) = \circ_{24}(2^{54}+2^{30}) = 2^{54} \neq 2^{54}+2^{31} = \circ_{24}(n)$.

Looking once more at the code generated by GCC for f32_u64 on PowerPC 64 bits, we observe another clever use of the round-to-odd technique:

$$\text{f32\_u64}(n) = \text{f32\_f64}(\text{f64\_u64}(\text{if } n < 2^{53} \text{ then } n \text{ else } n')), \qquad [7.15]$$

where $n' = (n \mid ((n\ \&\ \text{0x7FF}) + \text{0x7FF}))\ \&\ \sim\text{0x7FF}$.

When $n < 2^{53}$, the result of f64_u64(n) is exact and therefore a single rounding to *binary32* occurs. In the other case, unraveling the computation of n', we see that the least significant 11 bits of n' are 0; the most significant 52 bits of n' are identical to those of n; and the bit of n' of weight 2^{11} is equal to 0 if n is a multiple of 2^{12}, and is equal to 1 otherwise. Therefore, n' is n rounded to $q = \lceil \log_2 n \rceil - 11$ significant bits using round-to-odd mode. Since n' has only $q \leq 53$ significant bits, its conversion f64_u64(n') is exact. The correctness of [7.15] then follows from theorem 3.47, with $p = 24$ and $p + k = q \in [42; 53]$.

The same trick applies to the signed conversion f32_s64:

$$\text{f32\_s64}(n) = \text{f32\_f64}(\text{f64\_s64}(\text{if } |n| < 2^{53} \text{ then } n \text{ else } n')),$$

where n' is computed from n as in [7.15]. Owing to two's-complement representation of integers, the logical and arithmetic operations defining n' from n perform round-to-odd even if n is negative. Note that the then path is also correct if $|n| = 2^{53}$. GCC takes advantage of this fact by testing whether $-2^{53} \leq n < 2^{53}$, which can be done with only one conditional jump.

7.3.3. From FP numbers to integers

Conversions from FP numbers to integers are easier than the conversions described in sections 7.3.1 and 7.3.2. The general specification of FP-to-integer conversions, as given in the ISO C standards, is that they must round the given FP number f toward zero to obtain an integer. If the resulting integer falls outside the range of representable values for the target integer type (e.g. $[0; 2^{32})$ for target type u32), or if the FP argument is NaN, then the conversion has undefined behavior; it can produce an arbitrary integer result, but it can also abort the program.

CompCert's target architectures provide an instruction converting *binary64* FP numbers to signed 32-bit integers, rounding toward zero (the s32_f64 conversion). The behaviors of these instructions differ in the overflow case, but this does not matter because such overflow behavior is undefined by ISO C.

Conversion to unsigned 32-bit integers can be obtained from the signed conversion s32_f64 plus a case analysis:

$$\text{u32\_f64}(f) = \text{if } f < 2^{31}$$
$$\text{then s32\_f64}(f)$$
$$\text{else s32\_f64}(f \ominus 2^{31}) + 2^{31}.$$

The conversion u32_f64(f) is defined only if $f \in [0; 2^{32})$. In this case, the FP subtraction $f \ominus 2^{31}$ in the else branch is exact, and in either branch s32_f64 is applied to an argument in the $[0; 2^{31})$ range, where it is defined.

The same construction applies to 64-bit integers:

$$\text{u64\_f64}(f) = \text{if } f < 2^{63}$$
$$\text{then s64\_f64}(f)$$
$$\text{else s64\_f64}(f \ominus 2^{63}) + 2^{63}.$$

CompCert uses this implementation for the x86 platform, where s64_f64(f) is implemented using the x87 80-bit FP operations as s64_f80(f80_f64(f)). The only caveat is that the s64_f80 instruction of x87 rounds using the current rounding mode (to nearest even, by default); therefore, the rounding mode must be temporarily changed to round-toward-zero, which is costly.

The ARM and PowerPC 32-bit architectures provide no conversion instructions that produce 64-bit integers. For s64_f64 and u64_f64, CompCert uses a rather straightforward algorithm: it extracts the integer significand and shifts it appropriately based on the exponent. Algorithm 7.3 shows some pseudocode for u64_f64.

Algorithm 7.3 Pseudo-code for the u64_f64 conversion

```
u64 u64_f64(f64 f)
{
  int s = bit<63>(f);          // extract sign and
  int e = bits<62:52>(f) - 1023; // unbiased exponent
  if (s != 0 || e >= 64)       // f < 0 or f ≥ 2^64 ?
    return OVERFLOW;           // arbitrary result
  if (e < 0)                   // f < 1 ?
    return 0;
  u64 m = bits<51:0>(f) | 1<<52; // extract mantissa
  if (e >= 52)                 // and shift it
    return m << (e - 52);
  else
    return m >> (52 - e);
}
```

This completes the set of formally proved routines for converting FP numbers from and to integers, as mandated by the C standard, depending on the availability of some primitives in hardware. This also concludes this chapter on how to give precise semantics to FP operations for the C language, and how to implement and formally verify a compiler that complies to these semantics. This opens the way to verifying not just algorithms but also C programs, as will be done in the next chapter.

8

Deductive Program Verification

Chapters 5 and 6 have described several algorithms that perform FP computations and have shown how to formally verify their correctness (e.g. the computed value is close enough to the ideal value). Chapter 7 has presented a C compiler and the proof that it compiles FP operations in a way that does not invalidate the correctness of a C program. Now we are interested in the last piece of the puzzle, that is, proving the correctness of a C program and not just of the corresponding algorithm. Section 8.1 first motivates program verification and explains how it differs from algorithm verification as done in previous chapters. We rely here on deductive verification and the toolchain we use for the examples of this chapter is Frama-C/Jessie/Why3; more information on the corresponding methodology is given in section 8.2. Section 8.3 then presents several examples of C programs. Most of them have been previously described in Chapter 6, such as the area of a triangle or the average of two numbers. Finally, section 8.4 presents an alternative to Chapter 7 for ensuring that, whatever the compiler decisions about FP arithmetic, the compiled programs behave as specified.

8.1. Introduction

We have seen in the previous chapters theorems such as theorem 6.21 which states that algorithm 6.9 (Kahan's algorithm for an accurate discriminant) returns a value within two ulps of the correct result $b^2 - ac$. Let us now describe why we also want correct programs.

We first discuss the differences between an algorithm proof and a program proof. Our algorithm proofs are checked within a formal proof assistant, here Coq, using a formalization of FP arithmetic, here Flocq. The question of the underlying assumptions is important in order for the reader to trust the theorem enough to use it in a practical implementation. For example, all the theorems of this book rely on several axioms defining real numbers, but they are not mentioned in each theorem, only in the introductory section 2.3.2.

Even when no axioms are added, the formalization may not be exactly what the reader expects. In the particular case of Flocq, overflow is not taken into account. Consider for instance theorem 6.11 that states that for all binary FP numbers x and y, we have $\circ[(x + y)/2] = \circ((x + y)/2)$. This theorem holds in our formalization without overflow, but it is incorrect in the IEEE-754 standard as the addition may produce an overflow for sufficiently large values.

As for program proofs, they prevent some programming errors such as out-of-bound access or forgotten initialization. It can hardly happen when the algorithm is proved in Coq, or when the program is extracted from the algorithm (see section 2.1.5). However, a program written by hand may have mistakes, often simple ones, which are often detected when trying to verify it.

Moreover, programs written in a mainstream language such as C or Fortran or Python may be considered more readable than Coq theorems. Therefore, a verified C program with an explicit FP type such as double may be more convincing than the corresponding Coq theorem, even if the Coq theorem may be more general (it may hold for any radix and any precision).

The precise meaning of a program is determined by the language semantics. This is not simple, as seen in section 7.1: some computations may be done with extended precision, producing different results on the same program with different compiler options. Let us assume that we have a full semantics of the language and of the compiler. We choose the C language with CompCert's semantics for FP arithmetic described in section 7.2. In particular, this semantics comes with the parenthesizing (from left to right) and the precision of each operation from the static types. Note that no extended precision is ever used. This semantics gives us a formal and unique description of the executable code that is produced from the program.

In the verification process, there is a choice to be made about exceptional behaviors. A first possibility is to allow all exceptional values and computations on them [AYA 10]. This possibility can be useful in some cases as computations with infinities are quite reasonable, but it means more cases to verify for each operation. A second possibility is to allow infinities and to forbid NaNs. This removes the problems of NaN inputs in annotations described by Leavens [LEA 06]. He asserts that most FP specifications should have special cases for NaN inputs to be correct, but this is hardly readable and hardly interesting. He therefore claims that the *double* type, that is to say *binary64*, should not include NaNs, and that a special type *doubleWithNaN* should be added. The last possibility is the one we choose: we forbid both infinities and NaNs. It implies we have to prove that no exceptional behaviors occur, provided the inputs are finite. This means that every addition requires a proof that it does not overflow, and similarly for subtraction, multiplication, and division (with an additional required proof that the divisor is nonzero).

8.2. Our method and tools for program verification

Many methods exist for verifying a program, such as test [UTT 10], abstract interpretation [COU 77] (with many abstract domains such as octagons [MIN 06] or affine arithmetic [FIG 04]), deductive verification (see section 8.2.2), model checking [CLA 86], and symbolic execution [KIN 76]. Many tools based on one or several of these methods are available. Describing all of them is out of the scope of this book. We will focus only on the methods and tools we have been using for C program verification.

This section is organized as follows. Section 8.2.1 presents the ACSL specification language and how it supports FP arithmetic. Section 8.2.2 presents deductive verification, as well as the tools that will be used in section 8.3. Section 8.2.3 describes the verification process for a C program.

8.2.1. Specifications

What we want in this chapter is to link a program description with what we expect the program to do, such as computing an average or computing an FP number near $b^2 - ac$. This specification describes what the program does, or at least what the programmer expects the program to do. Specifications are common in deductive verification, but can also be used for documenting the program behavior, for guiding the implementation, and for facilitating the agreement between teams of programmers in modular development of software [HAT 12].

Specifications may be written using a formal language, that is, integrated with a programming language. The annotation language we use for the following C examples is named ACSL (for ANSI/ISO C Specification Language) [BAU 15] and has been defined by the developers of the Frama-C tool. In the ACSL language, annotations are written using first-order logic, contrary to Coq and its higher-order logic (see section 2.1). Generally, ACSL is quite similar to JML: it supports quantifiers and ghost code [FIL 14] (called model fields and methods in JML respectively).

Specifications include preconditions, postconditions, and loop invariants. For example, the specification of the following function states that it computes the square root of an integer x, or rather a lower bound on it:

```
//@ requires x >= 0;
//@ ensures \result * \result <= x;
```

```
int square_root(int x);
```

The precondition, introduced with **requires**, states that the argument x is nonnegative. Whenever some code contains a call to this function, the toolchain generates a verification condition stating that the input is nonnegative. The user has to prove it to ensure the program is correct. The postcondition, introduced with **ensures**, states the property satisfied by the return value \result. An important point is that, in ACSL, arithmetic operations are mathematical, not machine operations. In particular, the product \result * \result cannot overflow. Simply speaking, we can say that C integers are considered within specifications as mathematical integers.

In ACSL, an FP number f represents several values [BOL 07]:

– its FP value, that is, what is computed within the computer;

– its exact value, denoted by \exact(f). It is the value that would have been obtained if real arithmetic had been used instead of FP arithmetic for the computations. It means no rounding and no exceptional behavior. The execution path, however, is the one taken by the program based on FP values.

For instance, the following excerpt specifies the relative error on the content of the dx variable, which represents the grid step Δx (see section 9.3.2):

```
dx = 1./ni;
/*@ assert
  @    dx > 0. && dx <= 0.5 &&
  @    \abs(\exact(dx) - dx) / dx <= 0x1.p-53;
  @ */
```

The identifier dx represents the value actually computed (seen as a real number), while the expression \exact(dx) represents the value that would have been computed if mathematical operators had been used instead of FP operators. Note that 0x1.p-53 is a valid literal in ACSL (and in C as well) meaning $1 \cdot 2^{-53}$ (which is a *binary64* number).

Another meaningful example is that of the computation of $\exp(x)$ for a small x. The FP value is 1+x+x*x/2 computed in *binary64* and the exact value is $1 + x + \frac{x^2}{2}$. This allows a simple expression of the rounding error: $|f - \exact(f)|$. As done with integer operations, arithmetic operations on FP numbers inside annotations are mathematical, not machine operations. In particular, there is no rounding error inside the annotations.

8.2.2. Deductive verification

Deductive verification is the static analysis method that we rely on for our examples. Taking the definition from Filliâtre [FIL 11], deductive program verification is the art of turning the correctness of a program into a mathematical statement and then proving it. Deductive verification is modular: the specifications of the called functions are used to establish the proof without looking at their code. Note that the specification step is critical: if one puts too many preconditions (ultimately false), then the function cannot be reasonably used, if one puts too few postconditions (ultimately true), then one only proves there is no runtime error.

Deductive verification relies on Hoare's work [HOA 69] introducing the concept known today as Hoare triple. In modern notation, if P is a precondition, Q a postcondition, and s a program statement, then the triple $\{P\}s\{Q\}$ means that the execution of s in any state satisfying P ends in a state satisfying Q, assuming the execution does end. If the execution of s is required to terminate, this is total correctness, the other case being partial correctness. Given s and Q, it is possible to compute the weakest requirement over the initial state such that the execution of s will end up in a final state satisfying Q. Tools that compute these requirements are called verification condition generators and rely on weakest preconditions [DIJ 75]. It remains to check that P implies this weakest requirement.

For the C examples of this chapter, we use the deductive verification environment based on the following tools: Frama-C, Jessie, and Why3. Frama-C makes it possible to verify that the source code complies with an ACSL specification [KIR 15]. This architecture comes with several plugins for the static analysis of C code, such as value analysis using abstract interpretation, dataflow analysis, code transformation, and test-case generation. We use the Jessie plugin for deductive verification [MAR 07]. It internally uses the languages and tools of the Why3 platform.

This platform provides a programming language, a specification language, and a verification condition generator for programs written using these languages [BOB 11, FIL 13]. Why3 computes the verification conditions, using traditional techniques of weakest preconditions, and emits them to a wide set of existing automated and interactive theorem provers. The choice of the prover and some simplifications of the verification condition can be performed using a graphical interface. Why3 features quantification, polymorphism, a type for real numbers, and model fields and methods.

8.2.3. Our verification process

Let us now explain how to carry out the deductive verification of a C program. This process also described in Figure 8.1.

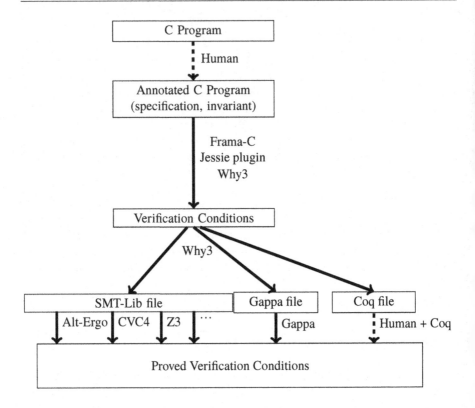

Figure 8.1. *Scheme for verifying a C program: when all the verification conditions are proved, the program is correct with respect to its specification.*

1) Annotate the program. Using C special comments, the programmer writes what the function expect (such as a positive parameter or an integer parameter smaller than 42) and what they ensure (such as a result within a small distance of the expected mathematical value). Other information may also be written: assertions, loop variants (to prove termination), and invariants (to handle what is happening inside loops).

2) Prove the corresponding verification conditions (VCs). This can be done for instance automatically using an SMT solver, or within a proof assistant. As explained in section 8.2.2, if all the theorems are proved, then the program is guaranteed to fulfill its specification (by proving the *behavior VCs*) and not to crash (by proving the *safety VCs*).

In practice, this is an iterative process: when the verification conditions are impossible to prove, one modifies the annotations (strengthening loop invariants, adding assertions or preconditions, and so on), until everything is proved.

Note that a single prover answering "yes" means that the verification condition is proved. An advantage of Why3 is its support for many provers (including some that do not appear in Figure 8.1 such as TPTP provers). It means that many automatic methods are available, so that the interactive proof burden is reduced.

8.3. Examples of annotated programs

We have seen many proved algorithms in previous chapters, but program verification presents a few specific difficulties. The first difficulty lies in writing a specification, since it needs to be correct and informative enough, as deductive verification is modular. Moreover, the specification may be long when there are many properties to be verified. The second difficulty is related to safety VCs, as these are additional theorems to be proved compared to algorithm proofs. These VCs are mostly about overflow, but some other may appear when a square root or a division are involved. The last difficulty is loop termination; especially when the termination condition is an FP test, it may be hard to ensure that the program will not loop forever.

The methodology described in section 8.2 has been applied to several C programs, some of which are given in this section. For all these examples, the fully annotated C code is given so that the specifications are comprehensive.[1] The programs are as follows:

– Section 8.3.1 shows a function performing an exact subtraction, which corresponds to Sterbenz' theorem. It comes with a very easy specification: a mathematical equality that states that the operation is exact. It leads to a single easy overflow VC.

– Section 8.3.2 shows the EFT for multiplication, which is based on Veltkamp's and Dekker's algorithms. The specification part is again quite easy, since the computed value is the rounding error of an FP multiplication, unless underflow occurs. Handling overflow is more complex as it relies on Veltkamp' algorithm properties.

– Section 8.3.3 shows the C function for computing the area of a triangle. Its specification only describes a relative error bound, but it exhibits an interesting safety VC. Indeed, we have to prove that the argument of a square root is nonnegative, while it is the result of a sequence of 11 FP operations.

– Section 8.3.4 gives two programs for computing the average of two FP numbers. One specification is trivial while the other one is very long (with properties such as error bound, symmetry, and sign). The safety VCs are the most interesting part, as we want to prove that neither program overflows on finite inputs.

---

1 All the C codes and their Coq proofs can be downloaded at http://www.lri.fr/~sboldo/research.html.

– Section 8.3.5 presents Malcolm's algorithm for determining the radix of an FP system. The specification is trivial and the safety VCs are easily proved. Nonetheless, this example illustrates a nontrivial proof of loop termination.

There is a larger verified program in the next chapter (in section 9.5), which deals with the discretization of the 1D wave equation described in section 9.1. Its specification is so long it is not given in full. There is no specific difficulty in the handling of safety VCs. The main difficulties lie in linking the partial differential equation to the algorithm and in proving the bound on the round-off errors.

Other examples verified using the same framework are given in [BOL 11a, MAR 14]. Note also an avionics example [GOO 13], linked to that of section 8.4.3.

8.3.1. *Exact subtraction*

The first example is a very simple program that refers to the property of section 5.1.1. The property states that, if x and y are FP numbers such that $\frac{y}{2} \le x \le 2y$, then $x - y$ is computed without error, because it fits into an FP number. Note that we implicitly assume that the inputs are finite FP numbers. The corresponding program is algorithm 8.1.

Algorithm 8.1 C program that depicts Sterbenz' theorem

```
/*@ requires  y/2. <= x <= 2.*y;
  @ ensures   \result == x-y;
  @*/

float Sterbenz(float x, float y) {
  return x-y;
}
```

As explained in section 8.2.1, all the computations inside annotations are exact. Therefore, the value x-y in the postcondition is the mathematical subtraction between the real numbers represented by x and y. Two VCs are generated: one for the postcondition (it corresponds to theorem 5.1) and one for safety corresponding to the fact that there will be no FP overflow. More precisely, the safety VC is that $\circ(x - y)$ is smaller than or equal to the largest finite FP number and it is proved by Gappa.

8.3.2. *Veltkamp's and Dekker's algorithms*

The computation of the EFT for the multiplication is described in section 5.2. When no FMA is available, it is based on two algorithms. The first one is Veltkamp's algorithm (algorithm 5.4) for splitting an FP number, described in section 5.2.1; we are here in a simple case as the radix is 2 and the rounding is to nearest, tie-breaking to even. The second one is Dekker's product (algorithm 5.5) for computing the FP multiplication error, described in section 5.2.2; we are also in a simple case as the radix is still 2.

Dekker's product (algorithm 5.5) returns two FP numbers r_h and r_ℓ such that $r_h + r_\ell = x \cdot y$. The following program only returns the error r_ℓ. It takes $\mathsf{xy} = r_h = \circ(x \cdot y)$ as a third input with the precondition at line 1 requiring this equality. As the operations are exact inside the annotations, we have to explicitly state that xy is the rounding to nearest with tie breaking to even, in the *binary64* format, of the real number $x \cdot y$.

For the sake of simplicity, a single program is given in algorithm 8.2, inlining Veltkamp's algorithm for both inputs and Dekker's product at lines 14–27. Underflow is not prevented, but the postcondition $x \cdot y = \mathsf{xy} + \mathsf{\backslash result}$ at line 6 only holds when no underflow occured. More precisely, we only consider the cases where the error is indeed an FP number. To ensure that, we require that $|x \cdot y| \geq 2^{-969}$ at line 5 with $-969 = e_{\min} + 2\varrho - 1$ (see section 5.2.2.1). As the algorithm also gives the exact error when either x or y is zero, we weaken the left-hand side of the postcondition to include this subcase.

As for the overflow, we require the inputs to be small enough at lines 2–4. Indeed, Veltkamp's algorithm may overflow when computing $\mathsf{px} = \circ(x \cdot C)$. If there is no overflow at this point, then there is no overflow in the subsequent FP computations, since $|\mathsf{qx}| \leq |\mathsf{px}|$ and since the other computed values are much smaller. In brief, this algorithm does not overflow if $x \cdot C$ does not overflow, which is ensured by the precondition $|x| \leq 2^{995}$. As for Dekker's product, the computations of $\circ(x \cdot y)$ and $\circ(\mathsf{hx} \cdot \mathsf{hy})$ are prevented from overflowing by the precondition $|x \cdot y| \leq 2^{1021}$. Proving that these preconditions are sufficient is complicated. In fact, some overflow VCs have to be proved in Coq as they rely on properties of Veltkamp's algorithm, e.g. $\mathsf{tx} \cdot \mathsf{ty}$ does not overflow when $\mathsf{hx} \cdot \mathsf{hy}$ does not.

8.3.3. *Accurate computation of the area of a triangle*

The next example implements the algorithm described in section 6.1. It computes the area of a triangle, given its side lengths as FP numbers. The common version of Heron's formula is $\sqrt{s\,(s-a)\,(s-b)\,(s-c)}$ where $s = \frac{a+b+c}{2}$, but it is inaccurate for needle-like triangles and the computations may lead to computing the square root of a negative number, due to round-off errors. Kahan's formula is mathematically equivalent to Heron's formula, but more accurate:

$$\tfrac{1}{4}\sqrt{(a+(b+c))\,(c-(a-b))\,(c+(a-b))\,(a+(b-c))}.$$

Algorithm 8.2 C program for Veltkamp's and Dekker's algorithms

---

```
1   /*@ requires xy == \round_double(\NearestEven,x*y);
2     @ requires \abs(x) <= 0x1.p995;
3     @ requires \abs(y) <= 0x1.p995;
4     @ requires \abs(x*y) <=  0x1.p1021;
5     @ ensures  ((x*y == 0 || 0x1.p-969 <= \abs(x*y))
6     @                     ==> x*y == xy + \result);
7     @*/
8   double Dekker(double x, double y, double xy) {
9
10    double C,px,qx,hx,py,qy,hy,tx,ty,r2;
11    C = 0x8000001p0;
12    /*@ assert C == 0x1p27 + 1; */
13
14    px = x * C;
15    qx = x - px;
16    hx = qx + px;
17    tx = x - hx;
18
19    py = y * C;
20    qy = y - py;
21    hy = qy + py;
22    ty = y - hy;
23
24    r2 = -xy + hx * hy;
25    r2 += hx * ty;
26    r2 += hy * tx;
27    r2 += tx * ty;
28    return r2;
29  }
```

As seen in section 6.1, its relative error in radix 2 is bounded by $\frac{19}{4}\varepsilon + 33\varepsilon^2$ with $\varepsilon = 2^{-53}$. Moreover, the absence of underflow can be detected *a posteriori* by checking whether the final result is large enough. The annotated program is in algorithm 8.3.

First, the square root is specified as an external function with a specification at lines 1–4: the function takes a nonnegative argument and produces the rounding to nearest of its exact square root. The mathematical exact value of the area of the triangle is computed with Heron's formula at lines 6–9 and denoted S. We then require the inputs of the function to be such that $0 \le c \le b \le a \le b + c$ at lines 11–12, as they must represent a valid triangle. We also require $a \le 2^{255}$ at line 13, in order to prevent overflows. This may seem too strong a hypothesis, but if we consider $a = b = c = 2^{256}$, then the last multiplication of Kahan's algorithm leads to rounding $3 \cdot 2^{1024}$ and thus overflows. The last annotation at lines 14–15 describes what the function guarantees: if the result is larger than 2^{-513}, then the relative error is smaller than $4.75 \cdot 2^{-53} + 33 \cdot 2^{-106}$.

Algorithm 8.3 C program for computing the area of a triangle

```
1   /*@ requires 0 <= x;
2     @ ensures \result == \round_double(\NearestEven,\sqrt(x));
3     @*/
4   double sqrt(double x);
5
6   /*@ logic real S(real a, real b, real c) =
7     @ \let s = (a+b+c)/2;
8     @         \sqrt(s*(s-a)*(s-b)*(s-c));
9     @ */
10
11  /*@ requires 0 <= c <= b <= a;
12    @ requires a <= b + c;
13    @ requires a <= 0x1p255;
14    @ ensures 0x1p-513 < \result
15    @    ==> \abs(\result-S(a,b,c)) <= (4.75*0x1p-53 + 33*0x1p-106)
16    @        * S(a,b,c);
17    @ */
18
19  double triangle(double a, double b, double c) {
20    return 0x1p-2 * sqrt((a+(b+c))*(a+(b-c))*(c+(a-b))*(c-(a-b)));
21  }
```

Let us now detail the VCs and their proofs. All of them were done either with Coq or with Gappa. Why3 is able to automatically extract a nice summary of the various VCs that tells which tool has proved them and the time it took (in seconds), as shown in Table 8.1. The last column has been added by hand and provides the number of lines of Coq proofs.

One VC (safety 12 in Table 8.1) is the precondition of the square root function, which requires the input to be nonnegative. This was already proved in Coq and we just had to plug the given proof. Another kind of VCs is about overflows (safety VCs 1–14 except 12 in Table 8.1). All those were automatic thanks to Gappa: the hypothesis $a \leq 2^{255}$ and the facts that $0 \leq c \leq b \leq a$ were sufficient for Gappa to prove that no exceptional behavior (infinities here) will occur.

Proof obligations		Gappa	Coq	
				Nb lines
Previous Coq proof			14.49	2,117
VC for behavior			23.51	82
VC for safety	1. floating-point overflow	0.01		
	2. floating-point overflow	0.01		
	3. floating-point overflow	0.01		
	4. floating-point overflow	0.01		
	5. floating-point overflow	0.01		
	6. floating-point overflow	0.00		
	7. floating-point overflow	0.00		
	8. floating-point overflow	0.01		
	9. floating-point overflow	0.00		
	10. floating-point overflow	0.01		
	11. floating-point overflow	0.01		
	12. precondition for call		3.85	13
	13. floating-point overflow	0.01		
	14. floating-point overflow	0.00		

Table 8.1. *VCs table for the triangle area program verification. Columns 3 and 4 are timings in seconds.*

Another VC is the function behavior. First we compare our algorithm to Heron's formula: for all real numbers a, b, and c, if $s = (a + b + c)/2$, then

$$\sqrt{s\,(s - a)\,(s - b)\,(s - c)}$$

is equal to

$$\tfrac{1}{4}\sqrt{(a + (b + c))\,(c - (a - b))\,(c + (a - b))\,(a + (b - c))}.$$

The proof is straightforward using the ring tactic (see section 4.1.1). It does not require any hypothesis on a, b, and c, as the square root of a negative number is 0.

For the round-off error, we use the radix-2 variant of theorem 6.1 (err_Δ_flt) of section 6.1. We have several hypotheses to satisfy concerning the precision, minimal exponent, and so on. The hypothesis $2^{-513} <$ result is of course exactly the inequality

$$2^{\left\lceil \frac{e_{\min} + p - 1}{2} \right\rceil - 2} < \tilde{\Delta}$$

required by the theorem, as $p = 53$ and $e_{\min} = -1074$ in *binary64*.

8.3.4. *Average computation*

This section describes two programs for computing approximations of the average of two FP numbers. The corresponding algorithms are described in section 6.5. When verifying the programs, an interesting part is the handling of overflow, since both algorithms were designed to prevent overflow. Therefore, the VCs created for each operation that guarantee that no overflow happens have to be dealt with, and may be nontrivial. Another interesting part is the annotations describing Sterbenz' average function. It is complicated as requirements [6.3]–[6.9] (see section 6.5) have to be described in ACSL. This section is organized as follows: section 8.3.4.1 describes an annotated absolute value. Section 8.3.4.2 describes Sterbenz' program and all its annotations. Section 8.3.4.3 describes the program computing the correctly-rounded average.

8.3.4.1. *Absolute value*

Both programs computing the average require an absolute value for tests. This function may come from a standard library and be implemented by testing or by playing with the first bit of the encoding of an FP number. As long as the specification is the same, any function will make the following programs work. We choose to define it using a test. The corresponding VC is automatically proved by Alt-Ergo.

Algorithm 8.4 C program for the absolute value

```
/*@ ensures \result == \abs(x); */
double abs(double x) {
  if (x >= 0) return x;
  else return (-x);
}
```

8.3.4.2. *Accurate Sterbenz' average program*

As explained in section 6.5.4, we have two algorithms for computing the average. The first one is algorithm 6.6 which corresponds to Sterbenz' hints. It is written as a C program in algorithm 8.6. The hard part here is to write the annotations of algorithms 8.5 and 8.6.

Here are some explanations about the annotations of algorithm 8.5. First, the floor function, which rounds down a real number to an integer, is specified at lines 1–4. The ulp function defined in section 3.1.3.2 is here denoted as ulp_d and defined in ACSL for *binary64* at lines 6–8 in a manner similar to what is done in Flocq.

Algorithm 8.5 C program for Sterbenz' average computation (part 1/2)

```
1   /*@ axiomatic Floor {
2     @ logic integer floor (real x);
3     @ axiom floor_prop: \forall real x; floor(x) <= x < floor(x)+1;
4     @ } */
5
6   /*@ logic real ulp_d (real x) =
7     @ \let e = 1 + floor (\log(\abs(x)) / \log(2));
8     @     \pow(2,\max(e-53,-1074)); */
9
10  /*@ logic real l_average (real x, real y) =
11    @ \let same_sign =
12    @   (x >= 0) ? ((y >= 0)) : (! (y >= 0));
13    @ (same_sign) ? ((\abs(x) <= \abs(y)) ?
14    @     \round_double(\NearestEven, x+\round_double(\NearestEven,
15    @         \round_double(\NearestEven, y-x)/2))   :
16    @     \round_double(\NearestEven, y+\round_double(\NearestEven,
17    @         \round_double(\NearestEven, x-y)/2)))  :
18    @     \round_double(\NearestEven,\round_double(\NearestEven, x+y)/2
19    @ */
20
21  /*@ lemma average_sym: \forall double x, y;
22    @         l_average(x,y) == l_average (y,x);
23    @ lemma average_sym_opp: \forall double x, y;
24    @         l_average(-x,-y) == - l_average (x,y);
25    @
26    @ lemma average_props: \forall double x, y;
27    @         \abs(l_average(x,y) - (x+y)/2) <= 3./2 * ulp_d((x+y)/2)
28    @ && (\min(x,y) <= l_average(x,y) <= \max(x,y))
29    @ && (0 <= (x+y)/2 ==> 0 <= l_average(x,y))
30    @ && ((x+y)/2 <= 0 ==> l_average(x,y) <= 0)
31    @ && ((x+y)/2 == 0 ==> l_average(x,y) == 0)
32    @ && (0x1p-1074 <= \abs((x+y)/2) ==> l_average(x,y) != 0);
33    @ */
```

The next block at lines 10–19 of algorithm 8.5 defines a logic function that computes the average following algorithm 6.6. In the ACSL syntax, it describes exactly what the program does. The reason for this duplication is that we want to prove $average(x,y) = average(y,x)$ (requirement (6.5)). In deductive verification, one only specifies a function with preconditions and postconditions on its result. Requirement (6.5) means *two* calls to the function. One cannot state a lemma with two calls to a C function, in particular as a generic C function may have side effects.

The solution is to have a logic function that has no side effects, and a lemma with two calls to this logic function. Therefore, we define a logic function called l_average. The postcondition at line 1 of algorithm 8.6 states that the function gives the same result as the real C program. After the definition of the l_average logic function come its properties at lines 21–33 corresponding to requirements [6.5], [6.6], [6.3], [6.4], [6.7], and [6.9] in that order.

Algorithm 8.6 C program for Sterbenz' average computation (part 2/2)

```
1   /*@   ensures \result == l_average(x,y);
2    @    ensures \abs(\result - (x+y)/2) <= 3./2 * ulp_d((x+y)/2);
3    @    ensures \min(x,y) <= \result <= \max(x,y);
4    @    ensures 0 <= (x+y)/2 ==> 0 <= \result;
5    @    ensures (x+y)/2 <= 0 ==> \result <= 0;
6    @    ensures (x+y)/2 == 0 ==> \result == 0;
7    @    ensures 0x1p-1074 <= \abs((x+y)/2) ==> \result != 0;
8    @ */
9
10  double average(double x, double y) {
11    int same_sign;
12    double r;
13    if (x >= 0) {
14      if (y >= 0) same_sign = 1;
15      else same_sign = 0; }
16    else {
17      if (y >= 0) same_sign = 0;
18      else same_sign = 1; }
19    if (same_sign) {
20      if (abs(x) <= abs(y)) r = x+(y-x)/2;
21      else r = y+(x-y)/2; }
22    else r = (x+y)/2;
23    //@ assert r == l_average(x,y);
24    return r;
25  }
```

The specification of the C function is given in algorithm 8.6. It has no precondition (any two finite FP numbers are handled). We guarantee there is no overflow without assuming any range for the inputs. The postconditions are first the equivalence with the logical function, and then the requirements that do not need two calls to the function, meaning Requirements [6.3], [6.4], [6.7], and [6.9] in that order at lines 2–7. Last is the C program at lines 10–25, with an assertion at line 23, which serves as a logical cut to ensure that the program is equivalent to its logical counterpart.

The behavioral VCs are proved either automatically or by calls to the Coq proofs described in section 6.5.4. The most interesting proofs of this example are the safety proofs, that is, the proofs related to overflow, as this is the only possible way for this program to fail. Almost all of them are proved automatically. Indeed, most operations do not overflow due to the case study of the signs of x and y, and this is handled automatically using Gappa. For a few operations, this is not sufficient and we need to rely on the fact that the result is between $\min(x, y)$ and $\max(x, y)$ (requirement [6.4]).

8.3.4.3. Correctly-rounded average program

As explained in section 6.5.5, we have an algorithm (algorithm 6.7) that computes the correctly-rounded average of two FP numbers. The corresponding program is algorithm 8.7. Both the annotations and the program are much smaller than in algorithms 8.5 and 8.6.

Algorithm 8.7 C program for correct average computation

```
1  /*@ requires 0x1p-967 <= C <= 0x1p970;
2    @ ensures \result == \round_double(\NearestEven, (x+y)/2);
3    @ */
4
5  double average(double C, double x, double y) {
6    if (C <= abs(x))
7      return x/2+y/2;
8    else
9      return (x+y)/2;
10 }
```

In algorithm 6.7, we had set $C = 2^{e_{\min}+2p+1}$. Here we allow larger values for C, but its minimal value is indeed $2^{e_{\min}+2p+1} = 2^{-967}$ in *binary64*. The requirement on C (minimal and maximal values) is the only precondition (line 1). Then comes the postcondition of the function at line 2: the result is the correct rounding of the average.

As before, the proofs of the behavior are quite simple as they are calls to previous Coq proofs. The difficult part, as expected, is overflow. All overflow VCs are handled automatically by Gappa, except for the VC that $x + y$ does not overflow, provided that $|x| < C \leq 2^{970}$. More precisely, even if y is the largest FP number, if $|x| < 2^{970}$, then $\circ(x + y)$ will not overflow as it will round to y.

8.3.5. Malcolm's algorithm

Malcolm's algorithm has not been described in previous chapters. It has been known since the 1970s and computes the radix of an FP format [MAL 72]. This may seem a strange idea as radix 2 is now ubiquitous, but this algorithm runs on any kind of architecture (including pocket calculators). If extended registers are used (see section 1.4.2), this algorithm may fail as explained by Gentleman and Marovich [GEN 74]. Note also that radix 10 is included in the IEEE-754 standard and a few CPU with radix-10 units have already appeared [SCH 09]. The termination is the most interesting feature of this program, in particular proving that a loop ending when A != A+1 actually ends.

This algorithm first computes one of the smallest positive values such that $x = \circ(x + 1)$. In radix 2, it is the smallest one, namely 2^ϱ, but that is not always the case in a generic radix. Then the algorithm computes the smallest FP value larger than x, which is $x + \beta$, with β the radix. The proof of this algorithm for any radix is rather complicated as the values 2^i may be inexact due to rounding on radices other than powers of 2. Our formalism for C programs only considers *binary32* and *binary64* numbers, hence only radix-2 computations. Therefore we can guess every value involved.

Let us describe both the annotations and the program of algorithm 8.8. The program postcondition (line 1) is that the result is the radix, that is 2.

Then, the annotations rely on a ghost variable i. It is a logical value, that is not used in the program, but used for more easily describing its behavior. More precisely, the variable A will always be a power of 2, so we add an invariant $A = 2^i$. This means that i is set and modified in order to keep this invariant correct. At line 2, A is set to 2, and then i is set to 1 at line 3. As i is a ghost variable, its modifications are done inside ACSL comments.

Now let us look into the main loop at lines 12–15. It is a while loop that will stop as soon as $A \neq \circ(A + 1)$. If the computations were infinitely precise, this would be an infinite loop. In the body of the loop, A is multiplied by 2, so we increase i by one, in order to keep the invariant $A = 2^i$. This invariant is explicitly stated in the annotations at line 8. Now let us see why this program terminates: $\circ(2^{53} + 1) = 2^{53}$. In *binary64*, there are 53 bits of significand, therefore the FP numbers closest to $2^{53} + 1$ are 2^{53} and $2^{53} + 2$. As ties break to even, 2^{53} is chosen. We now have an A such that $\circ(A+1) = A$. It is also the first one as $\circ(2^{52} + 1) = 2^{52} + 1 \neq 2^{52}$. We know that the loop will go on until i is 53 and A is 2^{53}. This gives us both the loop invariant: $1 \leq i \leq 53$ at line 9 and the variant. A variant is a nonnegative quantity that decreases at each iteration. As $1 \leq i \leq 53$ and i increases at each iteration, $53 - i$ is such a loop variant (line 10). We state that $i = 53$ and $A = 2^{53}$ at the end of the loop (line 17).

The program then includes another loop to determine the radix. We have an A such that $\circ(A+1) = A$. Moreover, it is the smallest such value in radix 2 (it is not always the smallest in any radix). The FP successor of this FP number A is $A + \beta$. The idea is to find the smallest integer B such that $\circ(A + B) = A + B$. The program first sets $B = 1$ at line 19. There is another loop at lines 25–28. In the body of the loop, B is increased

Algorithm 8.8 C program for computing the radix by Malcolm algorithm

```
1    /*@ ensures \result == 2.; */
2
3    double malcolm1() {
4      double A, B;
5      A = 2.0;
6      /*@ ghost int i = 1; */
7
8      /*@ loop invariant A == \pow(2.,i) &&
9        @              1 <= i <= 53;
10       @ loop variant (53-i); */
11
12     while (A != A+1) {
13       A *= 2.0;
14       /*@ ghost i++; */
15     }
16
17     /*@ assert i == 53 && A == 0x1.p53; */
18
19     B = 1;
20     /*@ ghost i = 1;*/
21
22     /*@ loop invariant B == i && (i == 1 || i == 2);
23       @ loop variant (2-i); */
24
25     while ((A+B)-A != B) {
26       B++;
27       /*@ ghost i++; */
28     }
29     return B;
30   }
```

by one. The interesting part is the loop condition: (A+B)-A != B. Indeed, we cannot test $\circ(A + B) = A + B$ as is. Malcolm's idea is to test $\circ[(A + B) - A]$. As B is between 1 and the radix, we easily have that $A \leq \circ(A + B) \leq 2A$. Sterbenz' theorem of section 5.1.1 applies, so $\circ[(A + B) - A] = \circ(A + B) - A$. Thus, testing whether $\circ[(A + B) - A]$ equals B is the same as testing whether $\circ(A + B)$ equals $A + B$. As we are in radix 2 and $A = 2^{53}$, we know exactly what happens: when $B = 1$, the loop condition is false and B is increased. When $B = 2$, the loop condition is true and we get out of the loop with a correct value for B (line 30). As before, we have to prove that the loop terminates. It also helps to give loop invariants. We reuse the ghost variable i that will be always equal to B. We cannot use B as is because variants must be integers, while B is an FP value. This equality and the possible values of B are put in the invariant (line 22). Thus, the value $2 - B = 2 - i$ is both positive and decreasing and can be set as a variant (line 23).

Overflow is not a problem as all the values are between 0 and 2^{54}. We can prove the postcondition and (more importantly) the termination of the program by proving 15 VCs, 10 of them being proved automatically.

In the proved program, we stop the loop when A equals $\circ(A + 1)$. In the original paper [MAL 72], the test was whether $\circ(\circ(A + 1) - A)$ does not equal 1. Our test is more efficient and both tests are equivalent in our context. Indeed as $A \geq 2$, we have $A \leq \circ(A + 1) \leq 2A$, therefore Sterbenz' theorem 5.1 can be applied and therefore, $\circ[(A+1) - A] = \circ(A+1) - A$. So $\circ[(A+1) - A] \neq 1$ if and only if $\circ(A+1) - A \neq 1$. So the original test is whether $\circ(A+1) \neq A+1$, which is equivalent to $\circ(A+1) = A$, as $\circ(A+1)$ is either A or $A+1$ due to the behavior of rounding to nearest, tie breaking to even.

8.4. Robustness against compiler optimizations

All the previous programs were formally proved correct with respect to their specifications. It should mean that they behave as formally described by the annotations. Unfortunately, this might not be the case due to compilation discrepancies. The same program may give several answers on several environments, as seen at the beginning of Chapter 7. This may be due to the availability on the processor of the fused multiply-add (FMA) operator (see section 1.1.4) or of 80-bit FP registers (see section 1.4.2), or to excessive optimizations.

As seen in section 7.1.2, the values of operations with FP operands are evaluated to a format whose range and precision may be larger than those of the type. This optimization opportunity applies to the use of an FMA operator for computing the expression $a \cdot b + c$, as the intermediate product is then performed with a much larger precision. This also means that the use of 80-bit registers and computations is allowed at will. In particular, for each operation, the compiler may choose to round the infinitely-precise result either to extended precision, or to *binary64* precision, or first to extended and then to *binary64* precision (double rounding). The question is then how to handle these discrepancies.

A first choice is to have a compiler with a clear and strict semantics as done in Chapter 7. A second choice is the direct analysis of the assembly code [YU 92, NGU 11]. Instead of analyzing the C code, we consider the assembly code after compilation, where all architecture-dependent information is known, such as the precision of each operation. We are left to check that the specifications, written at the C level, are valid with respect to the assembly code generated by the compiler. A drawback is that tools are not always able to interpret the compiler optimizations, a problematic one being inlining.

Another method is described in section 8.4.1, which covers all the compiler choices as far as extended registers and FMA are concerned. Section 8.4.2 is about program verification when the compiler is allowed to reorganize the additions.

8.4.1. *Extended registers and FMA*

As explained, a possible approach is to cover all the situations: the verification will guarantee that the specification will hold whatever the hardware, the compiler, and the optimization level [BOL 11d]. This is especially useful when the program will run on several platforms. Another advantage is that we stay at the source code level.

We rely on theorem 1.1 that covers the use of both extended precision, *binary64* precision, and double rounding. If f is any of those rounding of the real x, then we have both the following implications:

$$\text{if } |x| \geq 2^{-1022} \text{ then } \left| \frac{f - x}{x} \right| \leq 2050 \cdot 2^{-64}; \qquad [8.1]$$

$$\text{if } |x| \leq 2^{-1022} \text{ then } |f - x| \leq 2049 \cdot 2^{-1086}. \qquad [8.2]$$

If we rely on Chapter 7's semantics and if inputs and outputs are on 64 bits, subnormal additions and subtractions are correct (see section 5.1.2), so we may turn the second case into $f = x$. Unfortunately here, inputs may be 80-bit numbers from extended registers so equation [8.2] is not to be modified for additions and subtractions.

Theorem 1.1 gives rounding error formulas for a result f, which corresponds to various rounding denoted by $\square$ (64-bit, 80-bit, and double rounding). Now, we consider the FMA which computes $x \cdot y \pm z$ with a single rounding. The idea is simple: we see an FMA as a *rounded* multiplication followed by a rounded addition. So we only have to consider one more possible "rounding", which is the identity: $\square(x) = x$. This specific "rounding" magically covers all the FMA possibilities: the result of an FMA is $\square_1(x \cdot y + z)$, which can be seen as $\square_1(\square_2(x \cdot y) + z)$ with $\square_2$ being the identity. So we handle in the same way all the operations, with or without FMA, by considering one rounding for each basic operation (addition, subtraction, multiplication, division, square root, negation, and absolute value). Each rounding may be one of the four possible rounding ($\circ_{64}(x)$, $\circ_{80}(x)$, $\circ_{64}(\circ_{80}(x))$, x). Of course, the formulas of theorem 1.1 easily hold for the identity rounding. Some impossible cases are allowed: for example, all computations being exact. The important point is that *all* the actual possibilities are *included* in the considered possibilities. Furthermore, all of them have a rounding error bounded by theorem 1.1.

In practice, this needs to be applied to C programs. For this, we use a pragma that tells Jessie that the compiler may use extended registers and FMA. That pragma changes the definition of the FP operations. More precisely, it changes the operation postcondition, that is, how an operation result is defined in the VCs.

For all basic operations $x = a + b$, $a - b$, $a \cdot b$, a/b, $\sqrt{a}$, their definition is modified as follows. Chapter 7's semantics means that $f = \circ_{64}(x)$ is the only allowed result. When considering FMA and extended registers, our semantics means that the result f may be any real number such that equations [8.1] and [8.2] hold. Before, the value of each FP result was exactly known and was unique. Now, an FP result is characterized only by a small interval. Of course, only weaker properties can be proved, as only those that hold whatever the compilation choices are provable.

8.4.2. *Reorganization of additions*

To push the previous idea even further, we may also take into account that the compiler may reorganize additions. The idea here is that we will change the rounding error formula for the addition in order to guarantee that, even if $\circ[(a + b) + c]$ is transformed into $\circ[a + (b + c)]$ by the compiler, what is proved will still hold. For this, we use a formula of this form: $|\Box(a + b) - (a + b)| \le \varepsilon' \cdot (|a| + |b|) + \eta'$ (with given ε' and η'). Instead of an error proportional to $|a + b|$ as before, the error is proportional to $|a| + |b|$. This is a huge difference that handles the cancellations, but may increase a lot the proved bounds on the rounding error. Moreover, these ε' and η' depend upon the maximum number of reorganized additions [BOL 11d]. For instance, for a maximum of 16 additions, we may choose $\varepsilon' = 2051 \cdot 2^{-60}$ and $\eta' = 2049 \cdot 2^{-1082}$. Contrary to the previous method with only 80-bit computations and FMA, the obtained results taking associativity into account are not convincing. Even if the bounds are as tight as possible in the general case, they eventually give very coarse results, meaning large error bounds. This could have been expected as the worst case indeed produces a large rounding error. Moreover, the assumption bounding the number of reorganized additions (less than 16) is not checked in practice.

8.4.3. *The KB3D example*

In order to explain what these changes mean on the verification process and on what is provable, we focus on an example from avionics, described in section 8.4.3.1. It relies on a sign function to make a test. As there may have been FP rounding errors beforehand, a specific sign function with inaccuracy handling is presented in section 8.4.3.2. Our avionics example is formally proved as the previous examples assuming only 64-bit computations in section 8.4.3.3. An architecture-independent verification is done in section 8.4.3.4, with the possible use of FMA and extended registers.

8.4.3.1. Description of the KB3D example

The KB3D example takes as inputs the position and velocity vectors of two aircraft. The output is a conflict-prevention maneuver that involves the modification of a single parameter of the original flight path: vertical speed, heading, or ground speed [DOW 05, MUÑ 05, GOO 13]. This algorithm is distributed: each aircraft runs it simultaneously and independently. As any avionics application, the certification level is quite high. A formal proof in PVS of the whole algorithm has been done. It is rather long even if the program itself is rather short. But this formal proof was done assuming infinitely-precise computations. Unfortunately, FP inaccuracies may alter the decisions made by this algorithm.

Algorithm 8.9 C program, excerpt of KB3D – unannotated version

---

```
1  int sign(double x) {
2    if (x >= 0) return 1;
3    else return -1;
4  }
5  int eps_line(double sx, double sy, double vx, double vy) {
6    return sign(sx*vx+sy*vy) * sign(sx*vy-sy*vx);
7  }
```

We present in algorithm 8.9 a small part of KB3D (rewritten in C) that decides to go to the left or to the right by computing the sign of a quantity. This is a typical case where the result of a function may be entirely wrong due to FP inaccuracies. The algorithm was therefore modified in order to have the following specification: when it gives an answer, it is the correct one. It is similar to the geometric operators of section 6.6: the idea is to have an error bound on the value to be tested. If the magnitude of the value is larger than this error bound, it is sure to have the correct sign, regardless of the errors. In the other case, the result is of unknown sign and 0 is returned [BOL 10b, BOL 11d], and a common decision should then be taken or a more accurate result should be computed.

This example is especially interesting for two reasons. The first one is that it computes values looking like $a \cdot b + c \cdot d$. It is therefore reasonable that the compiler may use an FMA for improving performance and/or accuracy. Extended registers may also be used, like in any other FP program. The second reason is that this function makes a decision and this decision, when nonzero, should be the same on several aircraft running different hardware, which will be the case with algorithm 8.12.

8.4.3.2. Sign function

Our example relies on a particular sign function, whose code is in algorithm 8.10. There is first at line 1 a logic function for the sign called l_sign. It returns 1 when the value is nonnegative, and -1 otherwise.

Algorithm 8.10 C program, excerpt of KB3D, sign function

```
1   //@ logic integer l_sign(real x) = (x >= 0.0) ? 1 : -1;
2
3   /*@ ensures (\forall real y;
4   @                e1 <= x - y  <= e2 ==>
5   @                \result != 0 ==> \result == l_sign(y))
6   @                && \abs(\result) <= 1 ;
7   @*/
8   int sign(double x, double e1, double e2) {
9     if (x > e2)
10       return 1;
11     if (x < e1)
12       return -1;
13     return 0;
14   }
```

Let us look into the C code given at lines 8–14 before its specification. Instead of a test similar to that of l_sign, we have another kind of test. We have two additional inputs, e_1 and e_2, which stand for the minimal and maximal absolute error on the input x. More precisely, x is assumed to be some approximation of an ideal value with an absolute error interval of $[e_1; e_2]$. Therefore, when $x > e_2$, the ideal value is sure to be positive, hence the value 1 returned. When $x < e_1$, -1 is returned as the ideal value is negative. When $e_1 \leq x \leq e_2$, we do not have enough information to decide whether the ideal value is positive or negative. The function then returns 0, meaning "I do not know".

The specification is at lines 3–7. Line 6 is easy to understand: the result has an absolute value smaller than or equal to 1. This is easy as the result may be either 1, -1, or 0, and it is helpful to prove that no integer overflow may happen. Remember that the verification is modular, therefore the code of the sign function is not accessible when proving a function calling it, only the specification is available.

The most difficult part is at lines 4–5, where we consider any ideal value y. We suppose that the absolute error between x and y is between e_1 and e_2, that is to say

$e_1 \leq x - y \leq e_2$. We also suppose that the result is nonzero, meaning we are able to decide the sign of y. If both assumptions hold, then the function returns the correct sign (computed with the logical function l_sign) of the ideal value y.

This function is a means for deciding the sign of a real value, provided we have an approximation and an absolute bound on the quality of this approximation.

8.4.3.3. *KB3D – verification with no optimization*

We first explain the verification of the KB3D excerpt, assuming that each and every operation is done on 64 bits in algorithm 8.11.

Algorithm 8.11 C program, excerpt of KB3D

```
1   #define ME 0x1p-45
2
3   /*@ requires
4   @    \abs(sx) <= 100.0 && \abs(sy) <= 100.0 &&
5   @    \abs(vx) <= 1.0   && \abs(vy) <= 1.0;
6   @ ensures
7   @    \result != 0
8   @       ==> \result == l_sign(sx*vx+sy*vy)
9   @              * l_sign(sx*vy-sy*vx);
10  @*/
11
12  int eps_line(double sx, double sy, double vx, double vy) {
13    double t1, t2;
14
15    //@ assert \round_double(\NearestEven,ME) == ME;
16
17    t1 = sx*vx+sy*vy;
18    //@ assert -ME <= t1 - (sx*vx+sy*vy) <= ME;
19
20    t2 = sx*vy-sy*vx;
21    //@ assert -ME <= t2 - (sx*vy-sy*vx) <= ME;
22
23    return sign(t1, -ME, ME) * sign(t2, -ME, ME);
24  }
```

As explained in section 8.4.3.1, we have four inputs corresponding to the relative position (s_x, s_y) and relative velocity (v_x, v_y) of the other aircraft. The C code computes the sign of the real value $(s_x \cdot v_x + s_y \cdot v_y) \cdot (s_x \cdot v_y - s_y \cdot v_x)$. We have intermediate values for the two FP computations (t_1 and t_2) at lines 17 and 20. The result is the product of the signs at line 23.

Most of the proofs are automatic, using two different provers. We need some "communication" between provers, which is done by the use of assertions. In particular, we have an assertion on the absolute round-off error on t_1 at line 18 (and similarly for t_2 at line 21), to be be used for the postcondition of the sign function. Unfortunately, the postcondition of the sign function is a $\forall y, \ldots$ and this y does not seem to be correctly instantiated by SMT provers. We therefore need a small Coq proof to put things together: calling the sign function postcondition, the absolute error bound being available as it is proved by Gappa; This is done twice (for t_1 and t_2). We also have 11 VCs for overflow, all automatically handled by Gappa.

The FP error bound on both t_1 and t_2 is 2^{-45}, which is the smallest bound automatically proved by Gappa. It is defined as a macro at line 1 for the sake of readability. This value is due both to the maximal values of the inputs (100 for $|s_x|$ and $|s_y|$, and 1 for $|v_x|$ and $|v_y|$) and to the FP format $binary64$ (with gradual underflow). It helps in the Coq proof to assert at line 15 that 2^{-45} is an FP number.

The postcondition of this function is at lines 3–9. It begins with requirements on the values of the inputs, which must be reasonable values for the position and the speed of the aircraft. These bounds are useful to prove that no overflow occurs and to analyze the FP absolute round-off error. The specification then states a property on the result of the function when it is nonzero. When the result is zero, we know nothing (except that no exceptional behaviors occur). When the result is nonzero, it is the correct one, meaning it is the sign of the real number $(s_x \cdot v_x + s_y \cdot v_y) \cdot (s_x \cdot v_y - s_y \cdot v_x)$. In summary, this function may return either zero or the exact result.

8.4.3.4. KB3D – architecture-independent verification

Now let us assume that the compiler may use both the FMA operator and extended registers. The program is in algorithm 8.12. The first difference is at line 1. We have a pragma that tells the tools that extended registers and/or FMA may be used. As explained in section 8.4.1, this changes the postconditions of the basic operations into error bounds.

As expected, the bounds deduced in the previous section are not valid anymore, as in the worst case, we may have larger error bounds for one single operation (see sections 1.4.2 and 8.4.1). More precisely, we had a 2^{-45} bound that is now 0x1.90641p-45 (defined at line 3) which is more than 50% larger. This is due to the potential use of double rounding (a $binary80$ rounding followed by a $binary64$ rounding) at each operation. As before, nearly all the proofs are automatic, except for the postcondition, as a small Coq proof puts everything together.

The rest of the program is very similar to that of algorithm 8.11. In particular, the program preconditions and postconditions are exactly the same. The difference is that this program will answer 0 (meaning "I do not know") more often. But programs

---

Algorithm 8.12 C program, excerpt of KB3D – architecture-independent version

---

```
1   #pragma JessieFloatModel(multirounding)
2
3   #define ME      0x1.90641p-45
4
5   /*@ requires
6   @    \abs(sx) <= 100.0 && \abs(sy) <= 100.0 &&
7   @    \abs(vx) <= 1.0   && \abs(vy) <= 1.0;
8   @ ensures
9   @    \result != 0
10  @      ==> \result == l_sign(sx*vx+sy*vy)
11  @              * l_sign(sx*vy-sy*vx);
12  @*/
13  int eps_line(double sx, double sy,double vx, double vy) {
14    double t1,t2;
15
16    //@ assert \round_double(\NearestEven,ME) == ME;
17
18    t1 = sx*vx+sy*vy;
19    //@ assert -ME <= t1 - (sx*vx+sy*vy) <= ME;
20
21    t2 = sx*vy-sy*vx;
22    //@ assert -ME <= t2 - (sx*vy-sy*vx) <= ME;
23
24    return sign(t1, -ME, ME) * sign(t2, -ME, ME);
25  }
```

---

running on different hardware or compiled differently will give the good answer when they provide an answer (a nonzero value), which is the property we were looking for.

9

Real and Numerical Analysis

For the vast majority of FP users, FP arithmetic is not a research topic *per se*. It is just a tool to quickly get a result, hopefully accurate. This chapter describes experiments that extend beyond the realm of simple FP arithmetic toward real analysis and numerical analysis. Our running example comes from numerical analysis: it is a program to get an approximation to the solution of the 1D wave equation described in section 9.1. This application requires mathematical theorems that would have been next to impossible to prove using the real analysis formalization from Coq's standard library. This is why we use another formalization of real analysis detailed in section 9.2, in particular to prove the existence and regularity of the solution of the 1D wave equation using d'Alembert's formula. The properties of the numerical scheme, e.g. convergence, are proved in section 9.3. Then the round-off errors are bounded in section 9.4. Finally, the verification of the C program is explained in section 9.5.

9.1. Running example: three-point scheme for the 1D wave equation

The partial differential equation studied here is the one-dimensional wave equation. It models the propagation of waves along an ideal vibrating elastic string that is tied down at both ends. This is illustrated by Figure 9.1, the external forces being the gravity, the tree, and the child. The rope then oscillates as in Figure 9.2.

The partial differential equation is obtained from Newton's laws of motion. For the sake of simplicity, the gravity is neglected, so the string is supposed rectilinear when at rest. The string is also supposed to have a constant propagation velocity c. The wave equation is thus

$$\frac{\partial^2 p}{\partial t^2}(x, t) - c^2 \frac{\partial^2 p}{\partial x^2}(x, t) = s(x, t)$$

with given regular functions for the source term s, the initial position p_0, and the initial velocity p_1. See section 9.3.1 for more details.

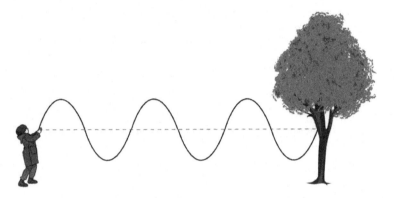

Figure 9.1. *Wave in a rope. The rope is tied down at both ends and rocked by the child. CC BY-SA 3.0.*

Figure 9.2. *Oscillating string, tied down at both ends.*

If we were not able to give an exact solution, it could have been approximated by a numerical scheme on a grid. We choose here to focus on the three-point scheme, a very simple scheme described in Figure 9.3: to compute a given position of the rope, we need its previous positions at three different space steps. The size of the grid is fixed at given constants $i_{\max}$ (width) and $k_{\max}$ (height). The main mathematical formula (for $k \in [1; k_{\max} - 1]$ and $i \in [1; i_{\max} - 1]$) is

$$\frac{p_i^{k+1} - 2p_i^k + p_i^{k-1}}{\Delta t^2} - c^2 \frac{p_{i+1}^k - 2p_i^k + p_{i-1}^k}{\Delta x^2} = s_i^k.$$

See section 9.3.2 for more details.

This scheme may then be turned into an algorithm and a C program. Here is its main loop assuming $s = 0$, with $i_{\max}$ being ni, $k_{\max}$ being nk, and a pre-computed constant a corresponding to $\frac{c^2 \Delta t^2}{\Delta x^2}$:

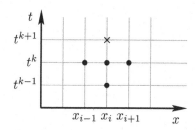

Figure 9.3. *Three-point scheme: p_i^{k+1} (at ×) depends on p_{i-1}^k, p_i^k, p_{i+1}^k, and p_i^{k-1} (at •).*

```
for (k = 1; k < nk; k++) {
    p[0][k+1] = 0.;
    for (i = 1; i < ni; i++) {
        dp = p[i+1][k] - 2. * p[i][k] + p[i-1][k];
        p[i][k+1] = 2. * p[i][k] - p[i][k-1] + a * dp;
    }
    p[ni][k+1] = 0.;
}
```

The main question we are interested in is then: does this program compute an accurate approximation to the mathematical solution? This requires two very different kinds of proofs. The first one bounds the method error, due to the convergence of the numerical scheme toward the mathematical solution. This is studied in section 9.3. The second one bounds the round-off errors and is studied in section 9.4. The program annotations and verification done in section 9.5 then rely on both.

The proofs about the convergence of the scheme rely on the existence and regularity of the mathematical solution. In this particular case, this is quite simple as we have a formula for this solution. This is d'Alembert's formula:

$$p(x,t) = \underbrace{\frac{1}{2}\left(p_0(x+ct) + p_0(x-ct)\right)}_{\alpha(x,t)} + \underbrace{\frac{1}{2c}\int_{x-ct}^{x+ct} p_1(\xi)\,d\xi}_{\beta(x,t)}$$

$$+ \underbrace{\frac{1}{2c}\int_0^t \int_{x-c(t-\tau)}^{x+c(t-\tau)} s(\xi,\tau)\,d\xi\,d\tau}_{\gamma(x,t)}$$

assuming p_0, p_1, and s are regular enough.

In order to prove that this formula satisfies the wave equation, one can compute the needed differentials and check the equality. Unfortunately, this is next to impossible with the standard library of real analysis of Coq. This is the reason why the Coquelicot library of real analysis was developed. This library conveniently copes with partial derivatives of parametric integrals. It is described in section 9.2.

9.2. Advanced formalization of real analysis

Several formal systems have libraries formalizing FP arithmetic. Most rely on real numbers, at least to compare FP computations with their real counterparts. The standard library of Coq has been described in section 2.3. It includes a classical axiomatization of real numbers. It also provides the usual definitions of standard real analysis such as finite limits of sequences and univariate functions, derivatives, Riemann integrals, and power series.

The main shortcoming we have to face here is that of dependent types described in section 2.1. The ubiquitous notion of differentiability is provided by the predicate called derivable_pt : (R → R) → R → Set, while the differentiation operator has the following type:

derive_pt : forall (f : R -> R) (x : R), derivable_pt f x -> R.

In other words, this operator takes a function f, a real number x, and a proof that f is differentiable at x, and it returns the value of $f'(x)$. Any derivative must therefore hold its proof of differentiability. Integrability follows the same pattern. This is cumbersome as these proofs pollute both statements and proof scripts. For instance, consider the lemma about the value of the integral of the constant function. Mathematicians would write $\int_a^b c\,dx = c \cdot (b - a)$. Using the standard library, this corresponds to

```
Lemma RiemannInt_P14 : forall a b c:R,
         Riemann_integrable (fct_cte c) a b.
(* 8 lines of proof *)
```

```
Lemma RiemannInt_P15 :
   forall (a b c:R) (pr:Riemann_integrable (fct_cte c) a b),
     RiemannInt (fct_cte c) a b pr = c * (b - a).
(* 31 lines of proof *)
```

with fct_cte being the constant function.

The statement of theorem RiemannInt_P15 is hard to read due to the dependent type (Riemann_integrable (fct_cte c) a b), meaning "the constant function equal to c is differentiable between a and b", which is required to write the statement.

Moreover, the dependent type may be complex if we integrate some differential or differentiate some integrals. As explained, given a differentiable function $p_1 : \mathbb{R} \to \mathbb{R}$, we consider

$$\beta(x,t) \overset{\text{def}}{=} \frac{1}{2c} \int_{x-ct}^{x+ct} p_1(\xi) \, d\xi. \tag{9.1}$$

For our running example, we need the values of $\frac{\partial^2 \beta}{\partial x^2}$ and $\frac{\partial^2 \beta}{\partial t^2}$.

The Coquelicot library[1] overcomes the deficiencies of Coq's standard library of real analysis [BOL 15d]. It extends it without introducing any new axiom and provides homogeneous definitions and lemmas for limits, derivatives, integrals, series, and power series with total functions (see section 9.2.1). These definitions are proved equivalent to those of the standard library, so that user developments can easily mix both libraries.

To ease the proof process, a tactic for automating proofs of differentiability is provided (see section 9.2.2). In addition to basic features of real analysis, several extensions are also provided: results about partial derivatives such as Schwarz and Taylor-Lagrange theorems (section 9.2.3), parametric integrals (section 9.2.4), and generalized limits (section 9.2.5).

Other features of the library, not described here, include filters, asymptotic behaviors, and a generalization of real analysis toward other spaces, such as complex numbers or matrices; see [BOL 15d, LEL 15] for more details.

9.2.1. *Total functions*

In pen-and-paper proofs, it is common to first write functions or formulas and prove that they are well-formed afterwards. Unfortunately, as seen in section 2.3.2, the derivative from the standard library is defined using dependent types, so one must first prove the integrability before being allowed to write the integral: see lemmas RiemannInt_P14 and RiemannInt_P15 above. Coquelicot builds total functions for limits, derivatives, integrals, series, and power series. They return the expected value in the case of convergence, differentiability, or integrability, and else return an arbitrary value. By avoiding dependent types throughout the library, theorem statements become much easier to write than with the standard library. Indeed, hypotheses of convergence, differentiability, and so on, can now be considered independently of the actual values. Without such a feature, it would be quite complex to state some theorems, such as the one stating that d'Alembert's formula is a solution

---

[1] http://coquelicot.saclay.inria.fr/.

of the wave equation. It also simplifies the proof process, as it makes rewriting rules usable in practice. This approach based on total functions is similar to what other proof assistants achieve due to Hilbert's ε operator but without having to assume the existence of such an operator in Coq. This operator can be seen as the following function:

```
epsilon: forall A : Type, inhabited A -> (A -> Prop) -> A
```

that chooses a satisfactory value when it exists. More precisely, we have the property:

```
epsilon_spec (A : Type) (i:inhabited A) (P : A->Prop) :
    (exists x, P x) -> P (epsilon i P).
```

There is no such function in Coq (without adding an axiom) but it may be constructed for some specific A, such as $\mathbb{R}$. The construction of the total functions relies on the limited principle of omniscience (LPO), which can be deduced from part of the axioms defining real numbers, namely completeness and total_order_T [BOL 12]. The functions that interest us here are the following ones: derivative and integral. More details about Coquelicot's other total functions can be found in [BOL 15d].

Besides the total function for derivative Derive, two predicates are available. The predicate (ex_derive f x) states that f is differentiable at point x and the predicate (is_derive f x y) states that f is differentiable at point x and its derivative at that point is y. This last predicate is proved to be equivalent to the predicate derivable_pt_lim from the standard library.

Likewise, we have a total function for the Riemann integral called RInt. Two predicates are also available: (ex_RInt f a b) states the Riemann integrability of f on the interval $[a;b]$; the predicate (is_RInt f a b If) states the Riemann integrability of f on the interval $[a;b]$ and that the real number If is indeed the integral. Equivalence with the standard library definitions is proved.

Let us go back to our running example. Given a function p_1, we define $\beta(x,t)$ as in equation 9.1. That is to say

```
Parameter p1 : R -> R.
Definition beta (x t : R) :=
    1/(2*c) * RInt (fun z => p1 z) (x - c * t) (x + c * t).
```

which looks like the corresponding mathematical statement.

Another example is the value of $\frac{\partial^2 \beta}{\partial t^2}(x,t)$, which is expected to be

Definition beta02 x t := c/2 *
    (Derive p1 (x + c * t) - Derive p1 (x - c * t)).

that is, $\frac{c}{2}\left(p_1'(x+ct)-p_1'(x-ct)\right)$.

The theorem we then want to prove is that β is twice differentiable and that its second derivative has the given value. A predicate for the nth differentiability and the nth derivative is available with the same semantic as is_derive: (is_derive_n f n x y) states that f is nth differentiable at point x and its nth derivative at that point is y. The Coq theorem is then

Lemma beta02_prop : forall x t,
    is_derive_n (fun u => beta x u) 2 t (beta02 x t).

Now that the theorem is stated, let us see how to prove it. More than the definition of predicates and functions for integrability/integral and differentiability/derivative, one needs numerous lemmas about them. In particular, their compatibility with the basic operations are proved in Coquelicot: composition, multiplication by a constant, opposite, and addition for function limits; derivatives and integrals; multiplication and multiplicative inverse for limits and derivatives. For limits, these theorems handle limits both at finite and infinite points and both finite and infinite limit values (see section 9.2.5). On top of these basic theorems, there is an automatic tactic described in section 9.2.2.

9.2.2. *Tactics for automating reasoning about differentiability*

When it comes to real analysis, some reasonings are mechanical, guided only by the structure of expressions, e.g. differentiability. It is thus important not only to provide tools that ease writing mathematics but also tools that automate reasoning whenever possible.

For this purpose, the Coquelicot library provides the auto_derive tactic. It is meant to help proving statements of the form (is_derive f x l), that is, f has a derivative at point x and it is equal to l. Variants of this tactic take care of equivalent statements from the standard library or statements where l is implicit. The tactic first reifies expression f, then performs symbolic differentiation on it, and finally compares the obtained derivative with l [BOL 12].

Along the differentiation computations, the tactic keeps track of all the side conditions needed for the final result to actually be a derivative of the original expression. The proofs of these properties are left to the user. For example, consider the proof of d'Alembert's formula properties: we want to prove Lemma `beta02_prop` from the end of section 9.2.1, which states that

$$\frac{\partial^2 \beta}{\partial t^2}(x,t) = \frac{c}{2}\left(p_1'(x+ct) - p_1'(x-ct)\right)$$

with β being defined in equation [9.1].

This statement is a second-order differentiability and a second derivative. For this example, the Coquelicot library provides a variant of the `auto_derive` tactic called `auto_derive_2`. This `auto_derive_2` tactic is built both upon `auto_derive` and upon another tactic called `auto_derive_fun`. This latest tactic takes a function as input and produces a hypothesis that gives an expression of its derivative provided some side conditions. For instance, calling

```
(auto_derive_fun (fun x => sqrt (exp x)))
```

gives a hypothesis:

```
forall x : R, 0 < exp x ->
  is_derive (fun t : R => sqrt (exp t)) x
    (1 * exp x * / (2 * sqrt (exp x)))
```

Given the `beta02_prop` goal, the `auto_derive_2` tactic requires the user to prove three goals:

– given a real number z, the function p_1 is integrable between $x - cz$ and $x + cz$, and is continuous around $x - cz$ and $x + cz$;

– the function p_1 is differentiable at $x - ct$ and $x + ct$;

– an automatically generated value is equal to (`beta02 x t`).

The first two goals are deduced from the fact that p_1 is differentiable on $\mathbb{R}$, therefore continuous and differentiable everywhere. The last goal is proved by unfolding `beta02` and a call to the `field` tactic (see section 4.1).

As seen with p_1, the `auto_derive` tactic can handle unknown function symbols. To do so, it builds the derivative using the `Derive` operator and it emits the appropriate differentiability goals. For known function symbols, such as sin or ln, a mechanism of type class instances makes it possible to indicate the derivatives and the differentiability domain. This mechanism supports all the elementary functions of Coq's standard library. The user can use the same mechanism to register more functions, if needed. Then, at the time of reification, whenever the tactic encounters a function symbol, it checks whether it is registered as being differentiable. If so, it tells the core tactic about its derivative, so that the core tactic does not use the `Derive` operator.

9.2.3. Partial derivatives

D'Alembert's formula is the solution to some systems of partial differential equations. This entails the need for multivariate functions and the ability to manipulate partial derivatives. Given a higher-order language, they can easily be defined from the standard one-dimensional derivative; the partial derivative of f with respect to the kth argument at point $(x_1, \ldots, x_n)$ is simply $\text{Derive}((u \mapsto f(x_1, \ldots, u, \ldots, x_n)), x_k)$. This was seen before at the end of section 9.2.1 in the theorem:

```
forall x t, is_derive_n (fun u => beta x u) 2 t (beta02 x t).
```

Things become more delicate when higher derivatives are needed. An example is Schwarz' theorem, which states that mixed second partial derivatives are equal (that is, the Hessian matrix is symmetric) under some continuity hypotheses. The Coquelicot formalism makes it easy to state, while it would have been nearly impossible with the standard library. Indeed, proof terms would creep every time the Derive total function appears in the statement below and they would be especially intricate since the derivatives only need to exist in a neighborhood of (x, y) and not on the whole $\mathbb{R}^2$ space. This is exactly what the predicate locally_2d means.

For the sake of readability, Schwarz' theorem is not written in Coq *in extenso*. The notation • means the variable considered for the differentiability or the derivative. That is to say (ex_derive f(•,v) u) means (ex_derive (fun z => f z v) u). Partial derivatives with respect to the first variable and to the second variables are written $\frac{\partial f}{\partial 1}$ and $\frac{\partial f}{\partial 2}$, respectively.

```
Lemma Schwarz : forall (f : R -> R -> R) (x y : R),
  locally_2d (fun u v =>
    ex_derive f(•,v) u /\ ex_derive f(u,•) v /\
    ex_derive ∂f/∂2(•,v) u /\ ex_derive ∂f/∂1(u,•) v) x y ->
  continuity_2d_pt ∂²f/∂1∂2 x y ->
  continuity_2d_pt ∂²f/∂2∂1 x y ->
  ∂²f/∂1∂2(x,y) = ∂²f/∂2∂1(x,y).
```

Coquelicot is not restricted to second partial derivatives; it also provides theorems about higher ones. An instance of such theorems is the Taylor-Lagrange approximation. It states that the following equality holds under some hypotheses on the bivariate function f:

$$f(x', y') = \text{Tpol}(f, n, (x, y), (x' - x, y' - y)) + \mathcal{O}(\|(x' - x, y' - y)\|^{n+1})$$

with

$$\mathsf{Tpol}(f, n, (x, y), (u, v)) \overset{\text{def}}{=} \sum_{p=0}^{n} \frac{1}{p!} \left(\sum_{m=0}^{p} \binom{p}{m} \cdot \frac{\partial^p f}{\partial x^m \partial y^{p-m}}(x, y) \cdot u^m \cdot v^{p-m} \right).$$

Again, stating such a theorem would have been impossible if not for the availability of total functions for defining partial derivatives. It would also have been cumbersome with Fréchet's higher derivative, since the nth such derivative at a given point is a multilinear map from $\mathbb{R}^{2n}$ to $\mathbb{R}$.

This Taylor-Lagrange theorem is needed by our running example as the approximation gives the regularity of the function p (d'Alembert's formula) described above. This regularity is the base of the convergence of the chosen numerical scheme (see sections 9.3.5 and 9.3.6).

9.2.4. Parametric integrals

Once one starts to manipulate analysis formulas mixing integrals and derivatives, there quickly comes a time when one has to compute the derivative of an integral (or the integral of a derivative). The fundamental theorem of calculus is sufficient to take care of the simpler case:

$$\left(x \mapsto \int_a^x f(t)dt \right)'(x) = f(x).$$

Unfortunately, integrals are hardly that simple and, as in our example, one might well have to compute the derivative of the following expression:

$$x \mapsto \int_{a(x)}^{b(x)} f(x, t)dt.$$

Not only do the bounds depend on x, the differentiation variable, but the integrand does too. In that case, the ability to commute differentials and integrals becomes crucial. Fortunately, Coquelicot provides the necessary tools. For instance, it proves that, under sufficient hypotheses, the derivative of the expression above is equal to

$$\int_{a(x)}^{b(x)} \frac{\partial f}{\partial x}(x, t)dt - f(x, a(x)) \cdot a'(x) + f(x, b(x)) \cdot b'(x).$$

An example that heavily manipulates derivatives and integrals is the proof of the relation between the elliptic integrals of the first and second kinds and the arithmetic-geometric mean [BER 13]. This was proved in Coq thanks to some theorems from Coquelicot.

9.2.5. Generalized limits

Since $\mathbb{R}$ is equipped with a linear order, people might want to compute limits of functions not only at finite values but also at $\pm\infty$. This kind of limit is unfortunately missing from the standard library of Coq; it is therefore impossible to state and prove that $\lim_{x \to -\infty} \exp(x) = 0$ without extending the formalism. Moreover, mathematicians are accustomed to writing equalities like:

$$\lim_{t \to x}(f(t) + g(t)) = \lim_{t \to x} f(t) + \lim_{t \to x} g(t)$$

regardless of the finiteness of x and of the actual limits. This may easily lead to too many cases to be studied if finite and infinite limits were handled separately.

Coquelicot encompasses all these cases by defining limits of real-valued functions as having values in the set $\overline{\mathbb{R}} = \mathbb{R} \cup \{-\infty, +\infty\}$. For instance, if we want to state that $\exp(x) \to 0$ when $x \to +\infty$, we may write (knowing that m_infty denotes $-\infty$):

Lemma is_lim_exp_m: is_lim (fun y => exp y) m_infty 0.

Coquelicot also extends the addition of two reals to a function from $\overline{\mathbb{R}}^2$ to $\overline{\mathbb{R}}$. As a result, the above equality about the sum of limits can be written. Moreover, this extended addition is proved to be continuous over $\overline{\mathbb{R}}^2 \setminus \{(+\infty, -\infty), (-\infty, +\infty)\}$ directly from the continuity of addition. The other usual arithmetic operators are handled in the same way by the library.

The intermediate value theorem is another example where its extension to $\overline{\mathbb{R}}$ is definitely useful. Given an open interval $(a; b)$, a function $f : \mathbb{R} \to \mathbb{R}$ continuous over $(a; b)$, and a real y, this theorem provides a real x such that $f(x) = y$ if $\lim_{t \to a} f(t) < y < \lim_{t \to b} f(t)$. Note that this theorem holds whether the bounds a and b and the two limits of f are finite or not. It is thus important that the user be given only one single theorem rather than 16 variants. Moreover, the theorem proved in the Coquelicot library is stated as a function rather than with an existential quantifier, so that it can be used to invert any continuous function.

In addition to d'Alembert's formula, the Coquelicot library has been exercised on two other applications [BOL 15d]. The first one illustrates the use of power series, by the definition and properties of Bessel functions [LEL 13]: convergence radius, differentiability, and numerous recurrence relations possibly including derivatives. The second application is an examination called the Baccalaureate. It is passed by most 18-year old French students at the end of high school and is required for attending university. Coquelicot was tried on the 2013 mathematics test of the scientific Baccalaureate [BAC 13], in real-life test conditions. This means it was done in a high school at the same time as the students. The two analysis exercises were done in about three hours out of four, showing the practicality of the library. Very few analysis questions were out of the scope of Coquelicot, showing that its breadth is at least comparable to that of the most scientifically-inclined 18-year old French students.

9.3. Method error of the 3-point scheme for the 1D wave equation

Now, let us go into the details of the numerical analysis scheme presented in section 9.1. This case study is a very simple one, as far as numerical analysis is concerned: it is the one-dimensional wave equation, discretized by a three-point scheme. Its verification encloses various proofs as different kinds of error may occur: the avoidable ones are mathematical errors in the pen-and-paper proof and programming errors in the C code; the unavoidable ones are method errors due to the numerical scheme and round-off errors due to the floating-point computations.

This section is dedicated to the formal proof of the convergence of the numerical scheme. More precisely, the goal of this section is to formally bound the method error. We are interested in the convergence error, which measures the distance between the continuous and discrete solutions. Precise definitions are given in section 9.3.1 about the continuous problem, in section 9.3.2 about the discretized problem, in section 9.3.3 about the properties of the scheme (consistency, stability, and convergence), and in section 9.3.5 about the required regularity of the continuous problem. The definition of the domination (big $\mathcal{O}$) is given in section 9.3.4. Then the proof is divided into three parts: consistency in section 9.3.6, stability in section 9.3.7, and convergence in section 9.3.8. Note also that the bound on the round-off errors in section 9.4 relies on a property of the numerical scheme, namely the fact that it does not blow up, which is also required for the convergence.

9.3.1. Description of the continuous problem

Let $x_{\min}$ and $x_{\max}$ be the abscissas of the endpoints of the string. Let $p(x, t)$ be the transverse displacement of the point of the string of abscissa x at time t from its equilibrium position; it is a (signed) scalar. Let c be the constant propagation velocity; it is a positive number that depends on the section and density of the string. Let $s(x, t)$ be the external action on the point of abscissa x at time t; it is a source term, such that $t = 0 \Rightarrow s(x, t) = 0$ (for the sake of initializations). Finally, let $p_0(x)$ and $p_1(x)$ be the initial position and velocity of the point of abscissa x, also called Cauchy data. As explained in section 9.1, our running example is the initial-boundary value problem

$$\forall t \geq 0, \ \forall x \in [x_{\min}; x_{\max}], \quad \frac{\partial^2 p}{\partial t^2}(x, t) + A(c)\, p(x, t) = s(x, t), \qquad [9.2]$$

$$\forall x \in [x_{\min}; x_{\max}], \quad \frac{\partial p}{\partial t}(x, 0) = p_1(x), \qquad [9.3]$$

$$\forall x \in [x_{\min}; x_{\max}], \qquad p(x, 0) = p_0(x), \tag{9.4}$$

$$\forall t \geq 0, \qquad p(x_{\min}, t) = p(x_{\max}, t) = 0, \tag{9.5}$$

where the differential operator $A(c)$ is defined by $-c^2 \frac{\partial^2}{\partial x^2}$.

The Coq counterpart is defined with a function taking a pair (x, t) as input. The function L corresponds to the left part of equation [9.2], it uses the Derive operator of section 9.2.1. Similarly, L1 is the differential with respect to the second variable, in order to correspond to equation [9.3]. Finally, L0 is the identity, in order to correspond to equation [9.3].

```
Definition is_solution (p: R*R -> R) :=
  (forall X:R*R, xmin <= fst X <= xmax -> L p X = s X) /\
  (forall x:R, xmin <= x <= xmax -> L1 p (x,0) = p1 x) /\
  (forall x:R, xmin <= x <= xmax -> L0 p (x,0) = p0 x) /\
  (forall t:R, p (xmin,t) = 0) /\
  (forall t:R, p (xmax,t) = 0).
```

As stated above in section 9.1, one can express the solution of this equation using d'Alembert's formula. Under some reasonable conditions on the Cauchy data p_0 and p_1 and on the source term s, this formula is the unique solution p to the initial-boundary value problem [9.2]–[9.5]. In this section, the external action on the rope s can be nearly any regular function (we only require that $s(x, 0) = 0$). In section 9.4, we will require $s = 0$ in order to bound the values taken by p.

For such a solution p, numerical analysts naturally associate with each time t the positive definite quadratic quantity called energy defined by

$$E(c)(p)(t) \overset{\text{def}}{=} \frac{1}{2} \left\| \left(x \mapsto \frac{\partial p}{\partial t}(x, t) \right) \right\|^2 + \frac{1}{2} \left\| (x \mapsto p(x, t)) \right\|^2_{A(c)},$$

where we denote:

$$\langle q, r \rangle \overset{\text{def}}{=} \int_{x_{\min}}^{x_{\max}} q(x) r(x) dx, \quad \|q\|^2 \overset{\text{def}}{=} \langle q, q \rangle, \text{ and } \|q\|^2_{A(c)} \overset{\text{def}}{=} \langle A(c) q, q \rangle.$$

The first term is interpreted as the kinetic energy and the second term as the potential energy, making E the mechanical energy of the vibrating string [JOH 86].

9.3.2. Description of the discretized problem

This continuous problem is then discretized, which means that approximate values are computed at points of a finite grid. Let $i_{\max}$ be the positive number of intervals of the space discretization. Let the space discretization step Δx and the discretization function $i_{\Delta x}$ be defined as

$$\Delta x \stackrel{\text{def}}{=} \frac{x_{\max} - x_{\min}}{i_{\max}} \quad \text{and} \quad i_{\Delta x}(x) \stackrel{\text{def}}{=} \left\lfloor \frac{x - x_{\min}}{\Delta x} \right\rfloor.$$

Let us consider the time interval $[0; t_{\max}]$. Let $\Delta t \in (0; t_{\max})$ be the time discretization step. Let us define

$$k_{\Delta t}(t) \stackrel{\text{def}}{=} \left\lfloor \frac{t}{\Delta t} \right\rfloor \quad \text{and} \quad k_{\max} \stackrel{\text{def}}{=} k_{\Delta t}(t_{\max}).$$

The compact domain $[x_{\min}; x_{\max}] \times [0; t_{\max}]$ is approximated by the regular discrete grid defined by

$$\forall k \in [0; k_{\max}], \forall i \in [0; i_{\max}], \quad \mathbf{x}_i^k \stackrel{\text{def}}{=} (x_i, t^k) \stackrel{\text{def}}{=} (x_{\min} + i\Delta x, k\Delta t). \qquad [9.6]$$

For a function q defined over $[x_{\min}; x_{\max}] \times [0; t_{\max}]$ (respectively $[x_{\min}; x_{\max}]$), we write q_{h} to denote any discrete approximation of q at the points of the grid, that is a discrete function over $[0; i_{\max}] \times [0; k_{\max}]$ (respectively $[0; i_{\max}]$). By extension, the notation q_{h} is also a shortcut to denote the matrix $(q_i^k)_{0 \le i \le i_{\max}, 0 \le k \le k_{\max}}$ (respectively the vector $(q_i)_{0 \le i \le i_{\max}}$). This is the common mathematical notation, so we use it in our mathematical formulas. Note that this notation, however, does not exist in our formalization: we have two different objects in Coq, one of type R -> R and the other of type nat -> R; there is no discretization operator.

As with the points of the discrete grid in equation 9.6, the exponent k denotes the kth time step and the subscript i the ith space step, beginning at $x_{\min}$. The notation $\bar{q}_{\mathrm{h}}$ is reserved for the approximation defined on $[0; i_{\max}] \times [0; k_{\max}]$ by

$$\bar{q}_i^k \stackrel{\text{def}}{=} q(\mathbf{x}_i^k) \quad (\text{resp. } \bar{q}_i \stackrel{\text{def}}{=} q(x_i)).$$

Let $p_{0\mathrm{h}}$ and $p_{1\mathrm{h}}$ be two discrete functions over $[0; i_{\max}]$. Let s_{h} be a discrete function over $[0; i_{\max}] \times [0; k_{\max}]$. Then, the discrete function p_{h} over $[0; i_{\max}] \times [0; k_{\max}]$ is said to be the solution of the three-point[2] finite difference scheme, as illustrated in Figure 9.3, when the following set of equations holds:

$$\forall k \in [2; k_{\max}], \forall i \in [1; i_{\max} - 1],$$

$$(L_{\mathrm{h}}(c)\,(p_{\mathrm{h}}))_i^k \stackrel{\text{def}}{=} \frac{p_i^k - 2p_i^{k-1} + p_i^{k-2}}{\Delta t^2} + (A_{\mathrm{h}}(c)\,p_{\mathrm{h}}^{k-1})_i = s_i^{k-1}, \qquad [9.7]$$

---

2 In the sense of "three spatial points", for the definition of the matrix $A_{\mathrm{h}}(c)$.

$\forall i \in [1; i_{\max} - 1],$

$$(L_{1h}(c)\,(\bar{p}_h))_i \stackrel{\text{def}}{=} \frac{p_i^1 - p_i^0}{\Delta t} + \frac{\Delta t}{2}(A_h(c)\,p_h^0)_i = p_{1,i}, \quad [9.8]$$

$\forall i \in [1; i_{\max} - 1], \quad (L_{0h}(\bar{p}_h))_i \stackrel{\text{def}}{=} p_i^0 = p_{0,i}, \qquad\qquad [9.9]$

$\forall k \in [0; k_{\max}], \quad p_0^k = p_{i_{\max}}^k = 0, \qquad\qquad\qquad [9.10]$

where the matrix $A_h(c)$, a discrete analog of $A(c)$, is defined for any vector q_h by

$$\forall i \in [1; i_{\max} - 1], \quad (A_h(c)\,q_h)_i \stackrel{\text{def}}{=} -c^2 \frac{q_{i+1} - 2q_i + q_{i-1}}{\Delta x^2}.$$

A discrete analog of the energy is defined by[3]

$$E_h(c)(p_h)^{k+\frac{1}{2}} \stackrel{\text{def}}{=} \frac{1}{2}\left\| \left(\frac{p_h^{k+1} - p_h^k}{\Delta t} \right) \right\|_{\Delta x}^2 + \frac{1}{2}\left\langle p_h^k, p_h^{k+1} \right\rangle_{A_h(c)} \qquad [9.11]$$

where, for any vectors q_h and r_h of size $i_{\max} + 1$, we define

$$\langle q_h, r_h \rangle_{\Delta x} \stackrel{\text{def}}{=} \sum_{i=0}^{i_{\max}} q_i r_i \Delta x, \qquad \|q_h\|_{\Delta x}^2 \stackrel{\text{def}}{=} \langle q_h, q_h \rangle_{\Delta x},$$

$$\langle q_h, r_h \rangle_{A_h(c)} \stackrel{\text{def}}{=} \langle A_h(c)\,q_h, r_h \rangle_{\Delta x}, \qquad \|q_h\|_{A_h(c)}^2 \stackrel{\text{def}}{=} \langle q_h, q_h \rangle_{A_h(c)}.$$

Note that the three-point scheme is parameterized by the discrete Cauchy data p_{0h} and p_{1h}, and by the discrete source term s_h. When these discrete inputs are respectively approximations of the continuous functions p_0, p_1, and s (e.g., when $p_{0h} = \bar{p}_{0h}$, $p_{1h} = \bar{p}_{1h}$, and $s_h = \bar{s}_h$), then the discrete solution p_h is an approximation of the continuous solution p.

9.3.3. Description of the scheme properties

In order to prove the convergence of the numerical scheme, the proof is composed of three steps. The first step, described in section 9.3.6, is called *consistency* and bounds the truncation error (defined below), that measures how much the continuous solution satisfies the numerical scheme. The second step, described in section 9.3.7, is called *stability* and bounds the growth of the discrete solution in terms of the input data (here p_{0h}, p_{1h}, and s_h) based on the energy (defined below). The last step,

---

[3] By convention, the energy is defined between steps k and $k + 1$, hence the notation $k + \frac{1}{2}$.

described in section 9.3.8, is called *convergence* and bounds the convergence error (defined below).

The convergence error e_h is defined by

$$\forall k \in [0; k_{\max}], \quad e_h^k \overset{\text{def}}{=} \bar{p}_h^k - p_h^k.$$

Note that when $p_{0h} = \bar{p}_{0h}$, we have $e_i^0 = 0$ for all i.

In contrast to FP errors, where the worst case is often considered, an average error is considered here as the quantity to be bounded is $\left\|e_h^{k_{\Delta t}(t)}\right\|_{\Delta x}$ and not $\max_{i,k}\left(\left|e_i^k\right|\right)$.

The numerical scheme is said to be *convergent* of order (m,n) uniformly on the interval $[0; t_{\max}]$ if the convergence error satisfies

$$\left\|e_h^{k_{\Delta t}(t)}\right\|_{\Delta x} = \mathcal{O}_{[0;t_{\max}]}(\Delta x^m + \Delta t^n). \tag{9.12}$$

As explained, the first step of the convergence proof is to bound the truncation error ε_h, which measures how much the continuous solution satisfies the numerical scheme. It is defined for $k \in [2; k_{\max}]$ and $i \in [1; i_{\max} - 1]$ by

$$\varepsilon_i^k \overset{\text{def}}{=} (L_h(c)\,(\bar{p}_h))_i^k - \bar{s}_i^{k-1},$$

$$\varepsilon_i^1 \overset{\text{def}}{=} (L_{1h}(c)\,(\bar{p}_h))_i - \bar{p}_{1,i},$$

$$\varepsilon_i^0 \overset{\text{def}}{=} (L_{0h}(\bar{p}_h))_i - \bar{p}_{0,i}.$$

Again, note that when $p_{0h} = \bar{p}_{0h}$ and $p_{1h} = \bar{p}_{1h}$, then for all i, $\varepsilon_i^0 = 0$ and $\varepsilon_i^1 = e_i^1/\Delta t$. Furthermore, when there is also $s_h = \bar{s}_h$, then the convergence error e_h is itself the solution of the same numerical scheme with inputs defined by, for all i, k,

$$p_{0,i} = \varepsilon_i^0 = 0, \quad p_{1,i} = \varepsilon_i^1 = \frac{e_i^1}{\Delta t}, \quad \text{and } s_i^k = \varepsilon_i^{k+1}.$$

The numerical scheme is said to be *consistent* with the continuous problem at order (m, n) uniformly on the interval $[0; t_{\max}]$ if the truncation error satisfies

$$\left\|\varepsilon_h^{k_{\Delta t}(t)}\right\|_{\Delta x} = \mathcal{O}_{[0;t_{\max}]}(\Delta x^m + \Delta t^n). \tag{9.13}$$

As explained, the second step of the convergence proof is about the limited growth of the numerical scheme. The numerical scheme is said to be *stable* if the discrete solution to the associated homogeneous problem (i.e., without any source term, $s(x,t) = 0$) is bounded independently of the discretization steps. More precisely, the numerical scheme is said to be uniformly stable on the interval $[0; t_{max}]$ if the discrete solution of the problem without any source term satisfies

$$\exists \alpha, C_1, C_2 > 0, \forall t \in [0; t_{max}], \forall \Delta x, \Delta t > 0, \quad \sqrt{\Delta x^2 + \Delta t^2} < \alpha \Rightarrow$$

$$\left\| p_h^{k_{\Delta t}(t)} \right\|_{\Delta x} \leq (C_1 + C_2 t) \left(\|p_{0h}\|_{\Delta x} + \|p_{0h}\|_{A_h(c)} + \|p_{1h}\|_{\Delta x} \right). \quad [9.14]$$

The Courant-Friedrichs-Lewy condition [COU 67], denoted by CFL(ξ) for ξ in $(0; 1)$, states that the discretization steps satisfy the relation

$$\frac{c \Delta t}{\Delta x} \leq 1 - \xi. \tag{9.15}$$

It means that the space step cannot go toward zero much faster than the time step.

The result to be formally proved states that if the continuous solution p is regular enough on $[x_{min}; x_{max}] \times [0; t_{max}]$ and if the discretization steps satisfy the CFL(ξ) condition, then the three-point scheme is convergent of order $(2, 2)$ uniformly on the interval $[0; t_{max}]$. Now let us describe the formal proof of this mathematical fact, including first some formalization details.

9.3.4. Domination (big $\mathcal{O}$)

A very first step is the definition of a big $\mathcal{O}$ for functions with two parameters. When considering a big $\mathcal{O}$ equality $a = \mathcal{O}(b)$, one usually assumes that a and b are two expressions defined over the same domain and its interpretation as a quantified formula comes naturally. Here, when manipulating Taylor expansions, the situation is a bit more complicated. Consider

$$f(\mathbf{x}, \mathbf{\Delta x}) = \mathcal{O}(g(\mathbf{\Delta x}))$$

when $\|\mathbf{\Delta x}\|$ goes to 0. If one were to assume that the equality holds for any $\mathbf{x} \in \mathbb{R}^2$, one would interpret it as

$$\forall \mathbf{x}, \exists \alpha > 0, \exists C > 0, \forall \mathbf{\Delta x}, \quad \|\mathbf{\Delta x}\| \leq \alpha \Rightarrow |f(\mathbf{x}, \mathbf{\Delta x})| \leq C \cdot |g(\mathbf{\Delta x})|,$$

which means that the constants α and C are in fact functions of $\mathbf{x}$. Such an interpretation happens to be useless, since the infimum of α may well be zero while the supremum of C may be $+\infty$.

As we wish to prove that the scheme converges when the size grids decrease, a solution is to bound the error by a finite C multiplied by a power of the discretization step size. A proper interpretation requires the introduction of a uniform big $\mathcal{O}$ relation with respect to the additional variable $\mathbf{x}$. Moreover, we consider that both $\mathbf{x}$ and $\mathbf{\Delta x}$ may be restricted to some sub-region $\Omega_\mathbf{x}$ and $\Omega_{\mathbf{\Delta x}}$ of $\mathbb{R}^2$; for instance $(0; +\infty)^2$ for the allowed grid size $\Omega_{\mathbf{\Delta x}}$.

$$\exists \alpha > 0, \exists C > 0, \forall \mathbf{x} \in \Omega_\mathbf{x}, \forall \mathbf{\Delta x} \in \Omega_{\mathbf{\Delta x}},$$

$$\|\mathbf{\Delta x}\| \leq \alpha \Rightarrow |f(\mathbf{x}, \mathbf{\Delta x})| \leq C \cdot |g(\mathbf{\Delta x})|. \quad [9.16]$$

This gives the following Coq definition for this uniform big $\mathcal{O}$. Its inputs include a property P that stands for $\Omega_{\mathbf{\Delta x}}$ and a type A that stands for $\Omega_\mathbf{x}$. Then the inequality must hold for all $\mathbf{\Delta x}$ (denoted by dX) small enough in the expected range.

```
Definition OuP (P : R * R -> Prop) (f : A -> R * R -> R)
              (g : R * R -> R) :=
  exists alp : R, exists C : R, 0 < alp /\ 0 < C /\
    forall X : A, forall dX : R * R,
      norm_l2 dX < alp -> P dX ->
        Rabs (f X dX) <= C * Rabs (g dX).
```

The implicit type A is generic in order for this definition and the dedicated theorems to be applicable both to $\mathbb{R}$, to $\mathbb{R}^2$, and to a dependent type representing a subset such as $[0; +\infty)$. The property P is also generic to share theorems; it handles $\mathbb{R}^2$ with (OuP (fun _ => True)) for Taylor approximation; it also handles the discretization steps for space and time that must be nonnegative with
(OuP (fun dX => 0 < fst dX /\ 0 < snd dX)).

To emphasize the dependency on both $\Omega_\mathbf{x}$ and $\Omega_{\mathbf{\Delta x}}$ (A and P), uniform big $\mathcal{O}$ equalities are now written

$$f(\mathbf{x}, \mathbf{\Delta x}) = \mathcal{O}_{\Omega_\mathbf{x}, \Omega_{\mathbf{\Delta x}}}(g(\mathbf{\Delta x})).$$

9.3.5. Differentiability and regularity

The notion of *sufficiently regular* functions is defined in terms of the full-fledged notation for the big $\mathcal{O}$. The convergence of the numerical scheme requires that the solution of the continuous equation is actually sufficiently regular. We rely on the differential operators $\frac{\partial}{\partial x}$ and $\frac{\partial}{\partial t}$ and on the Taylor polynomial Tpol defined in section 9.2.3:

$$\mathrm{Tpol}(f, n, (x, y), (u, v)) \overset{\text{def}}{=} \sum_{p=0}^{n} \frac{1}{p!} \left(\sum_{m=0}^{p} \binom{p}{m} \cdot \frac{\partial^p f}{\partial x^m \partial y^{p-m}}(x, y) \cdot u^m \cdot v^{p-m} \right).$$

Let $\Omega_{\mathbf{x}} \subseteq \mathbb{R}^2$. Let us say that the previous Taylor polynomial is a uniform approximation of order n of f on $\Omega_{\mathbf{x}}$ when the following uniform big $\mathcal{O}$ equality holds:

$$f(\mathbf{x} + \Delta \mathbf{x}) - \mathrm{Tpol}(f, n, \mathbf{x}, \Delta \mathbf{x}) = \mathcal{O}_{\Omega_{\mathbf{x}}, \mathbb{R}^2} \left(\| \Delta \mathbf{x} \|^{n+1} \right).$$

A function f is then said to be *sufficiently regular of order n uniformly on* $\Omega_{\mathbf{x}}$ when all its Taylor polynomials of an order smaller than n are uniform approximations of f on $\Omega_{\mathbf{x}}$. This regularity can be deduced using Taylor-Lagrange theorem from more common regularity hypotheses, such as f is of differentiability class C^{n+1}. But only this property, that the Taylor polynomial is a uniform approximation, is needed to establish the convergence of the scheme.

9.3.6. Consistency

The consistency of a numerical scheme expresses the fact that, for $\Delta \mathbf{x}$ small enough, the continuous solution taken at the points of the grid almost solves the numerical scheme (see the definition of the truncation error in section 9.3.3). More precisely, it is formally proved that when the continuous solution of the wave equation [9.2]–[9.5] is sufficiently regular of order 4 uniformly on $[x_{\min}; x_{\max}] \times [0; t_{\max}]$, the numerical scheme [9.7]–[9.10] is consistent with the continuous problem at order $(2, 2)$ uniformly on interval $[0; t_{\max}]$ (see equation [9.13] in section 9.3.3).

The proof is straightforward while involving long and complex expressions. The key idea is to always manipulate uniform Taylor approximations that are valid for all points of all grids when the discretization steps go down to zero.

For instance, to take into account the initialization phase corresponding to equation [9.8], one has to derive a uniform Taylor approximation of order 1 for the following continuous function (for any v sufficiently regular of order 3):

$$((x, t), (\Delta x, \Delta t)) \mapsto$$
$$\frac{v(x, t + \Delta t) - v(x, t)}{\Delta t} - \frac{\Delta t}{2} c^2 \frac{v(x + \Delta x, t) - 2v(x, t) + v(x - \Delta x, t)}{\Delta x^2}.$$

Note that the expression of this function involves both $x + \Delta x$ and $x - \Delta x$, meaning that one needs a Taylor approximation which is valid for both positive and negative growths. The proof would have been impossible if we had required $0 < \Delta x$ (as a space grid step) in the definition of the Taylor approximation.

A specific point about consistency is that it does not behave similarly with an infinite rope and a finite attached rope. A similar proof of convergence has been done considering an infinite rope and an infinite grid [BOL 10a]. It requires a lower bound for $c\frac{\Delta t}{\Delta x}$, which is useless in our case where the rope is tied down at both ends.

9.3.7. Stability

The stability of a numerical scheme expresses that the growth of the discrete solution is somehow bounded in terms of the input data (here, the Cauchy data p_{0h} and p_{1h}, and the source term s_h). For the proof of the round-off error (see section 9.4), a statement of the same form as equation [9.14] of section 9.3.3 is needed. Therefore, it was formally proved that, under the CFL(ξ) condition [9.15], the numerical scheme [9.7]–[9.10] is stable uniformly on interval $[0; t_{max}]$.

However, as the convergence of the numerical scheme was proved by using an energy-based technique,[4] it is more convenient to formulate the stability in terms of the discrete energy. More precisely, it is also formally proved that under the CFL(ξ) condition [9.15], the discrete energy [9.11] satisfies the following overestimation,

$$\sqrt{E_h(c)(p_h)^{k+\frac{1}{2}}} \le \sqrt{E_h(c)(p_h)^{\frac{1}{2}}} + \frac{\sqrt{2}}{2\sqrt{2\xi - \xi^2}} \cdot \Delta t \cdot \sum_{k'=1}^{k} \left\| s_h^{k'} \right\|_{\Delta x}$$

for all $t \in [0; t_{max}]$ and with $k = \lfloor \frac{t}{\Delta t} \rfloor - 1$.

The evolution of the discrete energy between two consecutive time steps is shown to be proportional to the source term. In particular, the energy is constant when the source is inactive. Thus, one obtains the following underestimation of the discrete energy:

$$\forall k, \quad \frac{1}{2} \left(1 - \left(c\frac{\Delta t}{\Delta x} \right)^2 \right) \left\| \frac{p_h^{k+1} - p_h^k}{\Delta t} \right\|_{\Delta x} \le E_h(c)(p_h)^{k+\frac{1}{2}}.$$

Therefore, the nonnegativity of the discrete energy is directly related to the CFL(ξ) condition. This is exactly the following Coq theorem:

```
Lemma minorationenergy: forall n, snd dX <> 0 -> 0 < fst dX ->
    1/2 * (1-Rsqr (c * (snd dX) / (fst dX))) *
    finite_squared_norm_dx (fst dX)
        (fun j => 1 / (snd dX) * (pnh dX j (n+1) - pnh dX j n))
    <= energy c (pnh dX) dX n.
```

Note that this stability result is valid for any input data p_{0h}, p_{1h}, and s_h. This property is used in the next section with input data other than those of the discrete solution.

---

4 An alternative is to use the Fourier transform, a more popular method among mathematicians.

9.3.8. *Convergence*

The convergence of a numerical scheme expresses the fact that the discrete solution gets closer to the continuous solution as the discretization steps decrease to zero. More precisely, it is formally proved that when the continuous solution of the wave equation [9.2]–[9.5] is sufficiently regular of order 4 uniformly on $[x_{min}; x_{max}] \times [0; t_{max}]$, and under the CFL($\xi$) condition [9.15], the numerical scheme [9.7]–[9.10] is convergent of order (2, 2) uniformly on interval $[0; t_{max}]$ (see definition (9.12) in section 9.3.3).

First, we prove that the convergence error e_h is itself the discrete solution to a numerical scheme of the same form but with different input data (as in section 9.3.3).[5] In particular, p_{0h} is zero and p_{1h} corresponds to some initialization error: $p_{1h} = e_h^1 / \Delta t$. The interesting value is the source term: it is the truncation error ε_h associated with the initial numerical scheme for p_h. Then, the previous stability result holds, and gives an overestimation of the square root of the discrete energy associated with the convergence error $E_h(c)(e_h)$ that involves a sum of the corresponding source terms, i.e., the truncation error. Finally, the consistency result also makes this sum a big $\mathcal{O}$ of $\Delta x^2 + \Delta t^2$ and a few more technical steps conclude the proof.

Theorem 9.1 (Convergence).

$$\left\| e_h^{k \Delta t(t)} \right\|_{\Delta x} = \mathcal{O} \left|_{\substack{t \in [0; t_{max}] \\ (\Delta x, \Delta t) \to 0 \\ 0 < \Delta x \wedge 0 < \Delta t \wedge \\ c \frac{\Delta t}{\Delta x} \le 1 - \xi}} (\Delta x^2 + \Delta t^2).$$

We have sketched the proof of the convergence of this simple numerical scheme. This proof means that formulas [9.7]–[9.10] compute something closer to the exact solution to the wave equation when the step sizes tend to zero [BOL 10a, BOL 13b]. This is a first step. The second step is to take into account the FP computations and the rounding errors.

9.4. Round-off error

The method error is proved as in the mathematical textbooks. Unfortunately, there is no such reference for the round-off error. Considering the program described in section 9.1, if we try naive forward error analysis, we get an error bound that is proportional to $2^k 2^{-53}$ for the computation of p_i^k. If this bound was sensible, it would cause the numerical scheme to compute only noise after a few steps. Fortunately, round-off errors actually compensate themselves and we end up with a bound proportional to $k^2 2^{-53}$. To take into account the compensations and hence prove a usable error bound,

---

5 There is no associated continuous problem.

one needs a precise statement of the round-off error [BOL 09a] to exhibit the error compensations made by the numerical scheme. We now assume a *binary64* FP format. To simplify the problem, we add two assumptions:

– There is no external force, that is $s = 0$.

– The initial position and velocity of the rope, p_0 and p_1, are such that the rope position is always between -1 and 1, that is $\forall x, t \; |p(x, t)| \leq 1$.

This last assumption is only a scaling of the inputs, as the rope is attached. Note that it only applies to the exact mathematical solution. Due to round-off and method errors, it does not mean that the computed position is between -1 and 1, but it will be possible to prove that the computed position is between -2 and 2. We could prove tighter bounds, say -1.1 and 1.1, but they do not bring anything to the rest of the proof.

To exhibit the compensations, we first study the round-off errors caused by one run of the body of the main loop (a local error) in section 9.4.1. Then we study the way the errors partly compensate in section 9.4.2 before bounding the global round-off error in section 9.4.3.

9.4.1. *Local round-off errors*

Let us remind ourselves of the core of the C program (see also section 9.1):

```
dp = p[i+1][k] - 2. * p[i][k] + p[i-1][k];
p[i][k+1] = 2. * p[i][k] - p[i][k-1] + a * dp;
```

with k ranging from 1 to $k_{\max}$ and i ranging from 1 to $i_{\max}$ and with a pre-computed constant a corresponding to $\frac{c^2 \Delta t^2}{\Delta x^2}$.

FP values as computed by the program have a tilde: $\tilde{a}$, $\tilde{p}_i^k$ to distinguish them from the discrete values of previous sections. They match the expressions a and p[i][k] in the program, while a and p_i^k are the exact mathematical values. The previous code means that

$$\tilde{p}_i^{k+1} = \circ \left[2\tilde{p}_i^k - \tilde{p}_i^{k-1} + \tilde{a} \cdot (\tilde{p}_{i+1}^k - 2\tilde{p}_i^k + \tilde{p}_{i-1}^k) \right].$$

Let δ_i^k be the (signed) FP error made in the previous two lines of C producing $\tilde{p}_i^{k+1}$. The δ_i^k are defined as follows:

$$\delta_i^{k+1} \stackrel{\text{def}}{=} \tilde{p}_i^{k+1} - (2\tilde{p}_i^k - \tilde{p}_i^{k-1} + a \cdot (\tilde{p}_{i+1}^k - 2\tilde{p}_i^k + \tilde{p}_{i-1}^k)).$$

To deduce a bound on δ_i^k, one has to estimate the range of $\tilde{p}_i^k$. For this, we rely on three properties. The first property is the assumption on the exact position of the rope:

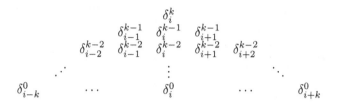

Figure 9.4. *Dependency pyramid: the global error of $\widetilde{p}_i^k$ depends upon the local error δ_i^k and upon all the previous local errors $\delta_{i+j}^{k-\ell}$ for $0 < \ell \leq k$ and $-\ell \leq j \leq \ell$.*

$|p(x,t)| \leq 1$. The second property is the bound on the values of the scheme: this is the stability property described in section 9.3.7. With reasonable assumptions on the grid size, one can prove that $\left|p_i^k\right| \leq 1.5$. The third property we exploit is the bound on the previous round-off errors. From the results of this section, one can prove that $\left|\widetilde{p}_i^k\right| \leq 2$ for $k \leq k_{\max}$, with a reasonable assumption on the maximal value of $k_{\max}$ (see below). That is to say, we have that the $\widetilde{p}_{i-1,i,i+1}^{k,k-1}$ used in the computation of $\widetilde{p}_i^{k+1}$ are in the range $[-2; 2]$.

The bound on δ_i^k is then easily deduced by the Gappa tool (see section 4.3) using a simple forward error analysis and interval arithmetic. It relies on another error bound for a, also computed by Gappa from the real code (see algorithm 9.5): $|\widetilde{a} - a| \leq 2^{-49}$. As the computations are done in *binary64*, the proved bound is that, for all i and k, $|\delta_i^k| \leq 78 \cdot 2^{-52}$.

9.4.2. Convolution of round-off errors

Note that the global FP error $\Delta_i^k = \widetilde{p}_i^k - p_i^k$ depends not only on δ_i^k, but also on all the $\delta_{i+j}^{k-\ell}$ for $0 < \ell \leq k$ and $-\ell \leq j \leq \ell$ as illustrated in Figure 9.4. Indeed round-off errors propagate along FP computations, and equation [9.7] shows that p_i^k is computed from $p_i^{k-1}, p_i^{k-2}, p_{i+1}^{k-1}$, and p_{i-1}^{k-1}.

The contributions of the δs to Δ_i^k, which are independent and linear (due to the structure of the numerical scheme), can be computed by performing a convolution with a well-chosen function $\lambda : (\mathbb{Z} \times \mathbb{N}) \to \mathbb{R}$. This function λ is called the fundamental solution [BOL 14] and represents the results of the same numerical scheme when fed with a single unit value:

$$\lambda_0^0 = 1, \qquad \forall i \neq 0, \ \lambda_i^0 = 0,$$

$$\lambda_{-1}^1 = \lambda_1^1 = a, \qquad \lambda_0^1 = 2(1-a), \qquad \forall i \notin \{-1,0,1\}, \ \lambda_i^1 = 0,$$

$$\forall k \geq 2, \ \lambda_i^k = a \cdot (\lambda_{i-1}^{k-1} + \lambda_{i+1}^{k-1}) + 2(1-a) \cdot \lambda_i^{k-1} - \lambda_i^{k-2}.$$

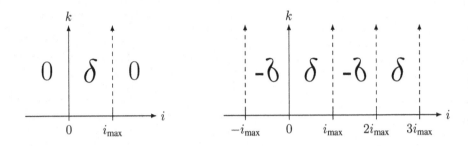

Figure 9.5. *Initial δ and its successive antisymmetric extension in space (its value is tossed and negated if $i \div i_{\max}$ is odd).*

Theorem 9.2. For all $i \in \mathbb{Z}$ and $k \in \mathbb{N}$, we have

$$\Delta_i^k = \widetilde{p}_i^k - p_i^k = \sum_{l=0}^{k} \sum_{j=-l}^{l} \lambda_j^l \, \delta_{i+j}^{k-l}.$$

For the sake of readability and simplicity of subtractions, we now consider $k \in \mathbb{Z}$. With some previous definitions, including that of λ, we can define Δ_i^k as in theorem 9.2 (denoted in Coq as Err)

Definition Err (i k : Z) :=
  sum_f_z 0 k
    (fun l => sum_f_z (-1) l (fun j => lambda a j l * delta (i+j)
      (k-1))).

Then, we can prove that it fulfills the desired recurrence relation:

Lemma Err_ok: forall i k:Z, (0 < k)%Z ->
Err i (k+1) = delta i (k+1) +
              a*(Err (i+1) k + Err (i-1) k) +
              2*(1-a)*Err i k - Err i (k-1).

Details of the proof can be found in [BOL 09a]. It mainly amounts to performing numerous tedious transformations of summations until both sides are proved to be equal. Another technical point is that this applies to an infinite rope with an unbounded set of i where p_i^k is computed. With a finite string, the ends of the rope ($i = 0$ or $i_{\max}$) imply that p_i^k and $\widetilde{p}_i^k$ are equal to zero, so Δ_i^k has to be zero too. This is solved by extending δ with its successive antisymmetric extension in space as explained in Figure 9.5 [JOH 86].

9.4.3. *Bound on the global round-off error*

The analytic expression of Δ_i^k can be used to obtain a bound on the round-off error. One needs two lemmas for this purpose.

Lemma 9.3. For all $k \geq 0$, we have $\displaystyle\sum_{i=-\infty}^{+\infty} \lambda_i^k = k + 1$.

Proof. By definition of λ, we have

$$
\begin{aligned}
\sum_{i=-\infty}^{+\infty} \lambda_i^{k+1} &= 2\widetilde{a} \sum_{i=-\infty}^{+\infty} \lambda_i^k + 2(1 - \widetilde{a}) \sum_{i=-\infty}^{+\infty} \lambda_i^k - \sum_{i=-\infty}^{+\infty} \lambda_i^{k-1} \\
&= 2 \sum_{i=-\infty}^{+\infty} \lambda_i^k - \sum_{i=-\infty}^{+\infty} \lambda_i^{k-1}.
\end{aligned}
$$

The sum by line satisfies a simple linear recurrence. As $\sum \lambda_i^0 = 1$ and $\sum \lambda_i^1 = 2$, we conclude that $\sum \lambda_i^k = k + 1$. ∎

Lemma 9.4. For all $i \in \mathbb{Z}$ and $k \geq 0$, we have $\lambda_i^k \geq 0$.

Proof. The demonstration, due to Kauers and Pillwein, was not formalized due to its large mathematical background.

$$
\lambda_i^k = \sum_{\ell=k}^{i} \binom{2\ell}{k+\ell} \binom{i+\ell+1}{2\ell+1} (-1)^{k+\ell} a^\ell = a^k \sum_{\ell=0}^{i-k} P_\ell^{(2k,0)}(1 - 2a)
$$

where P denotes the Jacobi polynomial.

Now the proof follows directly from the inequality of Askey and Gasper [ASK 72], which asserts that $\sum_{\ell=0}^{n} P_\ell^{(r,0)}(x) > 0$ for $r > -1$ and $-1 < x \leq 1$ (see theorem 7.4.2 in The Red Book [AND 99]). As $r = 2k \geq 0$ and as $0 < a < 1$ (so that $-1 < 1 - 2a < 1$), we have the required hypotheses that ensure that λ_i^k is nonnegative. ∎

Theorem 9.5. For all $i \in \mathbb{Z}$ and $k \geq 0$, we have

$$
|\Delta_i^k| = |\widetilde{p}_i^k - p_i^k| \leq 78 \cdot 2^{-53} \cdot (k + 1) \cdot (k + 2).
$$

Proof. According to theorem 9.2, Δ_i^k is equal to $\sum_{l=0}^{k} \sum_{j=-l}^{l} \lambda_j^l \, \delta_{i+j}^{k-l}$. It was previously proved that for all j and l, $|\delta_j^l| \leq 78 \cdot 2^{-52}$ (end of section 9.4.1) and that $\sum \lambda_i^l = l + 1$ (lemma 9.3). Since the λ_i^k are nonnegative (lemma 9.4), the error is easily bounded by $78 \cdot 2^{-52} \cdot \sum_{l=0}^{k} (l + 1)$. ∎

9.5. Program verification

We have described the proofs about both the method error and the round-off error. It remains to check that these formal proofs fit together and correspond to the C program. Following the methodology described in section 8.2, we annotate the C program with statements corresponding to theorems 9.1 and 9.5. It also remains to prove that this program never fails: this includes pointer access, division by zero, integer and floating-point overflow.

For the sake of simplicity, the program corresponds to the initial velocity p_1 and the source term s being zero. The computed values $\widetilde{p}_i^k$ are represented by a two-dimensional array p of size $(i_{\max} + 1) \times (k_{\max} + 1)$. This is a naive implementation; a more efficient implementation would store only two time steps.

The fully annotated program is six pages long, hence hardly readable. Selected excerpts are given and explained below.

Algorithm 9.1 Excerpt 1/5 of the C program for the 3-point scheme

---

```
1   /*@ logic real psol(real x, real t);
2   @
3   @ logic real psol_1(real x, real t);
4   @ axiom psol_1_def:
5   @   \forall real x; \forall real t;
6   @   \forall real eps; \exists real C; 0 < C &&
7   @   \forall real dx; 0 < eps ==> \abs(dx) < C ==>
8   @     \abs((psol(x + dx, t) - psol(x, t)) / dx
9   @       - psol_1(x, t)) < eps;
```

In algorithm 9.1, we first assume the existence of a solution psol to the 1D wave equation. This mathematical assumption is the goal of section 9.2. We also have to describe all the mathematical properties of psol, in particular the fact that it is the solution of the partial differential equation described above. Unfortunately, ACSL uses a first-order logic, where providing a differential operator is impossible. Therefore, for each derivative, we have to give its first-order definition, as the limit (using $\epsilon - \delta$ quantifications) of Newton's quotient. Only the first derivative of the solution with respect to its first variable psol_1 is given here (lines 3–9).

In the full C code, we then have all the assumptions [9.2]–[9.5] on psol. Only the first one is given in algorithm 9.2. The regularity of psol is another annotation, by assuming a Taylor-Lagrange approximation theorem of orders 3 and 4 (see section 9.2.3).

Algorithm 9.2 Excerpt 2/5 of the C program for the 3-point scheme

```
1   @ axiom wave_eq_2:
2   @   \forall real x; \forall real t;
3   @   0 <= x <= 1 ==>
4   @     psol_22(x, t) - c * c * psol_11(x, t) == 0;
```

Algorithm 9.3 Excerpt 3/5 of the C program for the 3-point scheme

```
1   @ predicate analytic_error{L} (double **p, integer ni,
2   @     integer i, integer k, double a, double dt)
3   @   reads p[..][..];
```

Algorithm 9.3 gives the declaration of the predicate analytic_error. This predicate is not defined in ACSL but directly in Coq. It states that the FP error can be stated with the formula of theorem 9.2 in section 9.4 with a double summation. It also states the bounds on the local round-off errors (see section 9.4.1).

Then come various notations for the method error and the specifications of some supplementary function (array2d_alloc for the matrix allocation and p_zero for an approximation of the mathematical p_0). For the sake of readability, we are skipping them. Then comes the function to be verified, forward_prop. The $\widetilde{p}_i^k$ (the FP versions of p_i^k) are computed and are all put in a matrix. All preconditions and postconditions are in algorithm 9.4.

At lines 3–6, we have most of the preconditions of this function. For instance, the size of the grid (ni for i_{max} and nk for k_{max}) must be greater than or equal to 2, but not too large (to prevent integer overflow) at line 3. Line 4 states that the value dt must be near enough Δt (relative error less than 2^{-51}). The Courant-Friedrichs-Lewy condition is at line 5 (see equation (9.15)). Line 6 requires that the grid is small enough so that the theorem about the method error, which is expressed as a big $\mathcal{O}$, can be applied.

The postconditions are at lines 8–16 of algorithm 9.4. The first one is the bound on the round-off errors of this numerical scheme, corresponding exactly to theorem 9.5 of section 9.4.3. The second one is the fact that this scheme is convergent of order (2, 2). It is based on theorem 9.1 of section 9.3.8, which involves a big $\mathcal{O}$. As said above, the fact that the size is small enough for the big $\mathcal{O}$ to be used is the requirement at line 6 of algorithm 9.4.

Finally, algorithm 9.5 presents the function forward_prop itself. It begins with its type and its variables (lines 1–5). Then dx is computed and its properties are asserted (lines 7–9). Then there is the first setting of p with the first initial condition p_0 so that p[$\cdots$][0] is correct at lines 18–21. Then there is the second setting of p so that p[$\cdots$][1] is also correct at lines 25–29. The main loop is at lines 40–41. Loop invariants have been kept while loop variants and assertions have been omitted for the sake of readability.

Algorithm 9.4 Excerpt 4/5 of the C program for the 3-point scheme

```
1    /*@ requires
2    [...]
3    @ ni>=2 && nk>=2 && ni <= 2147483646 && nk <= 7598581 &&
4    @ \abs(\exact(dt) − dt) / dt <= 0x1.p−51 &&
5    @ 0x1.p−500 <= \exact(dt) * ni * c <= 1 − 0x1.p−50 &&
6    @ \sqrt (1./( ni * ni)+\exact(dt)*\exact(dt)) < alpha_conv;
7    @
8    @ ensures
9    @ \forall integer i; \forall integer k;
10   @ 0 <= i <= ni ==> 0 <= k <= nk ==>
11   @    \round_error(\result[i][k]) <=
12   @         78./2 * 0x1.p−52 * (k + 1) * (k + 2);
13   @ ensures
14   @ \forall integer k; 0 <= k <= nk ==>
15   @ norm_dx_conv_err(\result, \exact(dt), ni, k) <=
16   @    C_conv * (1./( ni*ni) + \exact(dt)*\exact(dt ));
17   @ */
```

This example is the most complex C program of the book, it amounts to 33 lines of C code, 167 lines of annotations, for a total of 149 VCs; it also amounts to about 15,000 lines of Coq and 30 minutes of verification.

Let us comment on the results described here. First, the bound on the round-off error is quite reasonable: for k^2 computations, one has a bound proportional to $k^2 2^{-53}$. Note also that the bound on the method error is quite satisfactory: it is exactly the one from the mathematical textbooks, except for a few hypotheses which have been made explicit. For instance, the grid defined by $(\Delta x, \Delta t)$ must satisfy that $\Delta x > 0$ and $\Delta t > 0$, which is implicit for numerical analysts.

The example of this chapter combines formalizations of real analysis, numerical analysis and FP arithmetic. This combination is not artificial. Real analysis is required by numerical analysis for the existence and regularity of the exact solution. Numerical analysis is required as the stability of the scheme is used for the bounds on the local round-off errors. Both numerical analysis and FP arithmetic are required for the correctness of the program, as we want to prove that what the program actually computes is close to the exact solution to the wave equation, defined using real analysis.

Algorithm 9.5 Excerpt 5/5 of the C program for the 3-point scheme

```
1   double **forward_prop(int ni, int nk, double dt, double v,
2                         double xs, double l) {
3     double **p;
4     int i, k;
5     double a1, a, dp, dx;
6
7     dx = 1./ni;
8     /*@ assert dx > 0. && dx <= 0.5 &&
9       @ \abs(\exact(dx) − dx) / dx <= 0x1.p−53; */
10
11    a1 = dt/dx*v;
12    a = a1*a1;
13
14    /* Allocate space−time variable for the discrete solution. */
15    p = array2d_alloc(ni+1, nk+1);
16
17    /* 1st initial condition and boundary conditions. */
18    p[0][0] = 0.;
19    for (i=1; i<ni; i++) {
20      p[i][0] = p_zero(xs, l, i*dx); }
21    p[ni][0] = 0.;
22    /*@ assert analytic_error(p, ni, ni, 0, a, dt); */
23
24    /* 2nd initial condition (p_one=0) and boundary conditions. */
25    p[0][1] = 0.;
26    for (i=1; i<ni; i++) {
27      dp = p[i+1][0] − 2.*p[i][0] + p[i−1][0];
28      p[i][1] = p[i][0] + 0.5*a*dp;}
29    p[ni][1] = 0.;
30    /*@ assert analytic_error(p, ni, ni, 1, a, dt); */
31
32    /* Evolution problem and boundary conditions. */
33    /*@ loop invariant 1 <= k <= nk &&
34      @             analytic_error(p, ni, ni, k, a, dt); */
35    for (k=1; k<nk; k++) {
36      p[0][k+1] = 0.;
37      /*@ loop invariant 1 <= i <= ni &&
38        @             analytic_error(p, ni, i−1, k+1, a, dt); */
39      for (i=1; i<ni; i++) {
40        dp = p[i+1][k] − 2.*p[i][k] + p[i−1][k];
41        p[i][k+1] = 2.*p[i][k] − p[i][k−1] + a*dp;
42      }
43      p[ni][k+1] = 0.;
44      /*@ assert analytic_error(p, ni, ni, k + 1, a, dt); */
45    }
46    return p;
47  }
```

Bibliography

[AHM 15] AHMED Z., "Ahmed's integral: the maiden solution", *Mathematical Spectrum*, vol. 48, no. 1, pp. 11–12, 2015.

[AKB 10] AKBARPOUR B., ABDEL-HAMID A.T., TAHAR S. *et al.*, "Verifying a synthesized implementation of IEEE-754 floating-point exponential function using HOL", *The Computer Journal*, vol. 53, no. 4, pp. 465–488, 2010.

[AND 99] ANDREWS G.E., ASKEY R., ROY R., *Special Functions*, Cambridge University Press, 1999.

[ARM 10] ARMAND M., GRÉGOIRE B., SPIWACK A. *et al.*, "Extending Coq with imperative features and its application to SAT verification", in KAUFMANN M., PAULSON L.C. (eds), *Proceedings of the 1st Interactive Theorem Proving Conference (ITP)*, vol. 6172 of *Lecture Notes in Computer Science*, Springer, Berlin Heidelberg, 2010.

[ASK 72] ASKEY R., GASPER G., "Certain rational functions whose power series have positive coefficients", *The American Mathematical Monthly*, vol. 79, pp. 327–341, 1972.

[AYA 10] AYAD A., MARCHÉ C., "Multi-prover verification of floating-point programs", in GIESL J., HÄHNLE R. (eds), *Proceedings of the 5th International Joint Conference on Automated Reasoning (IJCAR)*, vol. 6173 of *Lecture Notes in Artificial Intelligence*, Springer, Edinburgh, Scotland, July 2010.

[BAC 13] "Baccalauréat général, Série S, Mathématiques, Session 2013", June 2013.

[BAR 89] BARRETT G., "Formal methods applied to a floating-point number system", *IEEE Transactions on Software Engineering*, vol. 15, no. 5, pp. 611–621, 1989.

[BAU 15] BAUDIN P., CUOQ P., FILLIÂTRE J.-C. *et al.*, ACSL: ANSI/ISO C Specification Language, version 1.9, 2015.

[BER 04] BERTOT Y., CASTÉRAN P., *Interactive Theorem Proving and Program Development. Coq'Art: The Calculus of Inductive Constructions*, Texts in Theoretical Computer Science, Springer, 2004.

[BER 13] BERTOT Y., "Proving the convergence of a sequence based on algebraic-geometric means to π", available at: http://www-sop.inria.fr/members/Yves.Bertot/proofs.html, 2013.

[BES 07] BESSON F., "Fast reflexive arithmetic tactics: the linear case and beyond", in ALTENKIRCH T., McBRIDE C., (eds), *Proceedings of Types for Proofs and Programs (Types'06)*, vol. 4502 of *Lecture Notes in Computer Science*, Springer, 2007.

[BOB 11] BOBOT F., FILLIÂTRE J.-C., MARCHÉ C. *et al.*, "Why3: shepherd your herd of provers", *Proceedings of the 1st International Workshop on Intermediate Verification Languages (Boogie 2011)*, Wrocław, Poland, pp. 53–64, August 2011.

[BOH 91] BOHLENDER G., WALTER W., KORNERUP P. *et al.*, "Semantics for exact floating point operations", *Proceedings of the 10th IEEE Symposium on Computer Arithmetic*, IEEE, pp. 22–26, June 1991.

[BOL 03] BOLDO S., DAUMAS M., "Representable correcting terms for possibly underflowing floating point operations", in BAJARD J.-C., SCHULTE M. (eds), *Proceedings of the 16th IEEE Symposium on Computer Arithmetic*, IEEE Computer Society Press, Los Alamitos, USA, 2003.

[BOL 04a] BOLDO S., Preuves formelles en arithmétiques à virgule flottante, PhD Thesis, École Normale Supérieure de Lyon, November 2004.

[BOL 04b] BOLDO S., DAUMAS M., "A simple test qualifying the accuracy of Horner's rule for polynomials", *Numerical Algorithms*, vol. 37, no. 1, pp. 45–60, 2004.

[BOL 05] BOLDO S., MULLER J.-M., "Some functions computable with a fused-mac", in MONTUSCHI P., SCHWARZ E. (eds), *Proceedings of the 17th IEEE Symposium on Computer Arithmetic*, IEEE, Cape Cod, USA, 2005.

[BOL 06a] BOLDO S., "Pitfalls of a full floating-point proof: example on the formal proof of the Veltkamp/Dekker algorithms", in FURBACH U., SHANKAR N. (eds), *Proceedings of the 3rd International Joint Conference on Automated Reasoning (IJCAR)*, Springer, Seattle, August 2006.

[BOL 06b] BOLDO S., DAUMAS M., KAHAN W. *et al.*, "Proof and certification for an accurate discriminant", in *12th IMACS-GAMM International Symposium on Scientific Computing, Computer Arithmetic and Validated Numerics*, Duisburg, Germany, September 2006.

[BOL 07] BOLDO S., FILLIÂTRE J.-C., "Formal verification of floating-point programs", in KORNERUP P., MULLER J.-M. (eds), *Proceedings of the 18th IEEE Symposium on Computer Arithmetic*, IEEE, Montpellier, France, June 2007.

[BOL 08] BOLDO S., MELQUIOND G., "Emulation of FMA and correctly-rounded sums: proved algorithms using rounding to odd", *IEEE Transactions on Computers*, vol. 57, no. 4, pp. 462–471, 2008.

[BOL 09a] BOLDO S., "Floats & Ropes: a case study for formal numerical program verification", in ALBERS S., MARCHETTI-SPACCAMELA A., MATIAS Y. *et al.* (eds), *Proceedings of the 36th International Colloquium on Automata, Languages and Programming (ICALP)*, vol. 5556 of *Lecture Notes in Computer Science*, Springer, Rhodes, Greece, July 2009.

[BOL 09b] BOLDO S., "Kahan's algorithm for a correct discriminant computation at last formally proven", *IEEE Transactions on Computers*, vol. 58, no. 2, pp. 220–225, IEEE Computer Society, February 2009.

[BOL 09c] BOLDO S., FILLIÂTRE J.-C., MELQUIOND G., "Combining Coq and Gappa for certifying floating-point programs", in CARETTE J., DIXON L., COEN C.S. *et al.* (eds), *Proceedings of the 16th Calculemus Symposium*, vol. 5625 of *Lecture Notes in Artificial Intelligence*, Grand Bend, Canada, 2009.

[BOL 10a] BOLDO S., CLÉMENT F., FILLIÂTRE J.-C. *et al.*, "Formal proof of a wave equation resolution scheme: the method error", in KAUFMANN M., PAULSON L.C. (eds), *Proceedings of the 1st Interactive Theorem Proving Conference (ITP)*, vol. 6172 of *Lecture Notes in Computer Science*, Springer, Edinburgh, Scotland, 2010.

[BOL 10b] BOLDO S., NGUYEN T.M.T., "Hardware-independent proofs of numerical programs", in MUÑOZ C. (ed.), *Proceedings of the 2nd NASA Formal Methods Symposium (NFM)*, NASA Conference Publication, Washington D.C., USA, April 2010.

[BOL 11a] BOLDO S., MARCHÉ C., "Formal verification of numerical programs: from C annotated programs to mechanical proofs", *Mathematics in Computer Science*, vol. 5, no. 4, pp. 377–393, 2011.

[BOL 11b] BOLDO S., MELQUIOND G., "Flocq: a unified library for proving floating-point algorithms in Coq", in ANTELO E., HOUGH D., IENNE P. (eds), *Proceedings of the 20th IEEE Symposium on Computer Arithmetic*, IEEE, Tübingen, Germany, 2011.

[BOL 11c] BOLDO S., MULLER J.-M., "Exact and approximated error of the FMA", *IEEE Transactions on Computers*, vol. 60, no. 2, pp. 157–164, 2011.

[BOL 11d] BOLDO S., NGUYEN T.M.T., "Proofs of numerical programs when the compiler optimizes", *Innovations in Systems and Software Engineering*, vol. 7, no. 2, pp. 151–160, 2011.

[BOL 12] BOLDO S., LELAY C., MELQUIOND G., "Improving real analysis in Coq: a user-friendly approach to integrals and derivatives", in HAWBLITZEL C., MILLER D. (eds), *Proceedings of the 2nd International Conference on Certified Programs and Proofs (CPP)*, vol. 7679 of *Lecture Notes in Computer Science*, IEEE, Kyoto, Japan, 2012.

[BOL 13a] BOLDO S., "How to compute the area of a triangle: a formal revisit", in *Proceedings of the 21st IEEE Symposium on Computer Arithmetic*, IEEE, Austin, USA, April 2013.

[BOL 13b] BOLDO S., CLÉMENT F., FILLIÂTRE J.-C. *et al.*, "Wave equation numerical resolution: a comprehensive mechanized proof of a C program", *Journal of Automated Reasoning*, vol. 50, no. 4, pp. 423–456, April 2013.

[BOL 14] BOLDO S., CLÉMENT F., FILLIÂTRE J.-C. *et al.*, "Trusting computations: a mechanized proof from partial differential equations to actual program", *Computers and Mathematics with Applications*, vol. 68, no. 3, pp. 325–352, 2014.

[BOL 15a] BOLDO S., "Formal verification of programs computing the floating-point average", in BUTLER M., CONCHON S., ZAÏDI F. (eds), *Proceedings of the 17th International Conference on Formal Engineering Methods (ICFEM)*, vol. 9407 of *Lecture Notes in Computer Science*, Springer International Publishing, November 2015.

[BOL 15b] BOLDO S., "Stupid is as stupid does: taking the square root of the square of a floating-point number", in BOGOMOLOV S., MARTEL M. (eds), *Proceedings of the 7th and 8th International Workshop on Numerical Software Verification*, vol. 317 of *Electronic Notes in Theoretical Computer Science*, Seattle, USA, pp. 50–55, April 2015.

[BOL 15c] BOLDO S., JOURDAN J.-H., LEROY X. *et al.*, "Verified compilation of floating-point computations", *Journal of Automated Reasoning*, vol. 54, no. 2, 2015.

[BOL 15d] BOLDO S., LELAY C., MELQUIOND G., "Coquelicot: a user-friendly library of real analysis for Coq", *Mathematics in Computer Science*, vol. 9, no. 1, pp. 41–62, 2015.

[BOL 17] BOLDO S., GRAILLAT S., MULLER J.-M., "On the robustness of the 2Sum and Fast2Sum algorithms", *ACM Transactions on Mathematical Software*, vol. 44, no. 1, June 2017.

[BOU 97] BOUTIN S., "Using reflection to build efficient and certified decision procedures", in ABADI M., ITO T. (eds), *Proceedings of the 3rd International Symposium on Theoretical Aspects of Computer Software (TACS)*, Springer, Sendai, Japan, 1997.

[BRI 08] BRISEBARRE N., MULLER J.-M., "Correctly rounded multiplication by arbitrary precision constants", *IEEE Transactions on Computers*, vol. 57, no. 2, pp. 165–174, 2008.

[BRI 12] BRISEBARRE N., JOLDEŞ M., MARTIN-DOREL É. *et al.*, "Rigorous polynomial approximation using Taylor models in Coq", in GOODLOE A., PERSON S. (eds), *Proceedings of 4th International Symposium on NASA Formal Methods (NFM)*, vol. 7226 of *Lecture Notes in Computer Science*, Springer, Norfolk, USA, 2012.

[BRU 11] BRUSENTSOV N.P., RAMIL ALVAREZ J., "Ternary computers: the Setun and the Setun 70", in IMPAGLIAZZO J., PROYDAKOV E. (eds), *Perspectives on Soviet and Russian Computing (SoRuCom)*, vol. 357 of *IFIP Advances in Information and Communication Technology*, 2011.

[CAR 95a] CARREÑO V.A., Interpretation of IEEE-854 floating-point standard and definition in the HOL system, Report no. 110189, NASA Langley Research Center, 1995.

[CAR 95b] CARREÑO V.A., MINER P.S., "Specification of the IEEE-854 floating-point standard in HOL and PVS", *1995 International Workshop on Higher Order Logic Theorem Proving and its Applications*, Aspen Grove, USA, 1995.

[CHL 13] CHLIPALA A., *Certified Programming with Dependent Types*, MIT Press, 2013.

[CLA 86] CLARKE E.M., EMERSON E.A., SISTLA A.P., "Automatic verification of finite-state concurrent systems using temporal logic specifications", *ACM Transactions on Programming Languages and Systems*, vol. 8, no. 2, pp. 244–263, April 1986.

[COD 80] CODY JR. W.J., WAITE W., *Software Manual for the Elementary Functions*, Prentice-Hall, Englewood Cliffs, NJ, 1980.

[COH 12] COHEN C., "Construction of real algebraic numbers in Coq", in BERINGER L., FELTY A. (eds), *Proceedings of the 3rd International Conference on Interactive Theorem Proving (ITP)*, Lecture Notes in Computer Science, Springer, Princeton, USA, August 2012.

[COR 00] CORNEA M., HARRISON J., IORDACHE C. *et al.*, "Divide, square root, and remainder algorithms for the IA-64 architecture", *Open Source for Numerics*, Intel Corporation, 2000.

[COU 67] COURANT R., FRIEDRICHS K., LEWY H., "On the partial difference equations of mathematical physics", *IBM Journal of Research and Development*, vol. 11, no. 2, pp. 215–234, 1967.

[COU 77] COUSOT P., COUSOT R., "Abstract interpretation: a unified lattice model for static analysis of programs by construction or approximation of fixpoints", *Proceedings of the 4th ACM SIGACT-SIGPLAN Symposium on Principles of Programming Languages (POPL)*, ACM, pp. 238–252, 1977.

[CRÉ 04] CRÉGUT P., "Une procédure de décision réflexive pour un fragment de l'arithmétique de Presburger", *Journées Francophones des Langages Applicatifs (JFLA)*, pp. 69–82, 2004.

[CRU 04] CRUZ-FILIPE L., GEUVERS H., WIEDIJK F., "C-CoRN: the constructive Coq repository at Nijmegen", ASPERTI A., BANCEREK G., TRYBULEC A. (eds), *Proceedings of the 3rd International Conference of Mathematical Knowledge Management (MKM)*, vol. 3119 of *Lecture Notes in Computer Science*, Springer, 2004.

[DAU 01] DAUMAS M., RIDEAU L., THÉRY L., "A generic library for floating-point numbers and its application to exact computing", *Proceedings of the 14th International Conference on Theorem Proving in Higher Order Logics (TPHOLs)*, Springer, Edinburgh, Scotland, pp. 169–184, 2001.

[DAU 10] DAUMAS M., MELQUIOND G., "Certification of bounds on expressions involving rounded operators", *ACM Transactions on Mathematical Software*, vol. 37, no. 1, pp. 1–20, 2010.

[DAV 72] DAVIS P.J., "Fidelity in mathematical discourse: is one and one really two?", *The American Mathematical Monthly*, vol. 79, no. 3, pp. 252–263, 1972.

[DEK 71] DEKKER T.J., "A floating-point technique for extending the available precision", *Numerische Mathematik*, vol. 18, no. 3, pp. 224–242, 1971.

[DEM 77] DEMILLO R.A., LIPTON R.J., PERLIS A.J., "Social processes and proofs of theorems and programs", *Proceedings of the 4th ACM SIGACT-SIGPLAN symposium on Principles of programming languages (POPL)*, ACM, pp. 206–214, 1977.

[DIJ 75] DIJKSTRA E.W., "Guarded commands, nondeterminacy and formal derivation of programs", *Communications of the ACM*, vol. 18, no. 8, pp. 453–457, 1975.

[DIN 07] DE DINECHIN F., LAUTER C.Q., MULLER J.-M., "Fast and correctly rounded logarithms in double-precision", *Theoretical Informatics and Applications*, vol. 41, pp. 85–102, 2007.

[DIN 11] DE DINECHIN F., LAUTER C.Q., MELQUIOND G., "Certifying the floating-point implementation of an elementary function using Gappa", *IEEE Transactions on Computers*, vol. 60, no. 2, pp. 242–253, 2011.

[DOW 05] DOWEK G., MUÑOZ C., CARREÑO V., "Provably safe coordinated strategy for distributed conflict resolution", *Proceedings of the AIAA Guidance, Navigation, and Control Conference and Exhibit*, San Francisco, USA, 2005.

[ERC 04] ERCEGOVAC M.D., LANG T., *Digital Arithmetic*, Morgan Kaufmann Publishers, 2004.

[FER 95] FERGUSON W.E., "Exact computation of a sum or difference with applications to argument reduction", *Proceedings of the 12th IEEE Symposium on Computer Arithmetic*, IEEE, Bath, England, pp. 216–221, 1995.

[FIG 95] FIGUEROA S.A., "When is double rounding innocuous?", *ACM SIGNUM Newsletter*, vol. 30, no. 3, pp. 21–26, 1995.

[FIG 04] DE FIGUEIREDO L.H., STOLFI J., "Affine arithmetic: concepts and applications", *Numerical Algorithms*, vol. 37, no. 1, pp. 147–158, 2004.

[FIL 11] FILLIÂTRE J.-C., Deductive program verification, Habilitation Thesis, University Paris-Sud, December 2011.

[FIL 13] FILLIÂTRE J.-C., PASKEVICH A., "Why3 — where programs meet provers", in FELLEISEN M., GARDNER P. (eds), *Proceedings of the 22nd European Symposium on Programming (ESOP)*, vol. 7792 of *Lecture Notes in Computer Science*, Springer, March 2013.

[FIL 14] FILLIÂTRE J.-C., GONDELMAN L., PASKEVICH A., "The spirit of ghost code", in BIERE A., BLOEM R. (eds), *Proceedings of the 26th International Conference on Computer Aided Verification (CAV)*, vol. 8559 of *Lecture Notes in Computer Science*, Springer, Vienna, Austria, July 2014.

[FOU 07] FOUSSE L., HANROT G., LEFÈVRE V. *et al.*, "MPFR: A multiple-precision binary floating-point library with correct rounding", *ACM Transactions on Mathematical Software*, vol. 33, no. 2, 2007.

[GAL 86] GAL S., "Computing elementary functions: a new approach for achieving high accuracy and good performance", in MIRANKER W.L., TOUPIN R.A. (eds), *Accurate Scientific Computations*, Springer, Berlin, 1986.

[GEN 74] GENTLEMAN W.M., MAROVICH S.B., "More on algorithms that reveal properties of floating point arithmetic units", *Communications of the ACM*, vol. 17, no. 5, pp. 276–277, 1974.

[GOL 91] GOLDBERG D., "What every computer scientist should know about floating-point arithmetic", *ACM Computing Surveys*, vol. 23, no. 1, pp. 5–48, 1991.

[GON 16] GONTHIER G., MAHBOUBI A., TASSI E., A small scale reflection extension for the Coq system, Report no. RR-6455, Inria, 2016.

[GOO 13] GOODLOE A.E., MUÑOZ C., KIRCHNER F. *et al.*, "Verification of Numerical Programs: From Real Numbers to Floating Point Numbers", *Proceedings of the 5th International Symposium on NASA Formal Methods*, Moffett Field, USA, pp. 441–446, May 2013.

[GOR 00] GORDON M., "From LCF to HOL: a short history", in PLOTKIN G., STIRLING C., TOFTE M. (eds), *Proof, Language, and Interaction: Essays in Honor of Robin Milner*, MIT Press, 2000.

[GRA 94] GRANLUND T., MONTGOMERY P.L., "Division by invariant integers using multiplication", *Proceedings of the ACM SIGPLAN Conference on Programming Language Design and Implementation (PLDI)*, ACM, pp. 61–72, 1994.

[GRA 15a] GRAILLAT S., LAUTER C., TANG P.T.P. *et al.*, "Efficient calculations of faithfully rounded L2-norms of n-vectors", *ACM Transactions on Mathematical Software*, vol. 41, no. 4, pp. 24:1–24:20, 2015.

[GRA 15b] GRANLUND T., THE GMP DEVELOPMENT TEAM, GNU MP: The GNU Multiple Precision Arithmetic Library, edition 6.0.0, 2015.

[GRÉ 05] GRÉGOIRE B., MAHBOUBI A., "Proving equalities in a commutative ring done right in Coq", in HURD J., MELHAM T. (eds), *Proceedings of the 15th International Conference on Theorem Proving in Higher Order Logics (TPHOLs)*, vol. 3603 of *Lecture Notes in Computer Science*, Springer, 2005.

[GRÉ 06] GRÉGOIRE B., THÉRY L., "A purely functional library for modular arithmetic and its application to certifying large prime numbers", in FURBACH U., SHANKAR N. (eds), *Proceedings of the 3rd International Joint Conference on Automated Reasoning (IJCAR)*, vol. 4130 of *Lectures Notes in Artificial Intelligence*, Springer, Seattle, USA, 2006.

[GRÉ 11] GRÉGOIRE B., POTTIER L., THÉRY L., "Proof certificates for algebra and their application to automatic geometry theorem proving", in STURM T., ZENGLER C. (eds), *Post-proceedings of the 7th International Workshop on Automated Deduction in Geometry (ADG)*, vol. 6301 of *Lecture Notes in Artificial Intelligence*, Springer, Shanghai, China, 2011.

[HAR 99a] HARRISON J., "A machine-checked theory of floating point arithmetic", in BERTOT Y., DOWEK G., HIRSCHOWITZ A. *et al.* (eds), *Proceedings of the 12th International Conference on Theorem Proving in Higher Order Logics (TPHOLs)*, Springer, Nice, France, 1999.

[HAR 99b] HARRISON J., KUBASKA T., STORY S., TANG P.T.P., "The computation of transcendental functions on the IA-64 architecture", *Intel Technology Journal*, vol. 4, 1999.

[HAR 00] HARRISON J., "Formal verification of IA-64 division algorithms", in AAGAARD M., HARRISON J. (eds), *Proceedings of the 13th International Conference on Theorem Proving in Higher Order Logics (TPHOLs)*, of *Lecture Notes in Computer Science*, vol. 1869, Springer, 2000.

[HAT 12] HATCLIFF J., LEAVENS G.T., LEINO K.R.M. *et al.*, "Behavioral interface specification languages", *ACM Computing Surveys*, vol. 44, no. 3, pp. 16:1–16:58, 2012.

[HIG 02] HIGHAM N.J., *Accuracy and Stability of Numerical Algorithms*, 2nd ed., SIAM, 2002.

[HIL 39] HILBERT D., BERNAYS P., *Grundlagen der Mathematik II*, Die Grundlehren der mathematischen Wissenschaften, 1939.

[HOA 69] HOARE C.A.R., "An axiomatic basis for computer programming", *Communications of the ACM*, vol. 12, no. 10, pp. 576–580, 1969.

[IBM 96] IBM, *The PowerPC Compiler Writer's Guide*, Warthman Associates, 1996.

[IEE 08] IEEE COMPUTER SOCIETY, IEEE Standard for Interval Arithmetic, Report no. 754-2008, August 2008.

[INT 06] INTEL CORPORATION, Intel Itanium Architecture Software Developer's Manual, edition 2.2, 2006.

[ISO 11] ISO, "International standard ISO/IEC 9899:2011, Programming languages – C", 2011.

[JAC 02] JACOBI C., Formal verification of a fully IEEE compliant floating point unit, PhD Thesis, Computer Science Department, Saarland University, Saarbrucken, Germany, 2002.

[JAC 05] JACOBI C., BERG C., "Formal verification of the VAMP floating point unit", *Formal Methods in System Design*, vol. 26, no. 3, pp. 227–266, 2005.

[JEA 11] JEANNEROD C.-P., KNOCHEL H., MONAT C. *et al.*, "Computing floating-point square roots via bivariate polynomial evaluation", *IEEE Transactions on Computers*, IEEE, vol. 60, no. 2, pp. 214–227, 2011.

[JEA 12] JEANNEROD C.-P., JOURDAN-LU J., "Simultaneous floating-point sine and cosine for VLIW integer processors", *Proceedings of the 23rd International Conference on Application-Specific Systems, Architectures and Processors (ASAP)*, pp. 69–76, July 2012.

[JEA 17] JEANNEROD C.-P., RUMP S.M., "On relative errors of floating-point operations: optimal bounds and applications", *Mathematics of Computation*, 2017.

[JOH 86] JOHN F., *Partial Differential Equations*, Springer, 1986.

[JOL 11] JOLDEŞ M., Rigorous polynomial approximations and applications, PhD Thesis, ENS de Lyon, France, 2011.

[KAH 83] KAHAN W., Mathematics Written in Sand — The HP-15C, Intel 8087, etc., 1983.

[KAH 86] KAHAN W., Miscalculating Area and Angles of a Needle-like Triangle, available at: http://www.cs.berkeley.edu/~wkahan/Triangle.pdf, 1986.

[KAH 04] KAHAN W., On the Cost of Floating-Point Computation Without Extra-Precise Arithmetic, available at: http://www.cs.berkeley.edu/~wkahan/Qdrtcs.pdf, November 2004.

[KAI 03] KAIVOLA R., KOHATSU K., "Proof engineering in the large: formal verification of Pentium®4 floating-point divider", *International Journal on Software Tools for Technology Transfer*, vol. 4, no. 3, pp. 323–334, 2003.

[KER 99] KERN C., GREENSTREET M.R., "Formal verification in hardware design: a Survey", *ACM Transactions on Design Automation of Electronic Systems*, vol. 4, no. 2, pp. 123–193, 1999.

[KET 04] KETTNER L., MEHLHORN K., PION S. *et al.*, "Classroom examples of robustness problems in geometric computations", in ALBERS S., RADZIK T. (eds), *Proceedings of the 12th European Symposium on Algorithms*, vol. 3221 of *Lecture Notes in Computer Science*, Springer, 2004.

[KIN 76] KING J.C., "Symbolic execution and program testing", *Communications of the ACM*, vol. 19, no. 7, pp. 385–394, 1976.

[KIR 15] KIRCHNER F., KOSMATOV N., PREVOSTO V. *et al.*, "Frama-C: a software analysis perspective", *Formal Aspects of Computing*, vol. 27, no. 3, pp. 573–609, 2015.

[KNU 98] KNUTH D.E., *The Art of Computer Programming*, 3rd edition, Addison-Wesley, Reading, 1998.

[KOR 05] KORNERUP P., "Digit selection for SRT division and square root", *IEEE Transactions on Computers*, vol. 54, no. 3, pp. 294–303, 2005.

[KRE 13] KREBBERS R., SPITTERS B., "Type classes for efficient exact real arithmetic in Coq", *Logical Methods in Computer Science*, vol. 9, no. 1, pp. 1–27, 2013.

[KUP 14] KUPRIIANOVA O., LAUTER C.Q., "Metalibm: a mathematical functions code generator", in HONG H., YAP C., (eds), *Proceedings of the 4th International Congress on Mathematical Software (ICMS)*, Springer, Seoul, South Korea, 2014.

[LEA 06] LEAVENS G.T., "Not a number of floating point problems", *Journal of Object Technology*, vol. 5, no. 2, pp. 75–83, 2006.

[LEC 35] LECAT M., *Erreurs de mathématiciens des origines à nos jours*, Castaigne, 1935.

[LEF 01] LEFÈVRE V., MULLER J.-M., "Worst cases for correct rounding of the elementary functions in double precision", in BURGESS N., CIMINIERA L. (eds), *Proceedings of the 15th IEEE Symposium on Computer Arithmetic*, IEEE, Vail, USA, 2001.

[LEL 13] LELAY C., "A new formalization of power series in Coq", *Proceedings of the 5th Coq Workshop*, Rennes, France, pp. 1–2, July 2013.

[LEL 15] LELAY C., Repenser la bibliothèque réelle de Coq : vers une formalisation de l'analyse classique mieux adaptée, PhD Thesis, École doctorale informatique de Paris-Sud (ED 427), June 2015.

[LER 09] LEROY X., "Formal verification of a realistic compiler", *Communications of the ACM*, vol. 52, no. 7, pp. 107–115, 2009.

[LET 04] LETOUZEY P., "A new extraction for Coq", in GEUVERS H., WIEDIJK F. (eds), *Proceedings of the International Workshop on Types for Proofs and Programs (TYPES)*, Lecture Notes in Computer Science, Springer, Berg en Dal, Netherlands, 2004.

[LIN 81] LINNAINMAA S., "Software for doubled-precision floating-point computations", *ACM Transactions on Mathematical Software*, vol. 7, no. 3, pp. 272–283, 1981.

[MAC 04] MACKENZIE D., *Mechanizing Proof: Computing, Risk, and Trust*, MIT Press, 2004.

[MAH 16] MAHBOUBI A., MELQUIOND G., SIBUT-PINOTE T., "Formally verified approximations of definite integrals", in BLANCHETTE J.C., MERZ S. (eds), *Proceedings of the 7th Conference on Interactive Theorem Proving (ITP)*, vol. 9807 of *Lecture Notes in Computer Science*, Springer, Nancy, France, 2016.

[MAL 72] MALCOLM M.A., "Algorithms to reveal properties of floating-point arithmetic", *Communications of the ACM*, vol. 15, no. 11, pp. 949–951, 1972.

[MAR 07] MARCHÉ C., "Jessie: an intermediate Language for Java and C Verification", in *Proceedings of the Workshop on Programming Languages meets Program Verification (PLPV)*, Freiburg, Germany, ACM, pp. 1–2, 2007.

[MAR 13] MARTIN-DOREL É., MELQUIOND G., MULLER J.-M., "Some issues related to double rounding", *BIT Numerical Mathematics*, vol. 53, no. 4, pp. 897–924, 2013.

[MAR 14] MARCHÉ C., "Verification of the functional behavior of a floating-point program: an industrial case study", *Science of Computer Programming*, vol. 96, no. 3, pp. 279–296, 2014.

[MAR 16] MARTIN-DOREL É., MELQUIOND G., "Proving tight bounds on univariate expressions with elementary functions in Coq", *Journal of Automated Reasoning*, vol. 57, no. 3, pp. 187–217, 2016.

[MAS 16] MASCARENHAS W.F., Floating point numbers are real numbers, available at: https://arxiv.org/abs/1605.09202, 2016.

[MAY 01] MAYERO M., Formalisation et automatisation de preuves en analyses réelle et numérique, PhD Thesis, University Paris VI, December 2001.

[MEL 07] MELQUIOND G., PION S., "Formally certified floating-point filters for homogeneous geometric predicates", *Theoretical Informatics and Applications*, vol. 41, no. 1, pp. 57–70, 2007.

[MEL 08] MELQUIOND G., "Proving bounds on real-valued functions with computations", in ARMANDO A., BAUMGARTNER P., DOWEK G. (eds), *Proceedings of the 4th International Joint Conference on Automated Reasoning (IJCAR)*, vol. 5195 of *Lecture Notes in Artificial Intelligence*, Springer, Sydney, Australia, 2008.

[MEL 12] MELQUIOND G., "Floating-point arithmetic in the Coq system", *Information and Computation*, vol. 216, pp. 14–23, 2012.

[MIN 95] MINER P.S., Defining the IEEE-854 floating-point standard in PVS, Report no. 110167, NASA Langley Research Center, 1995.

[MIN 06] MINÉ A., "The octagon abstract domain", *Higher-Order and Symbolic Computation (HOSC)*, vol. 19, no. 1, pp. 31–100, 2006.

[MON 08] MONNIAUX D., "The pitfalls of verifying floating-point computations", *ACM Transactions on Programming Languages and Systems*, vol. 30, no. 3, pp. 1–41, 2008.

[MOO 63] MOORE R.E., *Interval Analysis*, Prentice-Hall, Englewood Cliffs, 1963.

[MOO 98] MOORE J.S., LYNCH T., KAUFMANN M., "A mechanically checked proof of the AMD5K86 floating point division program", *IEEE Transactions on Computers*, vol. 47, no. 9, pp. 913–926, 1998.

[MOU 11] MOUILLERON C., REVY G., "Automatic generation of fast and certified code for polynomial evaluation", in ANTELO E., HOUGH D., IENNE P. (eds), *Proceedings of the 20th IEEE Symposium on Computer Arithmetic*, IEEE, Tübingen, Germany, 2011.

[MUL 10] MULLER J.-M., BRISEBARRE N., DE DINECHIN F. *et al.*, *Handbook of Floating-Point Arithmetic*, Birkhäuser, 2010.

[MUL 16] MULLER J.-M., *Elementary Functions, Algorithms and Implementation*, 3rd edition, Birkhäuser Boston, 2016.

[MUÑ 05] MUÑOZ C., SIMINICEANU R., CARREÑO V. *et al.*, KB3D Reference Manual - Version 1.a, Technical Memorandum no. NASA/TM-2005-213679, NASA Langley, Hampton, VA, 2005.

[NGU 11] NGUYEN T.M.T., MARCHÉ C., "Hardware-dependent proofs of numerical programs", in JOUANNAUD J.-P., SHAO Z. (eds), *Proceedings of the 1st International Conference on Certified Programs and Proofs (CPP)*, vol. 7086 of *Lecture Notes in Computer Science*, Springer, 2011.

[OGI 05] OGITA T., RUMP S.M., OISHI S., "Accurate sum and dot product", *SIAM Journal on Scientific Computing*, vol. 26, no. 6, pp. 1955–1988, 2005.

[OLE 99] O'LEARY J., ZHAO X., GERTH R. *et al.*, "Formally verifying IEEE compliance of floating point hardware", *Intel Technology Journal*, vol. 3, no. 1, pp. 1–14, 1999.

[PIE 17] PIERCE B.C., DE AMORIM A.A., CASINGHINO C. *et al.*, *Software Foundations*, University of Pennsylvania, 2017.

[POL 98] POLLACK R., "How to believe a machine-checked proof", *Twenty Five Years of Constructive Type Theory*, vol. 36, p. 205, Oxford University Press, 1998.

[ROU 14] Roux P., "Innocuous double rounding of basic arithmetic operations", *Journal of Formalized Reasoning*, vol. 7, no. 1, pp. 131–142, 2014.

[RUM 08] Rump S.M., Ogita T., Oishi S., "Accurate floating-point summation part I: faithful rounding", *SIAM Journal on Scientific Computing*, vol. 31, no. 1, pp. 189–224, 2008.

[RUS 98] Russinoff D.M., "A mechanically checked proof of IEEE compliance of the floating point multiplication, division and square root algorithms of the AMD-K7 processor", *LMS Journal of Computation and Mathematics*, vol. 1, pp. 148–200, 1998.

[RUS 99] Russinoff D.M., "A mechanically checked proof of correctness of the AMD-K5 floating point square root microcode", *Formal Methods in System Design*, vol. 14, no. 1, pp. 75–125, 1999.

[RUS 00] Russinoff D.M., "A case study in formal verification of register-transfer logic with ACL2: the floating point Adder of the AMD Athlon processor", in Hunt W.A., Johnson S.D. (eds), *Proceedings of the Third International Conference on Formal Methods in Computer-Aided Design (FMCAD)*, Springer, 2000.

[RUS 13] Russinoff D.M., "Computation and formal verification of SRT quotient and square root digit selection tables", *IEEE Transactions on Computers*, vol. 62, no. 5, pp. 900–913, 2013.

[SAW 02] Sawada J., Gamboa R., "Mechanical verification of a square root algorithm using Taylor's theorem", in Aagaard M.D., O'Leary J.W. (eds), *Proceedings of the 4th International Conference on Formal Methods in Computer-Aided Design (FMCAD)*, vol. 2517 of *Lecture Notes in Computer Science*, Springer, Berlin Heidelberg, 2002.

[SCH 09] Schwarz E.M., Kapernick J.S., Cowlishaw M.F., "Decimal floating-point support on the IBM System z10 processor", *IBM Journal of Research and Development*, vol. 53, no. 1, pp. 4:1–4:10, 2009.

[SHE 97] Shewchuk J.R., "Adaptive precision floating-point arithmetic and fast robust geometric predicates", *Discrete Computational Geometry*, vol. 18, pp. 305–363, 1997.

[SPI 11] Spitters B., van der Weegen E., "Type classes for mathematics in type theory", *Mathematical Structures of Computer Science*, vol. 21, no. 4, pp. 795–825, 2011.

[STE 74] Sterbenz P.H., *Floating Point Computation*, Prentice Hall, 1974.

[STE 05] Stehlé D., Zimmermann P., "Gal's accurate tables method revisited", in *Proceedings of the 17th IEEE Symposium on Computer Arithmetic*, IEEE, pp. 257–264, 2005.

[UTT 10] Utting M., Legeard B., *Practical Model-Based Testing: A Tools Approach*, Morgan Kaufmann, 2010.

[VEL 68] Veltkamp G.W., ALGOL procedures voor het berekenen van een inwendig product in dubbele precisie, RC-Informatie no. 22, Technische Hogeschool Eindhoven, 1968.

[VEL 69] Veltkamp G.W., ALGOL procedures voor het rekenen in dubbele lengte, RC-Informatie no. 21, Technische Hogeschool Eindhoven, 1969.

[YAP 95] YAP C.K., DUBÉ T., "The exact computation paradigm", in DU D.-Z., HWANG F.K. (eds), *Computing in Euclidean Geometry*, vol. 4, of *Lecture Notes Series on Computing*, World Scientific, vol. 4, pp. 452–492, 1995.

[YU 92] YU Y., Automated proofs of object code for a widely used microprocessor, PhD Thesis, University of Texas at Austin, 1992.

Index

Printed in the United States
By Bookmasters